Learning and Analytics in Intelligent Systems

Volume 13

The main aim of the series is to make available a publication of books in hard copy form and soft copy form on all aspects of learning, analytics and advanced intelligent systems and related technologies. The mentioned disciplines are strongly related and complement one another significantly. Thus, the series encourages cross-fertilization highlighting research and knowledge of common interest. The series allows a unified/integrated approach to themes and topics in these scientific disciplines which will result in significant cross-fertilization and research dissemination. To maximize dissemination of research results and knowledge in these disciplines, the series publishes edited books, monographs, handbooks, textbooks and conference proceedings.

More information about this series at http://www.springer.com/series/16172

Vishal Jain · Jyotir Moy Chatterjee
Editors

Machine Learning with Health Care Perspective

Machine Learning and Healthcare

Springer

Editors
Vishal Jain
Bharati Vidyapeeth's Institute of Computer Applications and Management
New Delhi, Delhi, India

Jyotir Moy Chatterjee
Lord Buddha Education Foundation
Kathmandu, Nepal

ISSN 2662-3447 ISSN 2662-3455 (electronic)
Learning and Analytics in Intelligent Systems
ISBN 978-3-030-40849-7 ISBN 978-3-030-40850-3 (eBook)
https://doi.org/10.1007/978-3-030-40850-3

This Springer imprint is published by the registered company Springer Nature Switzerland AG
The registered company address is: Gewerbestrasse 11, 6330 Cham, Switzerland

Preface

Health care is an important industry which offers value-based care to millions of people, while at the same time becoming top revenue earners for many countries. Today, the healthcare industry in the USA alone earns a revenue of $1.668 trillion. The USA also spends more on health care per capita as compared to most other developed or developing nations. Quality, value, and outcome are three buzzwords that always accompany health care and promise a lot, and today, healthcare specialists and stakeholders around the globe are looking for innovative ways to deliver on this promise. Technology-enabled smart health care is no longer a flight of fancy, as Internet-connected medical devices are holding the health system as we know it together from falling apart under the population burden. Machine learning in health care is one such area which is seeing gradual acceptance in the healthcare industry. Google recently developed a machine learning algorithm to identify cancerous tumors in mammograms, and researchers in Stanford University are using deep learning to identify skin cancer. Machine learning is already lending a hand in diverse situations in health care. Machine learning in health care helps to analyze thousands of different data points, suggest outcomes, and provide timely risk scores and precise resource allocation, and has many other applications. It is the era where we need to advance more information to clinicians, so they can make better decisions about patient diagnoses and treatment options, while understanding the possible outcomes and cost for each one. The value of machine learning in health care is its ability to process huge datasets beyond the scope of human capability and then reliably convert analysis of that data into clinical insights that aid physicians in planning and providing care, ultimately leading to better outcomes, lower costs of care, and increased patient satisfaction. Machine learning in medicine has recently made headlines. Machine learning lends itself to some processes better than others. Algorithms can provide immediate benefit to disciplines with processes that are reproducible or standardized. Also, those with large-image datasets, such as radiology, cardiology, and pathology, are strong candidates. Machine learning can be trained to look at images, identify abnormalities, and point to areas that need attention, thus improving the accuracy of all these processes. Long-term machine learning will benefit the family practitioner or internist at the bedside. Machine learning can offer an objective opinion to improve efficiency, reliability, and accuracy.

This book is a unique effort to represent a variety of techniques designed to represent, enhance, and empower multi-disciplinary and multi-institutional machine learning research in healthcare informatics. This book provides a unique compendium of current and emerging machine learning paradigms for healthcare informatics and reflects the diversity, complexity, and the depth and breadth of this multi-disciplinary area. The integrated, panoramic view of data and machine learning techniques can provide an opportunity for novel clinical insights and discoveries. Explore the theory and practical applications of machine learning in health care. This book will offer a guided tour of machine learning algorithms, architecture design, and applications of learning in healthcare challenges. One can discover the ethical implications of healthcare data analytics and the future of machine learning in population and patient health optimization. One can also create a machine learning model, evaluate performance, and operationalize its outcomes within a organization. This book will provide techniques on how to apply machine learning within your organization and evaluate the efficacy, suitability, and efficiency of machine learning applications. These are illustrated through leading case studies, including how chronic disease is being redefined through patient-led data learning. This book tried to investigate how healthcare organizations can leverage this tapestry of machine learning to discover new business value, use cases, and knowledge as well as how machine learning can be woven into pre-existing business intelligence and analytics efforts.

Healthcare transformation requires us to continually look at new and better ways to manage insights—both within and outside the organization today. Increasingly, the ability to glean and operationalize new insights efficiently as a by-product of an organization's day-to-day operations is becoming vital to hospitals and health care sector's ability to survive and prosper. One of the long-standing challenges in healthcare informatics has been the ability to deal with the sheer variety and volume of disparate healthcare data and the increasing need to derive veracity and value out of it.

Machine Learning with Health Care Perspective provides techniques on how to apply machine learning within your organization and evaluate the efficacy, suitability, and efficiency of machine learning applications. These are illustrated through how chronic disease is being redefined through patient-led data learning and the Internet of things. Explore the theory and practical applications of machine learning in health care. This book offers a guided tour of machine learning algorithms, architecture design, and applications of learning in health care. One will discover the ethical implications of machine learning in health care and the future of machine learning in population and patient health optimization. One can also create a machine learning model, evaluate performance, and operationalize its outcomes within organizations.

What You Will Learn?

- Gain a deeper understanding of various machine learning uses and implementation within wider health care.
- Implement machine learning systems, such as cancer detection and enhanced deep learning.
- Select learning methods and tuning for use in health care.
- Recognize and prepare for the future of machine learning in health care through best practices, feedback loops, and intelligent agents.

Who This Book Is For?

Healthcare professionals interested in how machine learning can be used to develop health intelligence—with the aim of improving patient health and population health and facilitating significant patient cost savings.

This book is a unique effort to represent a variety of techniques designed to represent, enhance, and empower multi-disciplinary and multi-institutional machine learning research in healthcare informatics. This book provides a unique compendium of current and emerging machine learning paradigms for healthcare informatics and reflects the diversity, complexity, and the depth and breadth of this multi-disciplinary area. The integrated, panoramic view of data and machine learning techniques can provide an opportunity for novel clinical insights and discoveries.

New Delhi, India | Vishal Jain
Kathmandu, Nepal | Jyotir Moy Chatterjee

Contents

Machine Learning for Healthcare: Introduction 1
Shiwani Gupta and R. R. Sedamkar

Artificial Intelligence in Medical Diagnosis: Methods, Algorithms and Applications 27
J. H. Kamdar, J. Jeba Praba and John J. Georrge

Intelligent Learning Analytics in Healthcare Sector Using Machine Learning 39
Pratiyush Guleria and Manu Sood

Unsupervised Learning on Healthcare Survey Data with Particle Swarm Optimization 57
Hina Firdaus and Syed Imtiyaz Hassan

Machine Learning for Healthcare Diagnostics 91
K. Kalaiselvi and M. Deepika

Disease Detection System (DDS) Using Machine Learning Technique 107
Sumana De and Baisakhi Chakraborty

Knowledge Discovery (Feature Identification) from Teeth, Wrist and Femur Images to Determine Human Age and Gender 133
K. C. Santosh and N. Pradeep

Deep Learning Solutions for Skin Cancer Detection and Diagnosis 159
Hardik Nahata and Satya P. Singh

Security of Healthcare Systems with Smart Health Records Using Cloud Technology 183
Priyanka Dadhich and Kavita

Intelligent Heart Disease Prediction on Physical and Mental Parameters: A ML Based IoT and Big Data Application and Analysis 199
Rohit Rastogi, D. K. Chaturvedi, Santosh Satya and Navneet Arora

Medical Text and Image Processing: Applications, Issues and Challenges 237
Shweta Agrawal and Sanjiv Kumar Jain

Machine Learning Methods for Managing Parkinson's Disease 263
Kunjan Vyas, Shubhendu Vyas and Nikunj Rajyaguru

An Efficient Method for Computer-Aided Diagnosis of Cardiac Arrhythmias 295
Sandeep Raj

Clinical Decision Support Systems and Predictive Analytics 317
Ravi Lourdusamy and Xavierlal J. Mattam

Yajna and Mantra Science Bringing Health and Comfort to Indo-Asian Public: A Healthcare 4.0 Approach and Computational Study 357
Rohit Rastogi, Mamta Saxena, Muskan Maheshwari, Priyanshi Garg, Muskan Gupta, Rajat Shrivastava, Mukund Rastogi and Harshit Gupta

Identifying Diseases and Diagnosis Using Machine Learning 391
K. Kalaiselvi and D. Karthika

Machine Learning for Healthcare: Introduction

Shiwani Gupta and R. R. Sedamkar

Abstract Machine Learning (ML) is an evolving area of research with lot many opportunities to explore. "It is the defining technology of this decade, though its impact on healthcare has been meagre"—says James Collin at MIT. Many of the ML industry's young start-ups are knuckling down significant portions of their efforts to healthcare. Google has developed a machine learning algorithm to help identify cancerous tumours on mammograms. Stanford is using a Deep Learning algorithm to identify skin cancer. US healthcare system generates approximately one trillion GB of data annually. Different academic researchers have come up with different number of features and clinical researchers with different risk factors for identification of chronic diseases. More data means more knowledge for the machine to learn, but these large number of features require large number of samples for enhanced accuracy. Hence, it would be better if machines could extract medically high-risk factors. Accuracy is enhanced if data is pre-processed in form of Exploratory Data Analysis and Feature Engineering. Multiclass classification would be able to assess different risk level of disease for a patient. In healthcare, correctly identifying percentage of sick people (Sensitivity) is of priority than correctly identifying percentage of healthy people (Specificity), thus research should happen to increase the sensitivity of algorithms. This chapter presents an introduction to one of the most challenging and emerging application of ML i.e. Healthcare. Patients will always need the caring and compassionate relationship with the people who deliver care. Machine Learning will not eliminate this, but will become tools that clinicians use to improve ongoing care.

Keywords Healthcare · Machine Learning · Feature Selection · Parameter Optimization · Diagnosis · Preprocessing

---

S. Gupta (✉) · R. R. Sedamkar
Computer Engineering, Thakur College of Engineering and Technology, Mumbai, India
e-mail: shiwani.gupta@thakureducation.org

R. R. Sedamkar
e-mail: rr.sedamkar@thakureducation.org

V. Jain and J. M. Chatterjee (eds.), *Machine Learning with Health Care Perspective*, Learning and Analytics in Intelligent Systems 13,
https://doi.org/10.1007/978-3-030-40850-3_1

1 Introduction

"Hiding within those mounds of data is knowledge that could change the life of a patient or change the world."—Atul Butte, Stanford University. More the data means better the accuracy and large number of features require even more number of instances making the computational complexity high. Hence dimensionality reduction is essential [1]. Poor clinical decisions due to lack of expertise can lead to loss of life which is unacceptable. Thus, Intelligent Decision Support System (DSS) is not to replace a doctor but to help him in order to perform decision making pertaining to particular disease. People around the globe are now more conscious of their health and undergo frequent health check-ups but the increase in work pressure and sedentary lifestyles has given rise to a greater number of youths getting mortal diseases particularly related to Heart, Diabetes, Cancer, Kidney, Parkinson, etc. The chapter focusses on recent literature reviewed to understand the developments in field of Machine Learning (ML) for Healthcare w.r.t missing values, class imbalance, binary/multiclass classification, multi typed features (nominal, ordinal), etc. According to "your total health" website; WHO estimates 11.1 million deaths from Cardiovascular Artery Disease (CAD) in 2020.

Similarly, multiple datasets available from different online sources (Kaggle, UCI ML repository, data.gov.in) for same disease classification have different number and names for features esp. Heart Disease Datasets [2–8]. Thus topics below demonstrate steps to build a robust machine learning model for healthcare.

2 Data Preparation (EDA)

Data preparation is to be done in order to clean and process collected data. According to a survey in Forbes, 80% time of data scientists is spent on **data preparation**. Data might have features in different scales; hence normalization is required [7] for k Nearest Neighbor (kNN), Support Vector Machine (SVM) and Neural Network (NN) type of parametric models. Data preparation takes a lot of time in a Machine Learning (ML) pipeline hence **Pandas profiling** in Python can be used to understand data. Below have been discussed almost all common practices as part of data preparation w.r.t. healthcare data:

2.1 Feature Cleaning

This step is important to identify missing values and outliers in numerical features. Histogram and Bar Chart can be plotted to understand the same.

2.1.1 Missing Value

Dealing with missing values is either to ignore them or treat them depending on the amount of missing data, variables dependent on missing values, and value missing in explanatory variable or predictor variable. Differing sources of missing values are required to be processed differently. Most obvious is to be able to get the actual value by repeating data collection process; but this is not practical. Missing data may be categorized as: missing completely at random (MCAR), missing at random (MAR) and missing not at random (MNAR) [9]. Let's look at techniques to treat the missing values:

I. **Deletion**: The best way to deal with missing values is deletion, which may not be applicable if values are missing at random.

   a. **Listwise**: Delete rows with missing values.
   b. **Pairwise**: Assumes missing data to be missing completely at random.

   Listwise deletion suffers more information loss than Pairwise deletion.

II. **Imputation: fancyimpute** package in python offers various robust ML models for imputing missing values. **SimpleImputer** and **IterativeImputer** class popularly known as **MICE** (Multiple Imputation by Chained Equations) from **sklearn.preprocessing** library of Python can be used for imputation. **Impyute** is also a library of missing data imputation algorithms. Following are several Missing value Imputation (MVI) techniques:

   a. **Popular Averaging Techniques**: Mean imputation works well with small numerical datasets, normally distributed and not skewed; median imputation for numeric variables may introduce bias and mode imputation (most frequently used) is preferred for categorical data. Mode can also be used for numeric variables. Thus, we are able to quickly fill missing values, but at the cost of variation in the dataset. Though a simple and computationally quick approach, it may lead to poor performance.
   b. **Predictive Techniques**: Predictive techniques are used to impute MCAR type of data by choosing such variables which have some relationship with missing observations. To impute; the dataset is divided into—set with no missing values for the feature called as training set and the other one with missing values called as test set, where the feature with missing values is treated as the target variable. There are various statistical methods and/or machine learning methods to impute such missing values.
   c. **KNN** (nonparametric) **Imputation** [4, 10]: Imputation is done using most frequent value among k neighbors for discrete variable and mean/mode for continuous variable. The similarity among two instances is computed using a distance function. If the features are similar in type, one should chose Euclidean distance e.g. width and height otherwise for dissimilar features, Manhattan distance can be chosen e.g. age, height, etc. Hamming distance is the number of categorical attributes for which the value is differing. This

method of imputation works well with both qualitative and quantitative features, even with multiple missing values. It is applicable to MCAR, MAR, MNAR missing data mechanisms with numeric (continuous, discrete), ordinal and categorical features. KNN algorithm takes large amount of time in analysing huge databases and is sensitive to outliers since it searches through the entire data in order to find similar instances. Moreover, there is a trade off in the value of k chosen. Greater k might include noisy attributes whereas lower k value would mean missing out significant attributes and increase the influence of noise. Thus, results are less generalizable. For binary classification problem, **odd value of k** would avoid ties.

Genetic Algorithm (GA), **Expectation Maximization (EM)** and **kMeans** [11] can also be used for imputation. EM works iteratively by using other features to impute a value termed as Expectation and checks this value to be the most likely i.e. termed as Maximization. If it is not the most likely, it iteratively re-imputes a more likely value until it is achieved. This procedure is followed preserving the relationship of imputed feature with other features. kMeans algorithm imputes missing value utilizing clustering with weighted distance.

Missingpy library in Python has **missForest** class that imputes missing values using **Random Forest** (RF) in an iterative manner. It begins imputing the column with smallest number of missing values. Initially missing values in non-candidate column are filled by mean for numerical and mode for continuous. Imputer fits a RF model with missing valued column as predictor variable. The imputer choses next column with the second smallest number of missing values in the first round. The process repeats itself for each column with a missing value, over multiple iterations for each column, until the stopping criterion is met.

2.1.2 Outliers in Data

"Observation which deviates so much from other observations as to arouse suspicion it was generated by a different mechanism"—Hawkins (1980). Outliers are the data points, three or more standard deviation (MAD) away from mean. We often tend to overlook outliers while building models; which tends to skew the data and reduce accuracy. Value < -1 or > 1 designates high skew. The boxplot () function [11] in **seaborn** library uses median and lower and upper quartiles (defined as the 25th and 75th percentiles) to identify outliers. InterQuartile Range or midspread/IQR = Q3 − Q1, where Q1 is lower quartile and Q3 is upper quartile. It detects outliers as any value beyond the range of $[Q1 - (1.5 \times IQR), Q3 + (1.5 \times IQR)]$.

Kurtosis is a measure of outliers (acceptable value is between −2 and 2). Visualization is used to detect outliers. Methods include **Box-plot**, **Histogram**, and **Scatter Plot** utilizing **plotly, bokeh, matplotlib, seaborn** and **pandas** plotting python libraries.

Following techniques may be used to deal with outliers:

a **Deleting observation**: If outlier values are through data entry or introduced during data processing or if the number of observations affected are very small, observations can be deleted to remove outliers.
b **Transforming and binning values**: Outliers tend to get eliminated if variables are transformed. One way of transforming variables is to use natural log which reduces the variation caused by extreme values.
c **Imputing:** Imputation utilising statistical averages works with artificial outliers.

One of the methods for outlier detection is Standard Score (**Z-Score**) [2] $z = \frac{(x-\mu)}{\sigma}$. Assuming your data to be drawn from Gaussian distribution, it computes the number of standard deviations a data point is from the sample's mean. Thus Z-score is a parametric method. To make the data fall into Gaussian distribution, transformation i.e. scaling can be applied. One can utilize **scipy.stats** module to compute the same.

2.1.3 Data Imbalance [3]

Data used in healthcare often has number of instances belonging to one class significantly low (<5%) as compared to instances belonging to the other class. Hence, most ML models e.g. Decision Tree, Logistic Regression could be biased towards majority class and hence return inaccurate performance measure. For patients who are healthy, the model would have very high accuracy and for patients who are diseased, the model would have extremely low accuracy. The overall accuracy would be high, not because the model is good but simply because the data is imbalanced, since the features of minority class being treated as noise are often ignored. Thus, following techniques can help one to train a classifier for detecting the minority class:

I. **Use the right metric for evaluating the model**: For a model generated using imbalanced data, applying inappropriate evaluation metrics can be dangerous. If accuracy is used to measure the quality of a model, a model which classifies all test samples into healthy class will have an excellent accuracy, but obviously, this model won't provide any valuable information for us. In such cases, other alternative evaluation metrics can be applied as:

   a. **Precision/Specificity** [7, 11, 12] is how many selected instances are relevant.
   b. **Recall/Sensitivity** [7, 11] is how many relevant instances are selected.
   c. **F1 score** [8] is harmonic mean of precision and recall. It gives balanced accuracy score.
   d. **MCC** [7] is Mathew's correlation coefficient between the observed and predicted binary classifications.
   e. **AUC_ROC** (Area Under the Curve_Receiver Operating Characteristics) is the relation between true-positive rate and false positive rate. Like precision and recall, accuracy is divided into sensitivity and specificity and models can be chosen based on the balance thresholds of these values. AUC should be equal to 1 for 100% correct predictions.

To compute performance, **confusion_matrix** [11] and **accuracy_score** can be used from **sklearn.metrics** module.

II. **Resample the training set**: Other than using different evaluation criteria to deal with imbalanced data, one can also work on getting different dataset. Two approaches to make a balanced dataset out of an imbalanced one are under-sampling and over-sampling.

   a. **Under-sampling** balances the dataset by reducing the size of the majority class by randomly selecting equal number of samples in the majority class as in the minority class. This method is used when no. of observations is large but it tends to discard potentially useful information which could be important for building rule classifiers. The sample chosen by random under sampling may be a biased sample not representative of the entire population. Thereby, resulting in inaccurate results with the actual test data set.
   b. **Over-sampling** [4] is used when the quantity of data is insufficient. It tries to balance dataset by increasing the size of minority class samples by using e.g. repetition, bootstrapping or Synthetic Minority Over-Sampling Technique (SMOTE) [5] and can be accomplished utilising **imblearn** class in Python. Over-sampling increases the likelihood of overfitting since it replicates the minority class instances. Moreover, SMOTE is not effective for high dimensional data since it increases Recall at cost of Precision. ADaptive SYNthetic Sampling (ADSYN) is an improvement over SMOTE since it adds randomness. Both generate new samples by interpolation.

A combination of over- and under-sampling is often successful as well.

III. **Use K-fold Cross-Validation (CV) appropriately**: Over-sampling takes observed samples of minority class and applies Bootstrapping to generate new random data based on a distribution function. If CV is applied after over-sampling, the model would overfit to the artificially bootstrapped data. Hence, CV should always be done before over-sampling the data, just as how feature selection should be implemented. Only by resampling the data repeatedly, randomness can be introduced into the dataset to prevent overfitting.
IV. **Ensemble different resampled datasets**: To generate more data, so that the model generalizes well can be done by building 'n' models with all samples of the minority class and n-random samples of the majority class. Thus, Ensemble models tend to generalize better [3].
V. **Design your own models**: There is no need for resampling, if the model is suited for imbalanced data. e.g. XGBoost internally takes care that the bags it trains on are not imbalanced. The resampling happens secretly. The cost function used penalizes misclassification of the minority class more than misclassification of majority class. Thus, it is possible to design models that generalize naturally in favor of the minority class.

3 Feature Engineering

Feature Engineering is extraction of features from data and transforming them into formats that are suitable for ML algorithms. "*The goal is to turn data into information, and information into insight.*"—Carly Fiorina. There are 3 types of data—Numerical (Discrete and Continuous), Categorical (qualitative), and Ordinal (numbers with mathematical meaning). The predictor variable in a classification task may be binary or multi [3]. To reduce the complexity due to increase in number of classes, multiclass classifier is simplified into a series of binary classification such as One-Against-One and One-Against-All [5].

3.1 Feature Transformation

I. **Normalizing**: To eliminate the effect of outlier that may negatively affect the accuracy of conclusion, different types of variables need to be brought in same order of magnitude. Feature Scaling benefits Gradient Descent with faster convergence. Distance calculation algorithms are greatly influenced by difference in scale among variables whereas Tree based algorithms are not affected by difference in magnitude. Normalization and Standardization can be used for scaling. Normalization can be achieved through Min-Max scaling using equation $xnormalized = \frac{(x - x_{\min})}{(x_{\max} - x_{\min})}$ bringing all numeric values in the range [0, 1] with zero mean and one standard deviation. The formula for standardization is $z = \frac{(x-\mu)}{\sigma}$ and is commonly known as **z-score**, where μ is mean and σ is standard deviation. Scaling can be accomplished through **StandardScaler** and **MinScaler** utility classes of **sklearn.preprocessing** package. It is advisable to remove extreme outliers before applying normalization else it would skew the values in your data to a small interval. **RobustScaler** class of Python can be used for skewed data.

II. **Feature encoding**: Major ML libraries work well with numerical variables. Nominal values can be misinterpreted by the learning algorithm. Missing values should be filled before encoding categorical features. Encoding allows algorithms which expect continuous features to use categorical features through **scikit-learn**'s **LabelEncoder** function. **OneHotEncoder** function is required for multilabelled features. OneHotEncoder converts n levels into n-1 new variables and can lead to **dummy variable trap** or **curse of dimensionality** (i.e. number of instances need to grow exponentially with number of features). It is not recommended to use OneHotEncoder with Tree based algorithms. Python provides **sklearn.preprocessing** package for the same. **get_dummies()** function of **pandas** package is a straightforward and easier way for the same.

III. **Discretisation/Binning**: Transforming continuous variables into discrete variables brings into non-linearity and thus improves the fitting power of model, minimizing the impact of extreme values and preventing overfitting possible with numerical variables.

IV **Skewed data:** It is necessary for the data distribution to be in range [−0.5, 0.5]. However, it is common for health care data to be distributed unsymmetrically with a long tail of high values. To handle right skewed data, a particular power of the data or log transform is used to bring it to near normal distribution.

3.2 Feature Extraction

When the data to be processed through an algorithm is too large, it's generally considered redundant. Analysis with large number of variables is computationally expensive, therefore we should reduce the dimensionality of these types of variables.

3.3 Feature Selection

In healthcare, accumulating data is a costly aspect [13]. Even, there is chance of data overfitting the model when number of observations is less and need for significant computation time when number of features is more. Hence, if machines could extract most informative features, the cost overhead on patients would reduce tremendously. Feature Selection is essential for simpler, faster, more reliable and robust ML models. Aim is to maintain accuracy and stability, improve runtime and avoid overfit. A feature selection technique benefits with redundant or irrelevant data which can be removed without much loss of information. Feature Selection algorithms are Filter based, Wrapper Based [2], Embedded [14] and Hybrid [14]. Python provides **feature_selection** module for feature selection.

I. **Filter based** methods are further categorized as **Basic**, Multivariate and Statistical. Filter methods rank features independent of the relationship among them. There are various measures in Filter based methods as Correlation based and Information Theory. The first step is to remove **constant** information which provides no / minimal information since it has same value for all instances of the feature. This can be checked by checking the variance of feature values, if the variance is 0 then the feature values are redundant. There is possibility of feature values being **quasi-constant** i.e. the variance < 0.01. Similarly, post one hot encoding of large datasets or dataset with lots of categorical values, there is chance of **duplicate** rows. Hence it is required to transpose the dataset and perform same operations as for constant and quasi constant columns.

   a. **Correlation** [6]: Next step should be to identify correlated features since effective results are appreciated if features correlate highly to the target and are uncorrelated to each other. Pearson Correlation Coefficient (PCC) [−1,

1] and Mathews Correlation Coefficient (MCC) perform the task. Coefficients closer to -1 designate strong negative relationship while coefficients closer to 1 designate strong positive relationship. Correlation is demonstrated through Correlation matrix. Correlated features do not necessarily affect the model performance (trees, etc.), but high dimensionality does and too many features hurt model interpretability. So, it's always better to reduce correlated features. Pearson Correlation is sensitive only to linear relationship and can be viewed through **Heat Map**. For Pearson correlation, it is recommended to drop features with values close to 0.

b. **Statistical** methods are fast but do not capture redundancy among features. These methods assign a score to each feature and thus rank them for inclusion or exclusion. Fisher score [7] and Chi square test can be used to measure the dependence among features. These methods often consider the feature independently and hence are univariate. Similarly, ANalysis of VAriance (ANOVA) parametric test identifies dependence among continuous variables.

c. **Information theory** is used extensively as it can measure nonlinear relationship among features. Mutual Information (MI) through Information Gain is used a lot in literature to identify how much knowing one variable reduces the uncertainty of other. MI measures similarity among features but is inconvenient to compute for continuous variables.

II. **Wrapper** methods are greedy, computationally intensive and exhaustive. They detect interaction among features by looking for subsets using Step Forward [7] or Step Backward [7]. Wrapper methods result in best feature subset for that particular type of model. Being exhaustive search, it builds a model and evaluates the subset for optimality through the score. This process is repeated until the performance of the model starts decreasing or increasing or till pre-determined number of features are extracted. Wrapper methods consider different combinations of selected features, and compare these to other combinations. Different combinations can be tried using **Relief**, **Gain Ratio** and **Entropy**. They detect the possible interactions between variables. The search process may be methodical, stochastic or may use heuristic such as **Best-First Search**, **Random Hill-Climbing** algorithm, or **Forward** and **Backward pass** respectively to add and remove features. **Boruta** works as wrapper around RF by finding all informative features than finding a subset of features on which some classifier has a minimal error. GA selects optimal features to fit in the objective function for SVM classifier [2]. GA wrapper around NN has shown accuracy of 93.85% on Z-Alizadehsani dataset [8]. GA wrapper around Adaboost onto Parkinson disease dataset identified 7 important features and gave 98.28% accuracy whereas GA on Bagging returned 10 features with accuracy 96.55% [9]. kMeans and GA have also been used for dimensionality reduction [11].

III. **Embedded** methods use a wrapper [14] to consider interaction between feature and model but doesn't build a different model each time a different feature subset is picked. Embedded method is faster and cheaper than Wrapper and more accurate than Filter. It uses importance of features by identifying node impurity. The method is constrained to limitation of the associated algorithm. Types include **Lasso** Regularization, **Decision Tree** (DT), Random Forest and Gradient Boosted Trees **derived importance**. Least absolute shrinkage and selection operator (LASSO) regularization adds penalty on different parameters to reduce their freedom making the model fit noise in training data and thus generalize well on test data. We can ascertain that if penalty is too high, important features are dropped and the performance of model drops. For regression, the coefficients of predictors are proportional to how much it contributes to the target variable. These methods work under the assumption that there is a linear relationship between explanatory and predictor variable, and explanatory variables are independent, normally distributed and scaled. Thus, the features whose coefficients are more than mean of all coefficients are selected. The variants include L1 (Lasso), L2 (Ridge) and L1/L2 (Elastic Net). L1 might shrink some parameters to zero but in L2, parameters never shrink to zero but approach it. **Bolasso**, an improvement to Lasso bootstraps samples. For Tree derived importance algorithms, a feature is more important if it reduces the impurity more, example being **Regularized Random Forest**.

IV. **Hybrid** method can be employed to utilize advantages of both filter and wrapper methods. **Recursive Feature Elimination** (RFE) is commonly used with SVM or RF to repeatedly construct a model and remove features with low weights. RFE is a greedy optimization algorithm which repeatedly creates models and aims to find the best performing feature subset by keeping aside the best or the worst performing feature at each iteration. RFE ranks features according to the order of their elimination.

It has been proven that Embedded methods have better accuracy than Filter [13].

4 Machine Learning Models to Classify Healthcare Data

Machine Learning (ML) algorithms build a mathematical model based on sample data. ML tasks are classified as Supervised Learning and Unsupervised Learning. In supervised learning, the training data is labelled and the response variable may be discrete/qualitative (for classification task) or continuous/quantitative (for regression task). Machine Learning for Healthcare diagnostics is a classification task where the dependent variable may be split into binary or multiple classes. Below are discussed several ML algorithms which have proven to give good diagnostics for Healthcare.

4.1 *Logistic Regression (LoR)*

Logistic Regression (LoR) [15] is a method for classifying data into dichotomous outcome. It computes the probability of an event occurrence utilizing logit function where **logit = log(p/1 − p) = log (probability of occurrence/probability of event not occurrence) = log (Odds)**. Estimation is done through Maximum Likelihood where coefficients of model provide information about importance of input features. Regularization prevents models from overfitting the training data. LoR requires quite large sample sizes. Figure below demonstrates sigmoidal curve separating data into 2 classes.

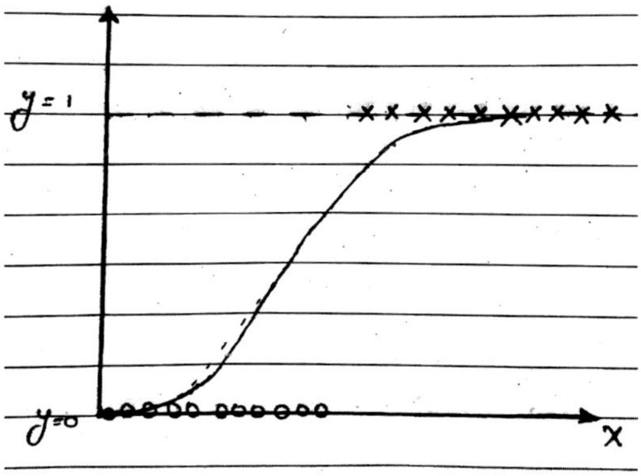

4.2 *Neural Networks (NN)*

Neural Networks (NN) are inspired by how the brain works [1, 4, 10, 12]. Artificial network of neurons (produce output by receiving input) allow machine to learn and fine tune itself by analysing new data. Activation function is used to turn input to output. e.g. sigmoid function. To compute the optimal values of parameters, a cost function is used. The network is run multiple times to make the model improve. Backpropagation (BPNN) algorithm helps to learn parameters for a NN. Gradient Descent algorithm or Adam optimizer is used to find optimal parameters [3, 12]. A NN with more number of parameters is prone to overfitting, thus regularization can prevent it. ANN can be built by importing the required **Keras** library and initializing the architecture of ANN to be sequential, which means stacking all layers, one on top of the other, sandwiching them and creating the ANN structure. Input layer has input = # of features, output = (# of features + 1)/2. Any hidden layer can have same # of input and output variables = (# of features + 1)/2. For the second layer, there is no need to add input nodes because the structure of the network is set to be feed forward, which means the outputs from one layer are used as the inputs for the next layer. Output layer has input = (# of features + 1)/2 and output = 1; a binary variable depicting diseased or not. Rectified Linear Unit (ReLU) activation function can be used in hidden layers and Sigmoid in output layer which provides a prediction between 0 and 1 where the value if greater than 0.5 means diseased else healthy. One can use **MultiLayerPerceptron** (MLPClassifier) class of **sklearn.neural_network** to build a supervised learning NN. Figure below demonstrates input, hidden and output layers of MultiLayerPercepetron neural network.

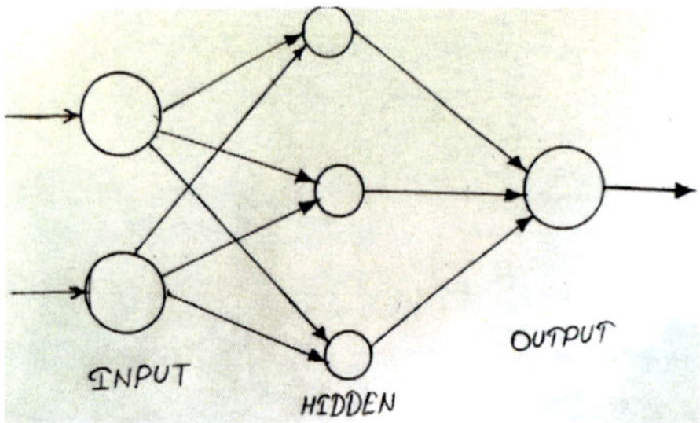

4.3 K Nearest Neighbor (KNN)

K Nearest Neighbor (KNN) or Instance Based Learner (IBL) is a lazy learner that can easily adapt onto unseen data [1, 6]. It is used in statistical estimation and pattern recognition as a non-parametric technique. For continuous variables, the distance measure used is Minkowski $= \left(\sum_{i=1}^{k} (|x_i - y_i|)^q\right)^{1/q}$ and for categorical variables, Hamming. Training set needs to be standardised before computing distance measure. Optimal value of k is between 3 and 10.

4.4 Support Vector Machine (SVM)

Support Vector Machine (SVM) [4, 6, 10, 12, 15] also termed as Large Margin Classifier constructs a hyperplane or decision boundary as shown in figure below. It transforms the original training data into higher dimension. For small number of features, Linear SVM can be used and for large number of features, Kernel trick is introduced. Training involves minimisation of the error function. Radial Basis Function (RBF) is by far the most popular choice of kernel types used in SVMs.

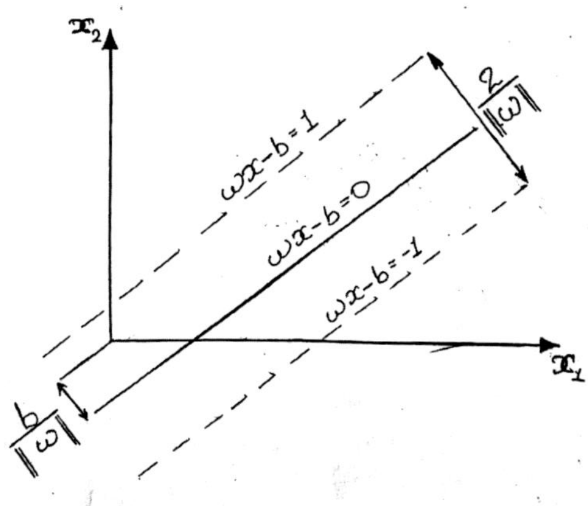

4.5 Decision Tree (DT)

Decision Tree (DT) [3, 4, 10] non parametric classifiers are easy to interpret and visualize and can easily capture non-linear patterns. It requires no normalization. DT are also biased to imbalanced dataset, so it is recommended to balance the dataset before creating the model. It divides the working area into subparts and checks for purity of class separation through Entropy and Information Gain or Gain Ratio or Gini Index. These metric are used in ID3, C4.5 (extension of ID3) and CART forms of Decision Tree respectively. It can be used for Missing Value Imputation and is also suitable for Feature Selection but DT can overfit noisy data which can be reduced by using Ensembles.

4.6 Ensemble

Learners [12] combine diverse set of algorithms to improvise stability and predictive power of the model. Most of the time, basic models do not perform well because they either have a high bias or they have too much variance to be robust. There is bias variance tradeoff if we alter the complexity of the same model. High bias, low variance model is consistently inaccurate whereas low bias, high variance model is accurate but inconsistent. What if we combine the models? The main challenge is to choose base models that make different kind of errors. There are 3 basic kind of ensembles:

I. **Bagging** (Bootstrap **Aggregation**) combines equally weighted results of homogenous base classifiers having high variance. Majority **Voting**/soft voting is used for combination in case of classification problems. Bootstrapping means every time one picks a sample; it is to be kept back before picking another sample. High variance for a model is not good, since its performance is sensitive to the training data provided. By increasing the size of the training set, one can decrease variance but can't improve the prediction power of model. So, even if more training data is provided, the model may still perform poorly. Hence Bagging improves stability and **overcomes overfitting**. Bagging samples with replacement whereas **Pasting** samples without replacement. All kinds of variables as categorical, interval-scaled, real, continuous, and binary can be handled. **Random Forest** (RF) an extension over bagging works on the principle of *wisdom of crowd*. RF is an ensemble of DT and is generated using a random selection of features at each node for determining the split. During classification, each tree votes and the most popular class is returned. Bagging is effective with limited data. **Bagged DT** and **Extra Tree** are other Bagging algorithms. Bagging uses same greedy algorithm to create each tree, thus very similar split points are chosen in each tree making them correlated even though each base model has been built independently. Bagging is suitable for complex models.

II. **Boosting** builds a model from training data and then creates the second model to correct errors of the first model by weighing it more in order to modify a weak learner to become better. Boosting trains weak classifiers with 50%–60% accuracy e.g. (DT with max_depth $= 1$, logistic regression linear classifier) and adds them to final strong classifier by weighing and thus is expensive, slow and unscalable. The misclassified samples gain weight and correctly classified ones lose weight. Algorithm learns from misclassified examples. It uses subsets of the original data to produce a series of averagely performing models and then "boosts" their performance by combining them together using a particular cost function (=majority vote). Thus, Boosting **decreases** model's **bias**. Boosting can lead to overfitting so we need to stop at right point. **Adaboost** [5, 16] / Additive modelling with base estimator as RF performs even better than RF and has linear complexity. The base model is trained on entire data. Model pays attention to training instances that previously underfitted and uses Stagewise Additive Modelling using a Multiclass Exponential Loss function. Difficult to classify instances are weighed more. AdaBoost is sensitive to noisy data and is highly affected by outliers because it tries to fit each point perfectly and is slow. **Gradient Boost Machine (GBM)** sequentially adds predictors each correcting its predecessor to minimize loss function (Mean Square Error (MSE) or F_1 score utilizing Gradient Descent. It sequentially modifies a weak learner to become better utilizing multiple core on CPU. Hyperparameter tuning is required to avoid overfitting. GBM are slow and hence not very scalable because of sequential model training hence e**X**treme **G**radient **B**oosting (**XGBoost**), an implementation of gradient boosted DT is designed for speed and performance [16]. It implicitly handles missing values. XGBoost is more regularized model to control overfitting than GBM. GBM and XGBoost eliminate variance but have overfitting issue. It is a se**quential** ensemble which tries to add new models that do well where previous models lack. It is suitable for low variance, high bias models which are consistently inaccurate. Variants include **Light GBM** (LGBM), **CatBoost** and **Regularized Greedy Forest** (RGF).

III. **Stacking** [5] combines heterogenous models and learns them in parallel to increase the predictive force of a classifier. **Bagging** and **Boosting** are normally used inside one algorithm (homogenous models), while **Stacking** is usually used to summarize several results from different algorithms (heterogenous models). **Stacking/**Stacked Generalization ensemble requires weighing the models. It is an Ensemble of ensembles. It performs better than individual models due to its smoothing nature. It is able to highlight each base model where it works best and discredit each base model where it performs poorly similar to Boosting. Stacking is fitting an ensemble with CV whereas **Blending** uses a holdout set from train set to make predictions.

Implementation can be done using **mlens** library of python for memory efficient parallelized ensemble learning.

5 Diagnosing Model Performance

When we try to predict the target variable using any ML technique, the main causes of difference in actual and predicted values are **noise, variance, and bias.** Bias means how far the prediction is from the target, which may be due to erroneous assumptions. $Bias = E\left[f(x) - \hat{f}[x]\right]$. Variance is how a random variable deviates from its mean in squared units. For high bias models, extracting more training data doesn't help as it causes both training and cross validation cost to be high and for high variance models; the cross-validation cost continues to decrease with increase in training data size. Thus, getting more training examples and trying smaller set of features fixes high variance whereas polynomial features fixes high bias. A **learning curve** plots the training and cross validation score as a function of the frequency of training examples taken into consideration. Gap between the two curves determines the interpretation of this model in the bias-variance landscape. In case, the training and cross validation score is well below the desired score, we can interpret the model to be having a low bias, hence underfit. In that case we might think of inducing new features and decrease the degree of regularization. In case, the two curves are symmetric around the desired score curve separated by a considerable gap, then it is the case of low bias, and high variance; the model is accurate but inconsistent. In this case we might think of adding more examples within the dataset and increase the order of regularization. Underfitting means high bias i.e. high train and test cost whereas overfitting has low train but high-test cost.

How to know if the model will predict the future well? Failing to carefully distinguish between performance on a training set and performance on a test set can have serious consequences since different learning machines will pay attention and ignore different characteristics of a data. A learning machine can be 99% accurate in memorizing the training data yet have very poor performance on a test data. One commonly used approach is called the "train-test" method. The model is trained on a number of different subsets of data and expected future performance is given by mean and standard deviation of the validation results on the validation data. A final check is done with the test data since test data results have *not* been used to tune the model. A typical breakdown for small data sets might be 70%:20%:10% for train:validation:test resulting in assessing model performance on small samples. **sklearn.model_selection** module can be utilised for the same. The performance of learning machine on the test set will generally tend to be lower than its performance on the training set since learning is not effectively extracting statistical regularities common to both the training data and test data. This problem can be addressed using **k fold CV** [7]. To reduce the variability, multiple rounds of CV averages performance, so that both memorization and generalization performance can be estimated. A variation to kfold CV is leave one out CV (LOOCV)/Jackknife. It is a procedure to obtain unbiased prediction and minimize overfitting risk [16]. Further, Stratified K Fold reduces both bias and variance.

The power of ML algorithms lies in appropriate allocation of the values of hyperparameters. Instead of assigning random values through trial and test or trying all possible combinations which may be exponentially expensive; Genetic Algorithm, a form of Evolutionary Computation has worked effectively for Hyperparameter Optimisation. GA choses fitness function to evaluate the quality of solution in terms of classification accuracy of the algorithm with varying parameters. If we can find best set of hyperparameter values then results could be appreciable. GA can be utilized with Roulette wheel selection and SVM for classification with single point crossover followed by mutation to preserve genetic diversity and elitism. Enhancement in accuracy is seen from 83.70% to 88.34% [2]. Researchers have worked towards increasing performance of NN from 84.62% to 93.85% through enhancing initial weights utilizing GA with 10-fold CV [8]. Feature vector chosen includes accuracy and number of selected features. GA-NN shows enhancement in accuracy when compared to Grid search. Other hyperparameters as no. of hidden layers, number of nodes per layer [12], choice of optimiser, learning rate and momentum can also be optimised [8]. Similarly, GA is able to optimize the number of base classifiers in an Ensemble [17]. Optimisation for both accuracy and diversity may naturally lead to small number of classifiers being selected. **TPOT** is a Python Automated Machine Learning tool that optimizes Machine Learning pipelines using Genetic Programming. Grid search and Randomized search fulfil the objective at the cost of computation time. Even in between combinations are not tested. Usually when deploying an algorithm, we make a range of choices in regards to selecting the hyper-parameters. As in the case of k-NN, this can be the value of k to configure; in case of a RF, it can be the density and depth of the ensemble of DT. **Validation curve** indicates how much a model generalizes as a function of the hyperparameter value.

6 Experimental Setups

To test above knowledge critical care disease datasets available online on UCI ML repository and Kaggle have been used. Several datasets have been chosen w.r.t. different types and number of independent variable, missing values, binary/multiclass prediction, etc.

Heart disease is a major health issue for humans. Following are several heart disease datasets available online:

Cleveland (303 instances) [2, 4, 7, 8], **Hungarian** (294 instances) [4, 7, 8], **Switzerland** (123 instances) [4, 7, 8] and **Long Beach V.A.** (199 instances) [4, 8] heart disease datasets have 76 attributes each of which 14 are useful. The predictor variable is multivalued in Cleveland, Switzerland and V.A. and has integer value from 0 (no presence) to 4. There are missing values in Hungarian, Switzerland and V.A.
Z-AlizadehSani [4, 8] dataset comprises of 56 attributes of 303 patients. The data has been collected from Cardiovascular Imaging Department, Rajaei Cardiovascular, Medical & Research Center, Iran University, Tehran, Iran. The classification is diagnosis of the patient as having Coronary Artery Disease (CAD) or not. The features are arranged in four groups: demographic, symptom and examination, ECG, and laboratory and echo.
Framingham [18] heart study dataset has 16 features of 4240 respondents and predicts Ten Year Coronary Heart Disease (CHD). It has 1% missing values.
Statlog [5] heart disease dataset has 270 instances with 13 attributes each and has no missing values.
Arrythmia [3, 6] heart disease dataset with 452 instances and 280 features has been used for testing on feature selection algorithms. It has missing values and is multivalued.

Pima Indian Diabetes dataset [11, 15, 17] from National Institute of Diabetes, Digestive and Kidney Diseases has 768 female patient record at least 21 years old with 9 features each. All features are numeric. It has been shown that 50% of patients with diabetes are not properly diagnosed.

Breast Cancer Wisconsin [6, 10] (Diagnostic) dataset with 10 numerical features describes characteristics of the cell nuclei present in the digitized image of a fine needle aspirate (FNA) of a breast mass in 3-dimensional space of 699 patients. It is the most common female cancer in the world with an estimated 1.67 million new cancer cases diagnosed in 2012 and with an annual incidence of approximately 144,000 new cases of breast cancers in India. It has missing data in 16 instances.

Lung Cancer [13] contains 22,401 samples with 164 features (87 lung adenocarcinomas and 77 adjacent normal tissues).

Chronic Kidney disease [16] is a menace that is affecting 10% of the world population.

Parkinson's disease (PD) [5, 6, 9] dataset from UCI has 22 features of 188 patients (107 men and 81 women) with ages ranging from 33 to 87 at the Department of Neurology, Istanbul University. It is estimated that 7–10 million are suffering from PD.

To experiment on above datasets, several Python libraries have been used. **NumPy** is core library for scientific computing. DataFrame is a two-dimensional object similar to a spreadsheet and the most commonly used **Pandas** object. One can use comma separated value (**csv**) file format to read/write data. describe () function of pandas library returns quick stats. **Matplotlib** visualization library in Python can be used to produce publication quality figures. Functions include bar chart, scatter chart, box-plot and histogram. **Scikit-learn** based on **Scipy** features number of supervised and unsupervised learning algorithms. SciPy is an optimization library. **Anaconda** is a third-party scientific distribution; providing tools (interactive python development environment) like **Spyder**, Jupyter notebook etc.

7 Results and Discussion

7.1 Data Preparation (EDA)

Data Wrangling task is the first block in Machine Learning block diagram as shown in Fig. 1 below. Exploratory Data Analysis (EDA) for Cleveland heart disease dataset shows 226 duplicate rows, 1 duplicate row for Hungarian and V.A. each. Cholesterol feature can be deleted in Statlog and Switzerland heart disease dataset and Exertional C.P. in Alizadeh Sani dataset. Cleveland heart disease dataset had missing values, so those rows were removed. Since this dataset is multiclass, patients with different risk factors were considered as diseased to convert multi class O/P to binary class. EDA for Cleveland heart disease dataset as shown in Fig. 2 demonstrates that the disease is not biased to certain age group or gender.

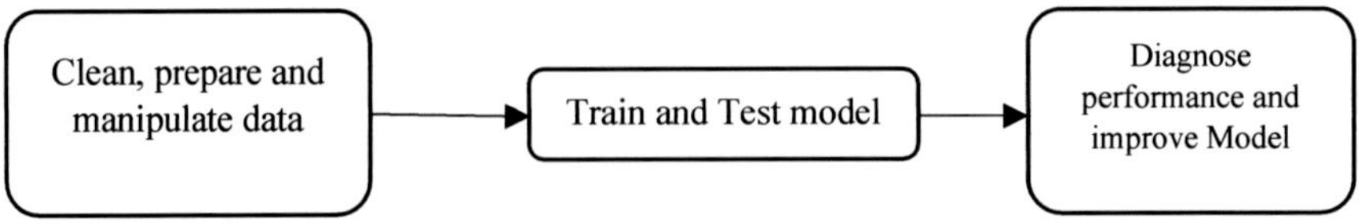

Fig. 1 Machine learning block diagram

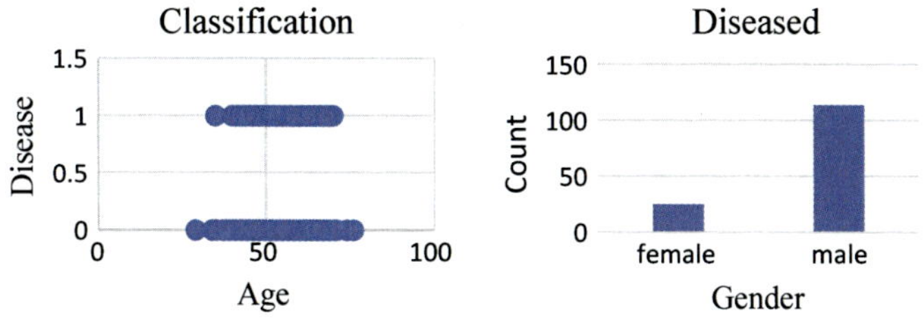

Fig. 2 EDA for Cleveland heart disease dataset

7.1.1 Feature Cleaning

On analysis of Pima Indian Diabetes dataset, missing values were found in skin thickness and insulin in 30% and 49% of rows. But domain knowledge tells that these two features can't have 0 value. Hence, these should be treated as missing values. Similarly, outliers can be detected through Box Plot/Whisker's Plot in almost all features (no. of pregnancy, glucose level, B.P., insulin level, BMI, diabetes pedigree function and age) and are shown in Fig. 3.

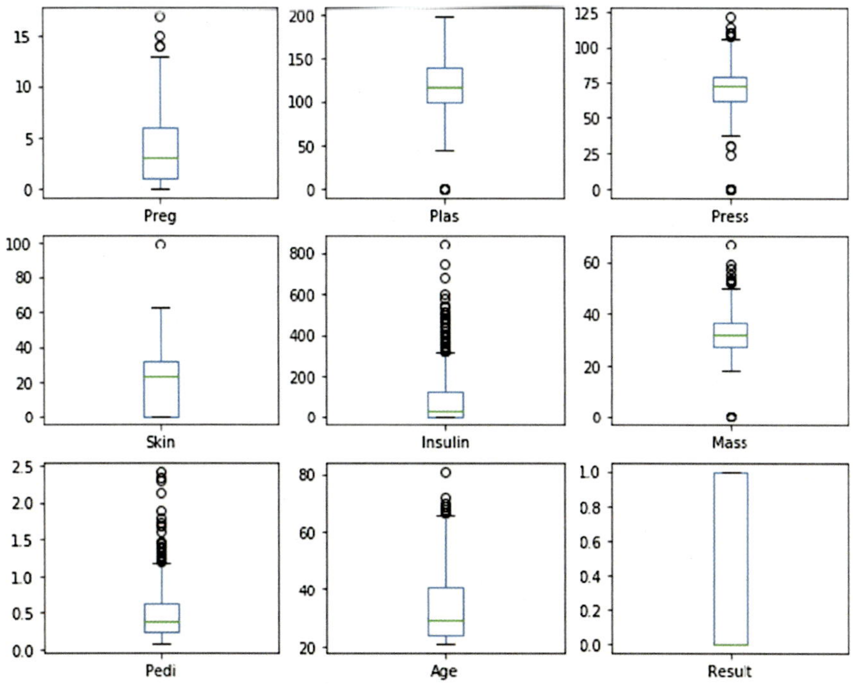

Fig. 3 Outlier detection for Pima Indian Diabetes dataset

Value	Count	Frequency (%)
0	3596	84.8%
1	644	15.2%

Fig. 4 Data Imbalance in Framingham Heart Disease dataset

7.1.2 Missing Data

The experimentation involved choosing Statlog heart disease dataset with no missing values, randomly deleting values and imputing through SimpleImputer, kNN and MICE. SimpleImputer utilises mean for numerical and most frequent for categorical and reduces accuracy considerably, no change in accuracy is observed with kNN and there is enhancement in accuracy with MICE.

7.1.3 Data Imbalance

Framingham Heart Disease dataset is imbalanced. The predictor variable is TenYearCHD. Figure 4 demonstrates class imbalance for Framingham heart disease dataset.

Ignoring imbalance gives results with low sensitivity and high specificity. Under sampling the majority class by deleting rows provides loss of relevant information whereas Oversampling minority class has risk of overfitting due to dependence in test and train data. Hence SMOTE (Synthetic minority oversampling technique) was employed which gave more than 20% enhancement in accuracy.

7.2 Feature Engineering

Scaling doesn't always enhance performance. kNN works with same scale data since uses Euclidean distance. SVM computes distance between separating hyperplane and support vector, hence same scale is required. NN requires scaling to equally distribute importance of each input. Scales of I/Ps do not matter much in LoR, DT, RF and NB. Experimentation was performed on Framingham dataset, balanced through under sampling and group imputation which prove the above statement.

7.2.1 Skewness

Figure 5 shows right skewed features through Histogram from Pima Indian Diabetes dataset. Histogram estimates distribution of data.

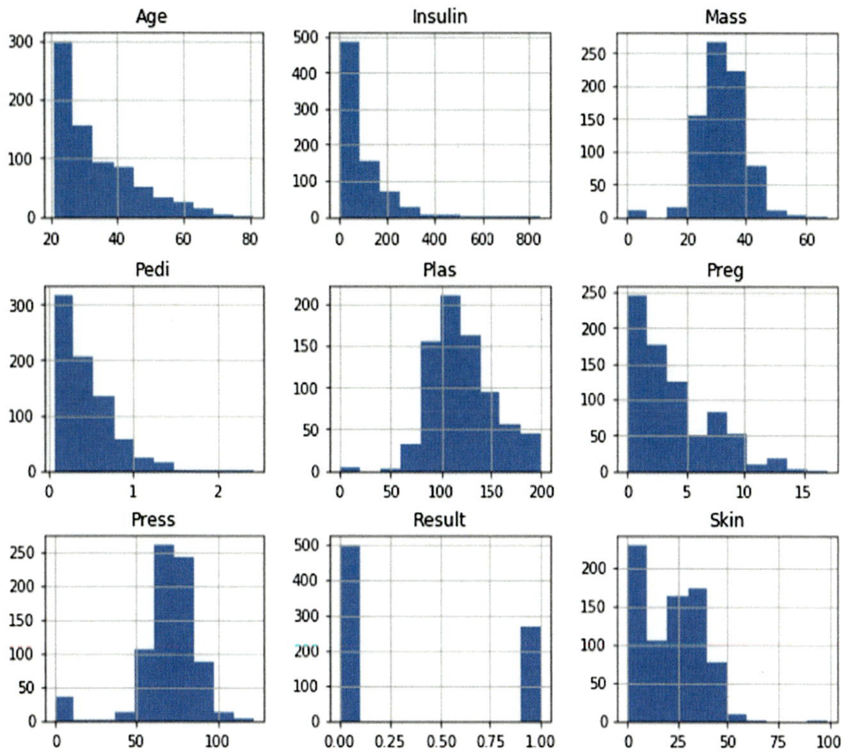

Fig. 5 Skewed features in Pima Indian Diabetes dataset

7.2.2 Feature Selection

Applying Feature selection (Filter followed by Embedded and followed by SFS Wrapper) onto Arrythmia dataset with 280 features reduced it to 7 with more than 15% enhancement in accuracy. Even for Z-AlizadehSani dataset with 55 features reduction came to 34 features with 3% enhancement in Area under the Curve (AUC) utilizing filter-based feature selection [13]. Feature selection through GA when applied to Framingham Heart Disease dataset provided enhancement in accuracy by 5% through SVM classifier and 16% through NN Classifier.

7.3 Building Model

Experimentation is done on Pima Indian Diabetes dataset with Bagged DT (CART), RF, ExtraTree, AdaBoost, GradientBoost and VotingEnsemble (of LoR, CART and

SVM). All achieve almost similar CV score. MLP Classifier and SVM show high sensitivity and F1 score onto Framingham heart disease dataset post oversampling with SMOTE, removing missing values through kNN imputation, scaling and splitting.

7.4 Evaluating Model Performance

Cross validation enhances performance. Overfitting occurs due to selection of model onto same data for training and testing. Thus, it is advisable to split the data into training and testing set. But a static split is not using your data efficiently. By varying what you learn on and what you're tested on, you generalize better. Experimentation was performed on Framingham dataset, balanced through under sampling and group imputation. Accuracy enhanced by 3%–6%. Figure 6 demonstrates Validation curve where the training and CV score with varying values of gamma hyperparameter of SVM classifier are shown onto Pima Indian Diabetes dataset. Results demonstrate an optimal value of the hyperparameter.

Figure 7 demonstrates Learning curve w.r.t. the train and CV score for SVM RBF kernel and NB onto Pima Indian Diabetes dataset. Results demonstrate an optimal size of training examples to achieve balance in train and validation scores. In first figure, for small amount of data, training score exceeds validation score. Adding more data leads to generalisation. In second figure, there is no point in adding more data, since curve converges to low score. **model_selection** class of **sklearn** package can be utilised for plotting learning curves.

Hyperparameter tuning of ensembles provide good results onto Breast Cancer dataset with Stratified KFold when applied with NN (KerasClassifier), SVM, XGBoost, CatBoost, LGBM, RGF and AdaBoost ensembles. Ensembles show much

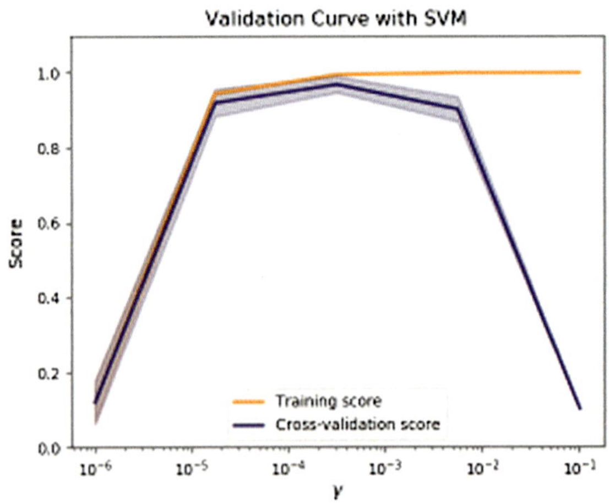

Fig. 6 Validation curve onto Pima Indian Diabetes dataset

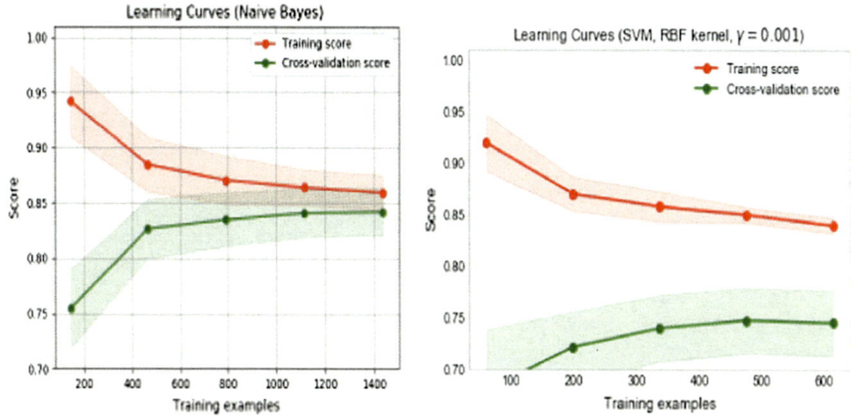

Fig. 7 Learning curve for Pima Indian Diabetes dataset

better performance than NN and corresponding hyperparameters are identified. Utilising GA for hyperparameter tuning SVM enhanced accuracy by 1% for RBF kernel parameters C and γ. Similarly, the accuracy enhanced through GA hyperparameter tuning by 2% utilising adam optimiser. No. of hidden layers, no. of nodes in hidden layers, learning rate and momentum were optimised.

8 Conclusions

The information used by healthcare professionals to diagnose a critical care patient can be used to create machine learning models that would further help medical practitioners better understand their patient's health to make better decisions. Thus, a disease prediction or diagnosis system will "learn" from labeled data to probabilistically predict the likelihood of a patient having the disease.

For diseased instances, precision is the percentage when the patient had disease and the machine diagnosed diseased whereas sensitivity/recall is the percentage when the machine was correct for predicting disease among all diseased patients. Thus, high value of sensitivity is important in comparison to accuracy in case of healthcare; which is no. of times the machine is right. Sensitivity enhancement happens with kNN imputation and SMOTE sampling. Thus, for healthcare, Type II errors which reflect Accuracy Paradox are not advisable.

Hence, there is need of reputed hospitals to electronically store data and use it for relevant analytics which can aid patients as well as practitioners in timely and accurate diagnosis. Thus, research prospect lie in enhancing the generalization ability of ML algorithms especially for imbalanced classes and multiclass data.

References

1. S. Gupta, R.R. Sedamkar, Apply Machine Learning for Healthcare to enhance performance and identify informative features, in *IEEE INDIACom; 6th International Conference on "Computing for Sustainable Global Development"*, BVICAM, New Delhi, India, 13–15 Mar 2019
2. C.B. Gokulnath, S.P. Shantharajah, *An Optimized Feature Selection Based on Genetic Approach and Support Vector Machine for Heart Disease* (Springer Nature, Iran, 2018)
3. E.R.Q. Fernanded, A.C.P.L.F. de Carvalho, X. Yao, Ensemble of classifiers based on multiobjective genetic sampling for imbalanced data. IEEE Trans. Knowl. Data Eng. **14**(8) (2015)
4. F. Babič, J. Olejár, Z. Vantová, J. Paralič, Predictive and descriptive analysis for heart disease diagnosis, in *FedCSIS*, vol. 11 pp. 155–163, IEEE Catalog Number: CFP1785N-ART c 2017, Slovakia, https://doi.org/10.15439/2017f219. ISSN 2300-5963
5. R. Pari, M. Sandhya, S. Sankar, *A Multitier Stacked Ensemble Algorithm for Improving Classification Accuracy* (IEEE, 2018)
6. S. Mahendru, S. Agarwal, *Feature Selection Using Metaheuristic Algorithms on Medical Datasets* (Springer Nature, Singapore, 2019)
7. S.M. Saqlain, M. Sher, F.A. Shah, I. Khan, M.U. Ashraf, M. Awais, A. Ghani, *Fisher Score and Matthews Correlation Coefficient-Based Feature Subset Selection for Heart Disease Diagnosis Using Support Vector Machines* (Springer, London, 2018)
8. Z. Arabasadi, R. Alizadehsani, M. Roshanzamir, H. Moosaei, A.A. Yarifard, Computer aided decision making for heart disease detection using hybrid neural network-Genetic algorithm. Comput. Methods Programs Biomed. **141**, 19–26 (2017). (Elsevier ScienceDirect)
9. N. Fayyazifar, M. Samadiani, *Parkinson's Disease Detection Using Ensemble Techniques and Genetic Algorithm* (IEEE, Pakistan, 2017)
10. I. Chlioui, A. Idri, I. Abnane, J.M.C. de Gea, J.L.F. Alemán, *Breast Cancer Classification with Missing Data Imputation* (Springer Nature, Switzerland, 2019)
11. T. Santhanam, M.S. Padmavathi, Application of K-Means and genetic algorithms for dimension reduction by integrating SVM for diabetes diagnosis. Procedia Comput. Sci. **47**, 76–83 (2015). (Elsevier ScienceDirect)
12. Y. Khan, U. Qamar, N. Yousaf, A. Khan, Machine learning techniques for heart disease dataset: a survey, in *ICMLC*, ACM, China, 22–24 Feb 2019
13. S. Gupta, R.R. Sedamkar, Feature Selection to reduce dimensionality of heart disease dataset without compromising accuracy. Int. J. Comput. Trends Technol. (IJCTT) **67**(6) (2019)
14. X.Y. Liu, Y. Liang, S. Wang, Z.Y. Yang, H.S. Ye, Hybrid genetic algorithm with wrapper embedded approaches for feature selection. IEEE Access **6**, 22863–22874 (2018)
15. Z. Yang, Y. Zhou, C. Gong, Diagnosis of diabetes based on improved Support Vector Machine and Ensemble Learning, in *ICIAI*, ACM, China, 15–18 Mar 2019
16. A. Ogunleye, Q.G. Wang, XGBoost model for Chronic Kidney Disease diagnosis. IEEE/ACM Trans. Comput. Biol. Bioinform. (2019)
17. S. Fletcher, B. Verma, Z.M. Jan, M. Zhang, The optimized selection of base-classifiers for ensemble classification using a multi-objective genetic algorithm, in *2018 IEEE International Joint Conference on Neural Networks (IJCNN)*, Australia
18. H.A.G. Elsayed, L. Syed, An Automatic early risk classification of hard coronary heart diseases using framingham scoring model, in *ICC* (ACM, Cambridge, UK, 2017)

Further Reading

19. https://www.analyticsvidhya.com
20. https://www.datacamp.com

21. https://machinelearningmastery.com
22. https://www.superdatascience.com
23. https://www.elitedatascience.com
24. https://towardsdatascience.com
25. https://datasciencecentral.com
26. https://medium.com
27. https://adataanalyst.com
28. https://kdnuggets.com
29. https://researchgate.net
30. https://knowledgehut.com
31. https://data.world
32. https://github.com
33. https://stts.stackexchange.com
34. https://linkedin.com
35. S. Raschka, Python Machine Learning (Packt Publishing, 2015)

Ms. Shiwani Gupta was born in India in 1981 and has completed her B.Tech. in C.S.E. in 2003 and M.Tech. in C.S.E. in 2009 from Lucknow, U.P., India. She is currently pursuing Ph.D. in Tech. from Mumbai University. Her area of interest includes Machine Learning, Artificial Intelligence, Biometrics, Algorithms, etc. She has over 16 years of teaching experience. She has over 60 publications in reputed journals and conferences.

Dr. R. R. Sedamkar was born in India in 1967. He holds Ph.D. in Engineering. M.E. and B.E. degree in C.S.E. His area of interest includes Networking and Data Compression. He has guided over 50 U.G. and over 15 P.G. projects. He is currently guiding 7 research scholars. He has over 25 years of teaching experience and is currently working as Dean Academics in TCET Mumbai.

Artificial Intelligence in Medical Diagnosis: Methods, Algorithms and Applications

J. H. Kamdar, J. Jeba Praba and John J. Georrge

Abstract Artificial intelligence (AI) has evolved rapidly since the late 1980s. Increasing of healthcare datasets and its performance, the past two decades have seen an exponential progress in publications on AI. However, with the advent of increased computational power, availability of AI devices was increased. There are two main devices in AI, machine learning, where structured data (i.e. images, EP and genetic data) are analyzed and natural language processing, where unstructured data are analyzed. Both AI devices have been improved in great detail over the past two decades for its methods, algorithms, and applications. However, various attempts and new methods of AI have been used in recent years and few diseases such as cancer, nervous system disease, cardiovascular disease, liver disease, congenital cataract disease, etc. were potentially analyzed using AI. Now a day an advanced method called deep learning has initiated a boom of AI and great modifications of diagnostic medical imaging systems like endoscopic diagnosis, pathology and dermatology will be expected in the near future. Herein, the authors give a basic technical knowledge about popular methods, algorithms and applications in medical diagnosis which emerged in the past years.

Keywords Medical diagnosis · Deep learning · Machine learning · Genetic algorithm · Cancer · Nervous system disease · Cardiovascular disease · Liver disease

---

J. H. Kamdar and J. Jeba Praba are considered as Joint first.

J. H. Kamdar
ICAR-Directorate of Groundnut Research, Junagadh, Gujarat, India

J. Jeba Praba
Department of Computer Applications, Christ College, Rajkot, Gujarat, India

J. J. Georrge (✉)
Department of Bioinformatics, Christ College, Rajkot, Gujarat, India
e-mail: johnjgeorrge@gmail.com

V. Jain and J. M. Chatterjee (eds.), *Machine Learning with Health Care Perspective*, Learning and Analytics in Intelligent Systems 13,
https://doi.org/10.1007/978-3-030-40850-3_2

1 Introduction

The term artificial intelligence (AI) is well-defined as a stream of science and engineering. It involves computational understanding with the creation of artefacts that reveal such behavior [41]. AI techniques and tools have been used in medical applications/diagnosis for over four decades. The overall objective was to improve healthcare by helping physicians in improving their effectiveness, efficiency and coherence. Improvements in accuracy and availability of large data of AI techniques have rapidly increased AI's choice for solving problems in medical diagnosis. AI software played a important role in further implementation for medical use. Applications of AI span a wide and diverse range of fields in clinical diagnosis that's why we are now in the fourth decade of AI. The core advantages of AI are accuracy, cost efficiency and replication. There are mainly two AI techniques/devices available, the primary one is Machine Learning (ML) followed by Natural Language Processing (NLP) which is the second one. In the primary technique, structured data were analyzed like genetic, imaging and electrophysiological (EP). While, in NLP techniques, unstructured data were analyzed like books, clinical notes and medical journal [23]. Further, the machine learning techniques are clustered into two methods, classical and deep learning. In this chapter, the authors give an overview of AI devices, algorithms and applications in medical diagnosis which emerged in the last years.

2 Machine Learning Techniques (ML)

ML is one of AI's major branches and most emerging subfields. AI software follows various features of human brainpower, such as planning, learning, logic, comprehension, perception, interaction, and the skill to relocate and exploit data/objects [3]. AI uses number of tools to mimic human intelligence in these areas.

ML techniques take empirical data as an input and anticipate the structures of the data. Empirical data includes age, gender, physical examination, imaging, gene expression, EP, etc. To incorporate these data sets, three algorithms can be used i.e. The fist algorithm is Supervised Algorithm, linked on to the second as Unsupervised Algorithm, while the last one is Semi-supervised Algorithm which promotes learning. In supervised ML, always have known classified data sets. It is mostly used for predicting relationships between trait of interest and outcome. Supervised algorithm used more often in medical diagnosis hence it gives more relevant outputs. Widely used supervised algorithms are (SVM) Support Vector Machine, Network of Neural, and the intense mode of algorithmizing through learning. One of the easiest ML algorithms to understand is SVM [8]. It is developed for binary (two-class) issues and can be applied to multi-class classification of data sets. The prediction is made by sketch the line between the two clusters and then. SVM has been widely used in medical diagnosis (Fig. 1).

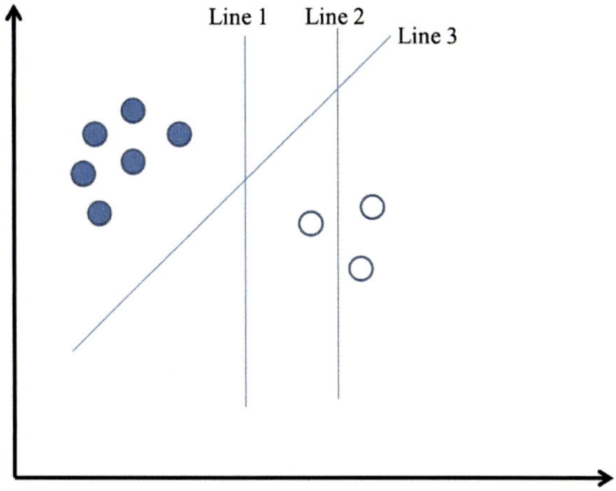

Fig. 1 Support vector machine (SVM) example

Neural networks [24] are one of the alternative ML methods which works finest when the line is not straight among two classes. A network that is neural denotes a trio of different layers: The dominant one is the layer of input, sheathed with another second layer of many, and finally displaying a layer of output (Fig. 2). Each node's inputs produce an output that is fed to the next node of network. The values of the traits as the initial inputs are fed into the network. The output defines classification of the traits. Ramesh et al. [37] reported that neural networks were the frequently used diagnostic tool. Further, another study shown that fuzzy logic–neural networks are the most frequently used AI technique. This can be intensively used in genetics, cardiology, radiology, and so forth [44].

Deep learning is an advanced version of the classical neural network technique. It has seen dramatic resurgence in the past six years due to increase in computational power and new data sets. More complex non-linear data sets can be analyzed by deep

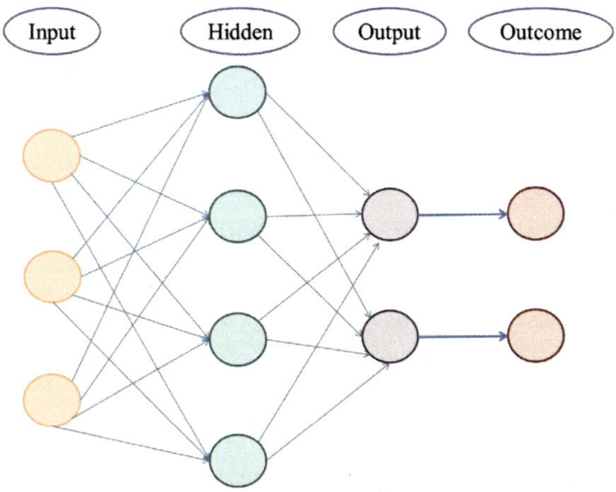

Fig. 2 Typical neural network configuration

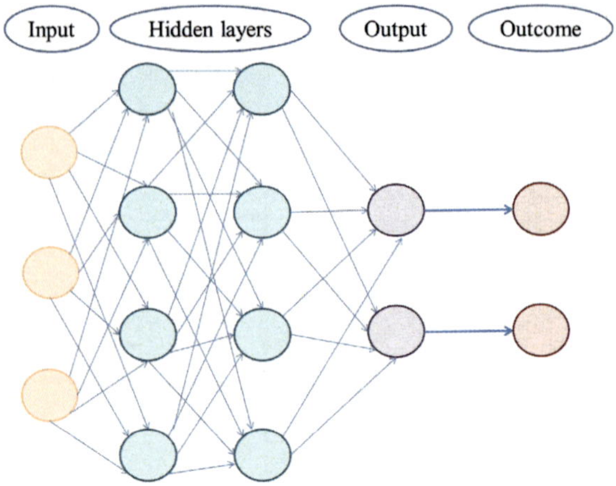

Fig. 3 Typical deep learning configuration with more than one hidden layer

learning. In recent years deep learning become very popular due to the growth of the size and ramification of data [38]. Compare to classical neural networks it uses more hidden layers (Fig. 3) and deployed more complex data with diverse structures [19]. In medical applications, there gives in many types of networks, and few vital one's are, Network of neural recurrent, accompanied by Network of neural deepness, Network of neural Convolution (CNN), and the Network of deep belief which was a most frequent methodology to develop deep learning of algorithms.

Recently, to assist disease diagnosis the CNN has been effectively deployed in the medical field with the help of popular software like, CNTK from Microsoft, TensorFlow from Google and Caffe from Berkeley AI Research. There are some cases in which the way of classifying the data is unacquainted. In this cases, unsupervised ML techniques can be used to conclude how to categorize the data. There are two popular unsupervised ML technique used to identify the most significant features that are used in classifying the data, (i) K-means clustering and (ii) Principal component analysis (PCA) [29, 35].

In K-means clustering, different clusters, or grouping can be used to discover the one where the points are nearest to each other in each group. While in case of PCA, it organized the data into a centered matrix. Individual row and column represent model of the data and value of a feature respectively. In centered matrix, the mean of the column is subtracted from the values in each column. It has been performed so that each column's mean become zero. Semi-supervised is a hybrid learning among unsupervised and supervised learning, which is appropriate for situations/data where the result for certain topics is incomplete [23].

3 Natural Language Processing (NLP)

The Natural language processing (NLP) is a platform of computational philology and AI promotiveness. It is the analytical process of creating a language that is natured to be realistic. In particular the medical forum and its literature contains enormous amount of printed study papers which apply NLP methods that lay interest in the usage of Electronic health records (EHR) system from which abundant information has been credited over the years. It includes electronic medical notes to advance the quality of health care through disease monitoring, evidence-based medicine and decision support. It is necessary to promote efficient NLP strategies to nurture texts clinically for tasks made rigid and reliable. In last 10 years, 1405 NLP based medical research publication was published with 18.39% average annual growth rate [6].

4 AI Resources

For those interested in using AI and its apps in a number of fields, a variety of resources are accessible online. These include internet software enabling investigators to perform AI experiments using their own information, statistical learning and logic inference software packages, open-source information mining systems, programming frameworks for AI methods, and businesses providing AI services for a free. Among them few selected tools are given in Table 1.

5 Application of AI in Medical Diagnosis

In last decade number of articles/books and data sets were available for diverse application of AI. Different AI algorithms either alone or with other methods were reviewed, and successfully deployed in medical diagnosis like, stroke, Alzheimer's disease, skin cancer, neurological diseases, acute ischemic stroke, etc.

The NLP was first introduced by Fiszman et al. [18] as a reading device with great flexibility for doctors to study the language of X-rays on the chest, and also used to treat victims of antiinfection. Followed by another remarkable medicionist Sweilam et al. [42] introduced least square support vector machine (LSSVM) as well as active set strategy by reviewing SVM in cancer diagnosis. In next two years, Orru et al. [33] used SVM to identify neurological as well as psychiatric disease imaging biomarkers. Swarm Optimized Wavelet Neural Network (PSOWNN) was used by Dheeba et al. [10] to identify breast defects in digital mammograms. Where, the inputs were images of mammographic that was finally dragonized and indicated as tumor. Thornhill et al. [43] recorded five quantitative shape descriptors of intraluminal carotid free-floating thrombus (FFT). This "signature" shape demonstrates the ability

Table 1 List of AI resources including the institution that developed the tools with source

Name of tool	Institute	Source
Alchemy: Open Source AI	Department of Computer Science and Engineering, UW, Seattle, WA, USA	alchemy.cs.washington.edu
AITopics	AAAI, Palo Alto, CA, USA	aitopics.org
Artificial Intelligence in Medicine Inc. (now INSPIRATA)	Artificial Intelligence in Medicine Inc., Toronto, Ontario, CND	www.aim.ca or www.inspirata.com
Artificial Medical Intelligence	Artificial Medical Intelligence, Eatontown, NJ, USA	www.artifcialmed.com
ELKI	Database and Information Systems, Ludwig Maximilians Universität München, Munich, GR	https://elki-project.github.io/
Encog Machine	Heaton Research, Inc., Chesterfield, MO, USA	www.heatonresearch.com/encog
Expertmaker	Expertmaker, San Francisco, CA, USA	www.expertmaker.com
Mathematica	Wolfram Research, Champaign, IL, USA	www.wolfram.com
MATLAB	MathWorks, Natick, MA, USA	www.mathworks.com
Neuroph	University of Belgrade, Belgrade, Serbia	neuroph.sourceforge.net
OpenCog	OpenCog Foundation, Rockville, MD, USA	opencog.org
Orange	Bioinformatics Laboratory, University of Ljubljana, Ljubljana, Slovenia	orange.biolab.si
RapidMiner	Rapid-I Inc., Burlington, MA, USA	www.rapid-i.com
Slicer	Brigham and Women's Hospital, HMS, Boston, MA, USA	www.slicer.org
Sourceforge	SourceForge.net, Geeknet Media. Dice Holdings, Inc., NY, USA	sourceforge.net/directory/scienceengineering/ai
Tools for Learning Artificial Intelligence	Laboratory for Computational Intelligence, UB, Columbia, Vancouver, CND	www.aispace.org
Video Games and Artificial Intelligence	Applied Games, Microsoft Research Cambridge, Cambridge, UK	research.microsoft.com/en-us/projects/ijcaiigames/
Weka 3	Machine Learning Group, UW, Hamilton, NZ	www.cs.waikato.ac.nz/ml/weka

to supplement standard lesion characterization with average precision in instances of suspected FFT for each method ranging from 65.2 to 76.4%.

In research it was Khedher et al. [27] who made use of the various combinations of SVM in company of different tools statistical in data, that detected the beginning stage of disease called Alzheimer. The following year, Hirschauer et al. [22] diagnosed Parkinson's disease that was based on types of motor, non-motor, and features of neuro imaging, through the help of an enhanced probabilistic neural network (EPNN). Gulshan et al. [21] used CNN to study the referable diabetic retinopathy (RDR) through the images of the retinal fundus in adults who were patients of diabetes. Rondina et al. [39] used Gaussian process regression to analyze anatomical stroke MRI images. They showed that better results were produced by extracting features through voxel patterns that represent lesion probability than by quantifying the lesion load per region.

In the last four years, Griffis et al. [20] acknowledged Naïve Bayes classification for a large group of stroke patients that observed and rectified heavy stroke lesions of people T1-weighted MRI scans. Farina et al. [16] defines and tests an offline man/machine interface that takes advantage of the spinal motor neuron and it's time to discharge. Esteva et al. [15] successfully conducted single CNN to identify how many people suffer from skin cancer? And the output was 129,450 data. In near course, an efficient establishment was done by Kamnitsas et al. [26] that featured a 11-layer deep, multi-scale, 3D CNN that was taken for the lesion segmentation in multimodal brain MRI and was given as a report of significance.

Long et al. [30] used DL with CNN to diagnose, stratify and treat congenital cataracts accurately. Miller et al. [31] used NLP to study EMR-based AE ascertainment and grading substantially for improved laboratory AE reporting. Castro et al. [5] applied NLP for implementing the EMR's power to obtain group of patients having intracranial aneurysms and its corresponding controls. In the year 2017, Afzal et al. [2], *put in action* the peripheral arterial disease (PAD) that included its major credentials from the notes of clinical narratives, through the aid of NLP. Further, these credentials were used to categorize the peripheral come ordinary arterial disease patients which resulted in more than 90% accuracy.

Zhong et al. [45] screened suicidal behavior of pregnant women using EMR based on diagnostic codes-structured and NLP-unstructured text. Afzal et al. [1] created an NLP-based algorithm to determine critical limb ischemia (CLI) from clinical narrative notes. For the identification of AIS patients, Kim et al. [28] used NLP and ML algorithms to classify the MRI radiology reports of the brain, into acute ischemic stroke (AIS) and non-AIS phenotypes. At present, Bacchi et al. [4] used combined approach of CNN and ANN for the prediction of Ischaemic Stroke Thrombolysis.

AI application is not restricted to human diagnosis only, few studies also reported in veterinary and agriculture sciences. The most important areas of veterinary and agriculture sciences are hybrid prediction, Cattle behaviors, disease identification, chewing patterns etc. ANN models for rumen fermentation pattern were developed in dairy cattle by Craninx et al. [9]. For medical database, Intelligent Rough Neural Network System (IRNNS) using Rough Set Theory (RST) and ANN was developed by Durairaj and Meena [12].

Santoni et al. [40] introduced and evaluated Gray Level Co-occurrence Matrix (GLCM) features. CNN learning for GLCMs can determine patterns with various variations, robustness to geometrical distortions and simple transformation of the cattle race in contrast, energy and homogeneity. Moving back in the past, Dongre et al. [11] developed a multiple layer, that brought into account, the neural network with back propagation. This network produced the mechanism of learning through ANN. Dutta et al. [13] reported a cattle behavior classification on models of ML that utilize a lot of sources, from magnetometer and tri-axis accelerometer collar sensors. To state, it is more a system that uses ML for data from swallowing signals of supplements like raw straw and weed that has been developed by Pegorini et al. [36] together with behavioral data, like rumination and idleness, for automated identification and classification of mastication patterns in calves. Morales et al. [32] reported an early warning SVM-based approach in commercial hens' egg production curves.

Disease detection is one of the most substantial concerns in agriculture. Few reports published/reported on the application of AI in agriculture indicate how recent and modern agriculture techniques are [25]. Chung et al. [7] reported automated and accurate detection in rice seedlings of pathogen Fusarium fujikuroi, which took less time than a naked eye examination. Pantazi et al. [34] developed a tool based on ANN/XY-Fusion, which detects the differentiated smut fungus Microbotyum silybum infected plants and healthy Silybum marianum plants during the growth of vegetation with 95.16% accuracy. For the real time control of thrips, Ebrahimi et al. [14] established a technique based on SVM image processing which differentiated parasites and thrips in strawberry. A strong classification was made to differentiate fresh and diseased leaves in the study of several plants and trees that was founded by Ferentinos [17] and was widely used as a CNN-based method.

6 Conclusion

In this chapter, we have performed a survey of AI-based research efforts applied in medical diagnosis. Our survey shows that deep learning is better than other techniques. Future research will focus on establishment of new algorithms, combinations, data sets to identified major diseases in early stages.

References

1. N. Afzal, V.P. Mallipeddi, S. Sohn, H. Liu, R. Chaudhry, C.G. Scott, I.J. Kullo, A.M. Arruda-Olson, Natural language processing of clinical notes for identification of critical limb ischemia. Int. J. Med. Inform. **111**, 83–89 (2018)
2. N. Afzal, S. Sohn, S. Abram, C.G. Scott, R. Chaudhry, H. Liu, I.J. Kullo, A.M. Arruda-Olson, Mining peripheral arterial disease cases from narrative clinical notes using natural language processing. J. Vasc. Surg. **65**(6), 1753–1761 (2017)

3. A. Agah, Introduction to medical applications of artificial intelligence, in *Medical Applications of Artificial Intelligence* (CRC Press, 2013), pp. 19–26
4. S. Bacchi, T. Zerner, L. Oakden-Rayner, T. Kleinig, S. Patel, J. Jannes, Deep learning in the prediction of ischaemic stroke thrombolysis functional outcomes: a pilot study. Acad. Radiol. (2019)
5. V.M. Castro, D. Dligach, S. Finan, S. Yu, A. Can, M. Abd-El-Barr, V. Gainer, N.A. Shadick, S. Murphy, T. Cai, G. Savova, Large-scale identification of patients with cerebral aneurysms using natural language processing. Neurology **88**(2), 164–168 (2017)
6. X. Chen, H. Xie, F.L. Wang, Z. Liu, J. Xu, T. Hao, A bibliometric analysis of natural language processing in medical research. BMC Med. Inform. Decis. Mak. **18**(1), 14 (2018)
7. C.L. Chung, K.J. Huang, S.Y. Chen, M.H. Lai, Y.C. Chen, Y.F. Kuo, Detecting Bakanae disease in rice seedlings by machine vision. Comput. Electron. Agric. **121**, 404–411 (2016)
8. C. Cortes, V. Vapnik, Support-vector networks. Mach. Learn. **20**(3), 273–297 (1995)
9. M. Craninx, V. Fievez, B. Vlaeminck, B. De Baets, Artificial neural network models of the rumen fermentation pattern in dairy cattle. Comput. Electron. Agric. **60**, 226–238 (2008)
10. J. Dheeba, N.A. Singh, S.T. Selvi, Computer-aided detection of breast cancer on mammograms: a swarm intelligence optimized wavelet neural network approach. J. Biomed. Inform. **49**, 45–52 (2014)
11. V.B. Dongre, L.S. Kokate, V.M. Salunke, S.M. Durge, V.N. Khandait, P.V. Patil, Artificial intelligence for prediction of standard lactation milk yield in Deoni cattle. Int. J. Livestock Res. **7**(11), 167–173 (2017)
12. M. Durairaj, K. Meena, A hybrid prediction system using rough sets and artificial neural networks. Int. J. Innov. Technol. Creative Eng. **1**, 16–23 (2011). ISSN: 2045-8711
13. R. Dutta, D. Smith, R. Rawnsley, G. Bishop-Hurley, J. Hills, G. Timms, D. Henry, Dynamic cattle behavioural classification using supervised ensemble classifiers. Comput. Electron. Agric. **111**, 18–28 (2015)
14. M.A. Ebrahimi, M.H. Khoshtaghaza, S. Minaei, B. Jamshidi, Vision-based pest detection based on SVM classification method. Comput. Electron. Agric. **137**, 52–58 (2017)
15. A. Esteva, B. Kuprel, R.A. Novoa, J. Ko, S.M. Swetter, H.M. Blau, S. Thrun, Dermatologist-level classification of skin cancer with deep neural networks. Nature **542**(7639), 115 (2017)
16. D. Farina, I. Vujaklija, M. Sartori, T. Kapelner, F. Negro, N. Jiang, K. Bergmeister, A. Andalib, J. Principe, O.C. Aszmann, Man/machine interface based on the discharge timings of spinal motor neurons after targeted muscle reinnervation. Nat. Biomed. Eng. **1**(2), 0025 (2017)
17. K.P. Ferentinos, Deep learning models for plant disease detection and diagnosis. Comput. Electron. Agric. **145**, 311–318 (2018)
18. M. Fiszman, W.W. Chapman, D. Aronsky, R.S. Evans, P.J. Haug, Automatic detection of acute bacterial pneumonia from chest X-ray reports. J. Am. Med. Inform. Assoc. **7**(6), 593–604 (2000)
19. I. Goodfellow, Y. Bengio, A. Courville, *Deep Learning* (MIT Press, Cambridge, 2016)
20. J.C. Griffis, J.B. Allendorfer, J.P. Szaflarski, Voxel-based Gaussian naïve Bayes classification of ischemic stroke lesions in individual T1-weighted MRI scans. J. Neurosci. Methods **257**, 97–108 (2016)
21. V. Gulshan, L. Peng, M. Coram, M.C. Stumpe, D. Wu, A. Narayanaswamy, S. Venugopalan, K. Widner, T. Madams, J. Cuadros, R. Kim, Development and validation of a deep learning algorithm for detection of diabetic retinopathy in retinal fundus photographs. Jama **316**(22), 2402–2410 (2016)
22. T.J. Hirschauer, H. Adeli, J.A. Buford, Computer-aided diagnosis of Parkinson's disease using enhanced probabilistic neural network. J. Med. Syst. **39**(11), 179 (2015)
23. F. Jiang, Y. Jiang, H. Zhi, Y. Dong, H. Li, S. Ma, Y. Wang, Q. Dong, H. Shen, Y. Wang, Artificial intelligence in healthcare: past, present and future. Stroke Vasc. Neurol. **2**(4), 230–243 (2017)
24. J.H. John, Neural network and physical systems with emergent collective computational abilities. Proc. Natl. Acad. Sci. U.S.A. **79**, 2554–2558 (1982)
25. A. Kamilaris, F.X. Prenafeta-Boldú, Deep learning in agriculture: a survey. Comput. Electron. Agric. **147**, 70–90 (2018)

26. K. Kamnitsas, C. Ledig, V.F. Newcombe, J.P. Simpson, A.D. Kane, D.K. Menon, D. Rueckert, B. Glocker, Efficient multi-scale 3D CNN with fully connected CRF for accurate brain lesion segmentation. Med. Image Anal. **36**, 61–78 (2017)
27. L. Khedher, J. Ramírez, J.M. Górriz, A. Brahim, F. Segovia, Alzheimer's Disease Neuroimaging Initiative, Early diagnosis of Alzheimer's disease based on partial least squares, principal component analysis and support vector machine using segmented MRI images. Neurocomputing **151**, 139–150 (2015)
28. C. Kim, V. Zhu, J. Obeid, L. Lenert, Natural language processing and machine learning algorithm to identify brain MRI reports with acute ischemic stroke. PLoS ONE **14**(2), e0212778 (2019)
29. S. Lloyd, Least squares quantization in PCM. IEEE Trans. Inf. Theory **28**(2), 129–137 (1982)
30. E. Long, H. Lin, Z. Liu, X. Wu, L. Wang, J. Jiang, Y. An, Z. Lin, X. Li, J. Chen, J. Li, An artificial intelligence platform for the multihospital collaborative management of congenital cataracts. Nat. Biomed. Eng. **1**(2), 0024 (2017)
31. T.P. Miller, Y. Li, K.D. Getz, J. Dudley, E. Burrows, J. Pennington, A. Ibrahimova, B.T. Fisher, R. Bagatell, A.E. Seif, R. Grundmeier, Using electronic medical record data to report laboratory adverse events. Br. J. Haematol. **177**(2), 283–286 (2017)
32. I.R. Morales, D.R. Cebrián, E. Fernandez-Blanco, A.P. Sierra, Early warning in egg production curves from commercial hens: a SVM approach. Comput. Electron. Agric. **121**, 169–179 (2016)
33. G. Orru, W. Pettersson-Yeo, A.F. Marquand, G. Sartori, A. Mechelli, Using support vector machine to identify imaging biomarkers of neurological and psychiatric disease: a critical review. Neurosci. Biobehav. Rev. **36**(4), 1140–1152 (2012)
34. X.E. Pantazi, D. Moshou, R. Oberti, J. West, A.M. Mouazen, D. Bochtis, Detection of biotic and abiotic stresses in crops by using hierarchical self-organizing classifiers. Precis. Agric. **18**, 383–393 (2017)
35. K. Pearson, LIII. On lines and planes of closest fit to systems of points in space. Lond. Edinb. Dublin Philos. Mag. J. Sci. **2**(11), 559–572 (1901)
36. V. Pegorini, L.Z. Karam, C.S.R. Pitta, R. Cardoso, J.C.C. da Silva, H.J. Kalinowski, R. Ribeiro, F.L. Bertotti, T.S. Assmann, In vivo pattern classification of ingestive behavior in ruminants using FBG sensors and machine learning. Sensors **15**, 28456–28471 (2015)
37. A.N. Ramesh, C. Kambhampati, J.R. Monson, P.J. Drew, Artificial intelligence in medicine. Ann. R. Coll. Surg. Engl. **86**(5), 334 (2004)
38. D. Ravì, C. Wong, F. Deligianni, M. Berthelot, J. Andreu-Perez, B. Lo, G.Z. Yang, Deep learning for health informatics. IEEE J. Biomed. Health Inform. **21**(1), 4–21 (2016)
39. J.M. Rondina, M. Filippone, M. Girolami, N.S. Ward, Decoding post-stroke motor function from structural brain imaging. NeuroImage Clin. **12**, 372–380 (2016)
40. M.M. Santoni, D.I. Sensuse, A.M. Arymurthy, M.I. Fanany, Cattle race classification using gray level co-occurrence matrix convolutional neural networks. Procedia Comput. Sci. **59**, 493–502 (2015)
41. S.C. Shapiro, Artificial intelligence, in *Encyclopedia of Artificial Intelligence*, vol. 1, 2nd edn., ed. by S.C. Shapiro (Wiley, New York, 1992)
42. N.H. Sweilam, A.A. Tharwat, N.A. Moniem, Support vector machine for diagnosis cancer disease: a comparative study. Egypt. Inform. J. **11**(2), 81–92 (2010)
43. R.E. Thornhill, C. Lum, A. Jaberi, P. Stefanski, C.H. Torres, F. Momoli, W. Petrcich, D. Dowlatshahi, Can shape analysis differentiate free-floating internal carotid artery thrombus from atherosclerotic plaque in patients evaluated with CTA for stroke or transient ischemic attack? Acad. Radiol. **21**(3), 345–354 (2014)
44. A. Yardimci, A survey on use of soft computing methods in medicine, in *Proceedings of the 17th International Conference on Artificial Neural Networks*, Porto, Portugal (2007), pp. 69–79
45. Q.Y. Zhong, E.W. Karlson, B. Gelaye, S. Finan, P. Avillach, J.W. Smoller, T. Cai, M.A. Williams, Screening pregnant women for suicidal behavior in electronic medical records: diagnostic codes vs. clinical notes processed by natural language processing. BMC Med. Inform. Decis. Making **18**(1), 30 (2018)

Mr. J. H. Kamdar is a research scholar at ICAR-Directorate of Groundnut Research, Junagadh, Gujarat, India. He is pursuing Ph.D. degree from Saurashtra University, Rajkot, Gujarat, India. He has nearly 8 years of experience in biotechnology and allied areas. He has published more than 16 research papers, book chapter, and popular articles in the internationally and nationally reputed journals and magazines. He has also presented his research work in more than 17 International/national conferences and also awarded for best oral/poster presentation.

Mrs. J. Jeba Praba One of a few researchers in the field of cloud security in the Gujarat province of India, Mrs. Jeba Praba, J., (Assistant Professor, Department of Computer Science & Applications, Christ College, Rajkot, Gujarat, India), is a Ph.D. research scholar of Marwadi University, Rajkot, Gujarat, India has more than 15 years of experience in teaching, research and academic administration. Under her supervision, more than 100 students have completed their B.Sc. & M.Sc. dissertation works. She has been an organizing committee member for various International and National Symposia, Workshops, seminars, etc. She has published many research papers and book chapters in the internationally reputed journals and books. She has also presented her research work in numerous International and national conferences and has attended several international and national conferences.

Dr. John J. Georrge (Assistant Professor & Head, Department of Bioinformatics, Christ College, Rajkot), has more than 15 years of experience in teaching, research and academic administration. He is currently holding an additional charge as a Bioinformatics Nodal officer of Gujarat State Biotechnology Mission (GSBTM, Govt. of Gujarat) and coordinator of Department of Biotechnology (DBT), Govt. of India sponsored Postgraduate diploma in the Computational Biology course. He has received the Best Teacher Award by the Gujarat Science Academy. Dr. Georrge is an elected Member of Royal Society of Biology, United Kingdom and the Bentham Ambassador of Bentham Science Publishers, United Arab Emirates. He is a Visiting Researcher at the Indian Institute of Science (IISc), Bangalore.

Intelligent Learning Analytics in Healthcare Sector Using Machine Learning

Pratiyush Guleria and Manu Sood

Abstract Machine Learning and its role in the health care sector is the area of research in emerging times. There are learning types in Machine Learning which involves Supervised, Unsupervised and Reinforcement Learning. These techniques become important to unearth the hitherto unknown relationship from data which become useful for society. In the proposed chapter, the author has discussed the intelligent learning analytics achieved using Machine Learning and predicted the patient prognosis based on the input dataset values using python. Here, predictive modeling is done that uses historical data to predict an output variable. The Machine Learning applications in healthcare are becoming boon to patients for identifying diseases and diagnostics. The Healthcare sector can benefit from the ability of technologies such as Machine Learning to support them in the intelligent analysis of vast amount of data. Machine Learning in Healthcare sector helps to analyze the data and predict the outcome. The intelligent learning analytics achieved through Machine Learning can break down information to enable it to make predictions.

Keywords Emerging · Health · Learning · Predicted · Reinforcement · Supervised · Unsupervised

1 Introduction

With the enormous growth of data on a large scale in the medical field has stressed upon the urgent requirement of Machine Learning. Machine Learning is becoming helpful in extracting meaningful information from voluminous data. The Machine Learning algorithm can help in classification and clustering of data, but the major

P. Guleria (✉)
NIELIT Shimla, Shimla, Himachal Pradesh, India
e-mail: pratiyushguleria@gmail.com

M. Sood
Department of Computer Science,
Himachal Pradesh University, Shimla, Himachal Pradesh, India
e-mail: soodm_67@yahoo.com

V. Jain and J. M. Chatterjee (eds.), *Machine Learning with Health Care Perspective*, Learning and Analytics in Intelligent Systems 13,
https://doi.org/10.1007/978-3-030-40850-3_3

challenge in the Healthcare sector is of an accurate dataset. The predictions obtained using Machine Learning algorithms would be accurate only when the training datasets are free from malicious data.

In the Healthcare sector, Machine Learning techniques like deep learning are generating effective results in image classification and predictions. The sources of data collection have also increased with recent advancements which involve data collection in SRL Laboratories, mobile-based technologies, CT Scan, MRI images, etc. but along with it, the computational research challenges for extracting useful information has also increased a lot. The challenges in data collection involve (a) the continuous growth of data, (b) storage cost, (c) heterogeneous data, (d) different data types, (e) regions, (f) diverse data sources, (g) complexity in data, etc. In such a scenario, effective implementation of Machine Learning algorithms can lead to a breakthrough in the healthcare sector and will provide an opportunity for Intelligent Learning clinical insights and discoveries. In [1], the author has proposed a neural network-based algorithm for disease risk prediction using structured and unstructured data from the hospital. Machine learning utilization in health sector helps to discover patterns of diseases from large datasets. Authors [2] have classified datasets of poisoning attacks using machine learning to find targeted errors.

The data in the electronic form provides a major opportunity for researchers in the healthcare sector for knowledge data discovery to improve the health sector. Machine learning with the help of computational techniques handles complex datasets to obtain meaningful results [3]. There is a large scope of Machine Learning algorithms in the processing of big data in the health care sector [4]. Authors in [5] have proposed the Machine Learning classifiers to assist healthcare-related decisions. It acts as a decision support system for the electronic classification of patient records. The Machine Learning techniques help in analyzing (a) unstructured, structured data, (b) clustering of patients having similar medical symptoms, (c) prediction of diseases [6].

The Machine Learning tools which help in predictive analytics of Healthcare activities are as follows (a) Apache Mahout, (b) Skytree (c) karmasphere, (d) BigML. These tools perform data mining and big data analytics [7]. In [8], authors have reviewed the challenges and opportunities of image analysis and machine learning in digital pathology. The deep learning techniques in confluence with artificial neural networks are capable of performing unsupervised learning from unstructured and complex data in the form of images. The unsupervised and supervised learning techniques extract attributes in voluminous data to train and build the ML models to develop predictive models.

In [9], authors have done the literature survey on deep learning in healthcare sector. The deep learning technology results in improved healthcare and resolves the challenges of transforming health data which are as follows: (a) heterogeneous data attributes in medical data, (b) unstructured data format, (c) missing values in the dataset, (d) data from multiple domains, (e) complexity in data. The data mining and statistical approaches helps in building predictive models.

2 Motivation

The utilization of Artificial Intelligence and Machine Learning in healthcare is the need of an hour. In Machine Learning, the ability of python like languages can be explored effectively for the benefit of society. Machine Learning consists of classification and clustering algorithms which learn from voluminous dataset to perform predictive analytics. The predictive results help the physicians to assist them in getting updated information related to healthcare and immediate relief to patients. Machine learning in healthcare sector needs to develop analytical engines/product, transforming the way healthcare ecosystem accesses and derives value from unified clinical datasets to deliver best-in-class/cutting edge patient care and thereby improve health outcomes.

3 Machine Learning

Machine learning consists of Unsupervised and Supervised learning categories. In Unsupervised learning, clustering techniques are there whereas, in supervised learning, classification and regression techniques are there. The Machine learning categories are shown in Fig. 1.

3.1 Advantageous Sectors of Machine Learning

There are certain sectors where Machine Learning is playing an important role in intelligent learning analytics and provide a lot of opportunities in research and development. Machine Learning field helps in (a) finding patterns in data, (b) automate data analysis, (c) communication, security surveillance, speech and audio processing, (d) object detection, healthcare, (e) localization and tracking, (f) image processing, security, and forensics.

Apart from it, other sectors which are advantageous to Machine Learning are as follows:

(a) Machine learning algorithms help in the prediction of prognosis in the Healthcare sector.
(b) ML helps in image, video and text recognition.
(c) ML can improve from past experiences and learn from them.
(d) Computational image analysis for predictive modeling in the Healthcare sector.
(e) Fraud detection in Financial and Banking sectors.
(f) Real-time data analysis in the online shopping industry.
(g) Machine learning helps in proper customer care and relationship management approach. The ML predicts the trends of the market from the voluminous dataset, following effective customer feedbacks, surveys.

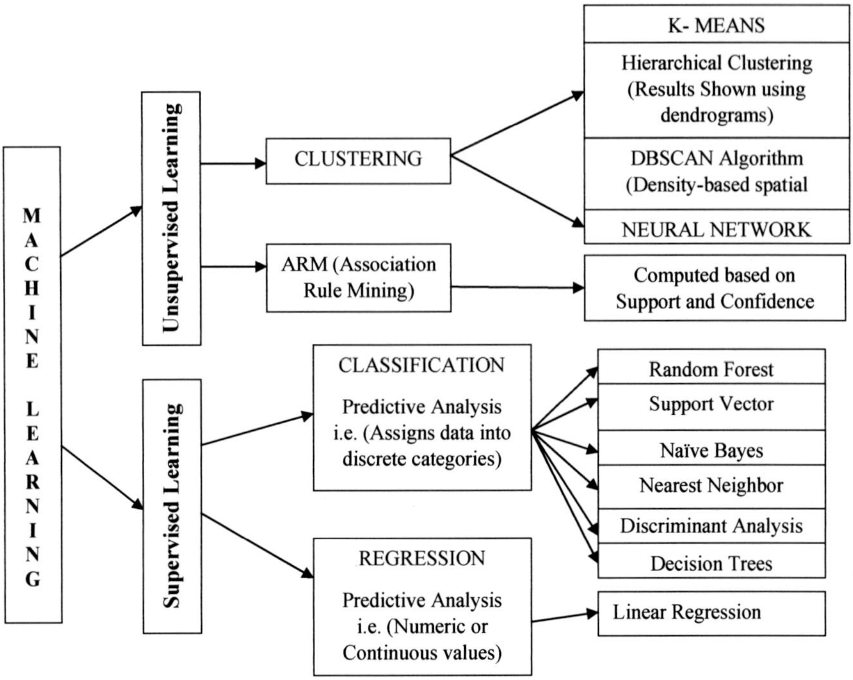

Fig. 1 Machine learning categories

(h) Machine Learning applications in speech and character recognition; face detection, biometric, iris data analysis.
(i) Natural language processing using Machine Learning.
(j) Signal processing and sensor data analytics.
(k) Intelligent semantic analysis, semantic web, and ontologies.
(l) Smart farming.
(m) Bio-technology for drug development and medical therapy.

3.2 *Machine Learning Challenges*

There are certain challenges in Machine Learning with Healthcare Sector which are as follows:

(a) The existing state of technology and datasets for healthcare providers, who continue to struggle to streamline data from legacy infrastructures.
(b) Attain actionable insights from disparate systems.

(c) Data needs to be standardized and optimized before it can be used to improve scientific knowledge.
(d) The data pipeline also needs to be continuously updated to reflect the real world, instead of a single snapshot in time.

4 Machine Learning Tools

The popular Machine learning tools and framework are shown in Table 1 and Fig. 2. The Machine learning libraries helpful in Intelligent data analytics are (a) Scikit-learn, (b) PyTorch, (c) NLTK toolkit and (d) Tensorflow, etc. whereas the Machine learning algorithms are prebuilt in some of the popular tools which are (a) Weka, (b) KNIME, (c) RapidMiner, (d) SPSS, (e) KEEL, (f) Orange and (g) MATLAB etc.

Table 1 Machine learning tools and libraries

S. No.	Library/tools	Description
1.	Scikit-learn	It is a library in python for data analysis and mining. It provides models for classification, clustering, and regression
2.	PyTorch	PyTorch is an open-source python library. The PyTorch library is used for natural language processing, neural networks
3.	Tensorflow	It is a JavaScript library in machine learning to build and train the model. It is a library for numerical computations using dataflow graphs
4.	Keras	Keras is a Python deep learning library. It runs on top of the TensorFlow
5.	Weka	Weka is known as Waikato environment for knowledge analysis. It is a machine learning algorithm for performing data mining tasks
6.	KNIME	KNIME is Konstanz information miner is free open source software for a machine learning data analytics and mining techniques
7.	RapidMiner	RapidMiner is software for (a) data preprocessing, (b) preparation, (c) machine learning, (d) text mining and (e) predictive analytics
8.	NLTK	NLTK is a natural language toolkit for natural language processing in python
9.	KEEL	KEEL means knowledge extraction based on evolutionary learning. It performs classification and clustering using data mining algorithms
10.	Orange	It is the tool for interactive data analysis and visualization. Orange provides the toolkit for (a) data mining, (b) machine learning and (c) data visualization. It can be used as a python library also
11.	MATLAB	MATLAB is a computing environment for machine learning, data analysis, regression analysis, classification, clustering approaches etc. It is a programming language for mathematical computing
12.	SPSS	It is the statistical package for social sciences. SPSS is used for statistical data analysis

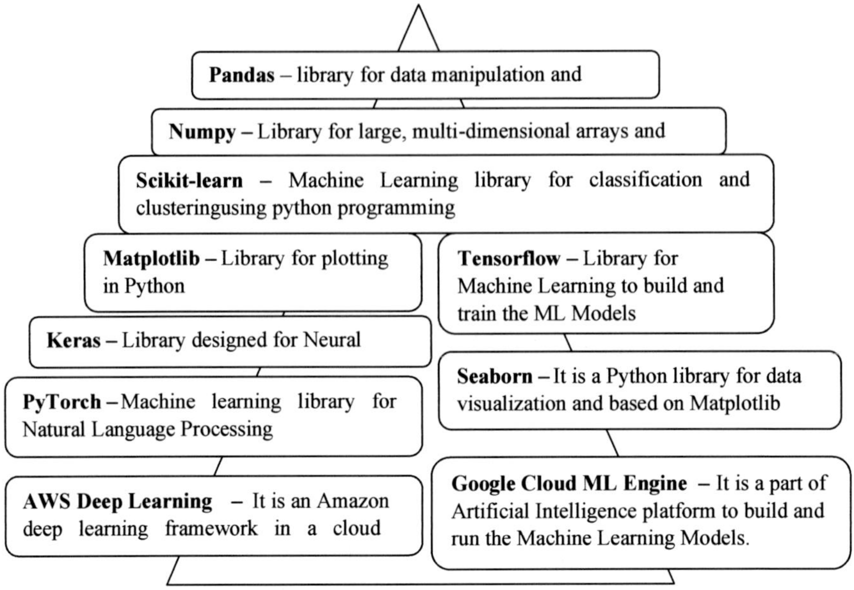

Fig. 2 Machine learning framework

5 Proposed Machine Learning Model in the Healthcare Sector

A Patient-centric Machine Learning model is proposed in Fig. 3. The proposed model shows the Machine Learning process. The model is divided into a phased manner. In the first phase, the patients' symptoms are recorded as input dataset and then the appropriate Machine learning model is selected to train the system. The system after training will be enabled to predict patient diagnosis. The crucial phase of the proposed model is the Knowledge representation which will regularly update itself for further training. The patient diagnostic results predicted by the system will be updated into the knowledge system as patient history for further precise analytical results and research.

With the help of Machine Learning model, authors have predicted the diagnosis for patients based on the symptoms shown in Table 3.

5.1 Objective

To predict the prognosis of patients based on the symptoms from the dataset shown in Table 3 using python programming.

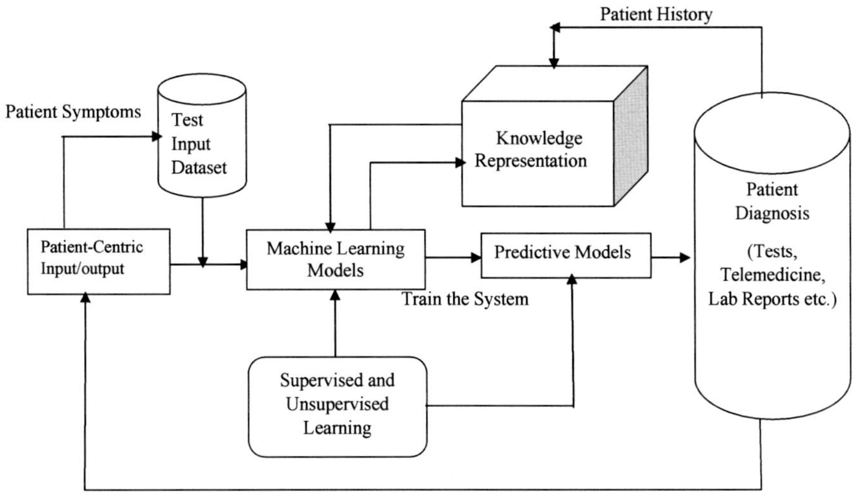

Fig. 3 Patient-centric machine learning model

Table 2 Python programming libraries

Name of library	Purpose of the library
NumPy	Library to work on numerical mathematics, arrays and matrices
Matplotlib	Library to work on an extension of NumPy and for plotting in Python. It provides a Matlab like an Interface in python
Pandas	Library for data manipulation and analysis

Input Dataset: The training dataset consists of the attributes categorized in Table 3. The training dataset having attributes shown in the table is the symptoms of different diseases. The sample training dataset is shown in Table 3.

In this application, the patient symptoms are predicted through patient-centric data learning model shown in Table 4. The results shown in Table 4 are helpful for (a) patients who may not avail hospital facility at nearest place, (b) can take telemedicine, (c) medical practitioners to work on personal health record of patients, (d) hospital operators for collecting patients feedback, (e) pharmaceutical researchers, (f) data generated from IOT enabled devices.

5.2 Pre-processing of Data

The dataset is slice and dice to separate features from predictions. The next step is to do dimensionality reduction for removing redundancies. In Machine Learning, the dataset is split into a training set and test set. The libraries imported for generating the results are shown in Table 2.

Table 3 Training dataset for predictive prognosis

Vomiting, Digginess	Fatigue	Nausea, Unsteadiness	Weightloss	Restlessness	Spinning movement, Loss of Balance	Irregularsugarlevel	Breathlessness	Sweating	Vision Problem	Chest Pain	Obesity	Excessive Hunger	Increased Appetite	Polyuria	Prognosis
0	1	0	1	1	1	1	0	0	1	0	1	1	1	1	Diabetes
0	1	0	1	1	1	1	0	0	1	0	1	1	1	1	Diabetes
0	0	0	1	1	1	1	0	0	1	0	1	1	1	1	Diabetes
0	1	0	0	1	1	1	0	0	1	0	1	1	1	1	Diabetes
0	1	0	1	1	1	0	0	0	1	0	1	1	1	1	Diabetes
1	0	0	0	0	0	0	1	1	0	1	0	0	0	0	Heart attack
1	0	1	0	0	1	0	1	0	0	0	0	0	0	0	Vertigo
1	0	0	0	0	0	0	1	1	0	1	0	0	0	0	Heart attack
1	0	1	0	1	1	0	0	0	0	1	0	0	0	0	Vertigo
0	1	0	1	1	0	1	0	0	1	0	1	1	1	1	Diabetes
⋮	⋮	⋮	⋮	⋮	⋮	⋮	⋮	⋮	⋮	⋮	⋮	⋮	⋮	⋮	⋮
1	0	0	0	0	0	0	0	1	0	1	0	0	0	0	Heart attack
n	n	n	n	n	n	n	n	n	n	n	n	n	n	n	Heart attack/ Diabetes/ Vertigo?

The training data is the learning dataset which is used to train an algorithm. The training dataset helps the algorithm to predict accurate decisions with new input value based on certain parameters. The testing dataset is just to evaluate the performance of the algorithm. In the testing dataset, only the input data is given whereas, the expected output is not shown. The testing dataset validates the results of training data and also prevents from conditions of over fitting and under fitting of results.

5.3 *Results and Discussions*

The prognosis results are shown in Table 4. Here, the symptoms are matched with the training dataset and accordingly, the patient diagnoses are predicted. The dataset is trained using predefined libraries in python programming language.

Table 4 Patient diagnostic results

Please reply with yes/Yes or no/No for the following symptoms
Breathlessness?
No
Vomiting?
Yes
Weight loss?
No
Spinning Movement and Loss of Balance?
Yes
Unsteadiness?
Yes
Digginess?
Yes
Output: ['You may have Vertigo']
Symptoms present: ['unsteadiness', 'vomiting', 'digginess', 'spinning movement', 'loss of balance']
Symptoms given ['vomiting', 'headache', 'nausea', 'spinning movements', 'lossofbalance', 'unsteadiness']

6 Machine Learning Approaches

6.1 Supervised Learning Approach

In the supervised learning approach, with labeled input, the desired output is there. With the help of the supervised learning approach, Machine learning helps in developing data-driven insights which lead to improved designs and decisions. Machine learning applications implemented using Software like MATLAB and Python follow the following steps:

(a) import dataset,
(b) preprocessing and clean the data by removing outliers and missing data,
(c) explore,
(d) train the dataset,
(e) test to check the model adequacy, and finally deploy.

6.1.1 The Solution to Healthcare Problems Using Linear Regression Model

Here, one of the supervised learning approaches are implemented i.e. Linear Regression. The Linear Regression technique uses predictive modeling to predict an event or outcome. It uses a technique of learning a continuous value from a set of features. The dataset for the experimentation purpose is obtained from MATLAB. The dataset description is shown in Table 5. The predictor names for this regression model are: {sex, age, wgt and smoke}.

Objective: The objective is to predict the Blood Pressure values which a continuous target variable i.e. known to be a Regression task. To model the systolic pressure as a function of patients, the attributes classified are mentioned as below:

(a) Age
(b) Smoker
(c) Gender
(d) Weight

Preprocess data: Gender and Smoke attributes have two choices each. Therefore, there is a need to change these attributes into a categorical form: Gender = {'M', 'F'}, Smoke = {1, 0}.

Table 5 Dataset Description

Sex	Age	Weight	Smoke	Systolic blood pressure	Diastolic blood pressure	Remarks
Predictor variable	Predictor variable	Predictor variable	Predictor variable			The attribute values are represented as follows:
m	38	176	1	124	93	**Sex** = {Male—'m', Female—'f' **Smoke** = {1—'yes' 0—'no'}
m	43	163	0	109	77	
f	38	131	0	125	83	
f	40	133	0	117	75	
f	49	119	0	122	80	
f	46	142	0	121	70	
f	33	142	1	130	88	
m	40	180	0	115	82	
m	38	176	1	124	93	
m	43	163	0	109	77	
…	…	…	…	…	…	
m	48	177	0	114	86	

6.1.2 Results and Discussions

The Linear Regression Model applied is shown in Eq. 1 i.e.

$$\text{sys} \sim 1 + \text{age} + \text{smoke} + \text{sex} * \text{wgt} \tag{1}$$

The plot main effects are shown in Fig. 4. This plot displays the main effects. The small circles represent the magnitude of the effect and the blue lines displays the upper and lower confidence limit values for the main effect. The predicted systolic blood pressure values are shown in Figs. 5 and 6. The predictors for which the results are displayed in Figs. 5 and 6 are:

Sex = {m}, age = {43}, wgt = {159} and smoke value = {y}.

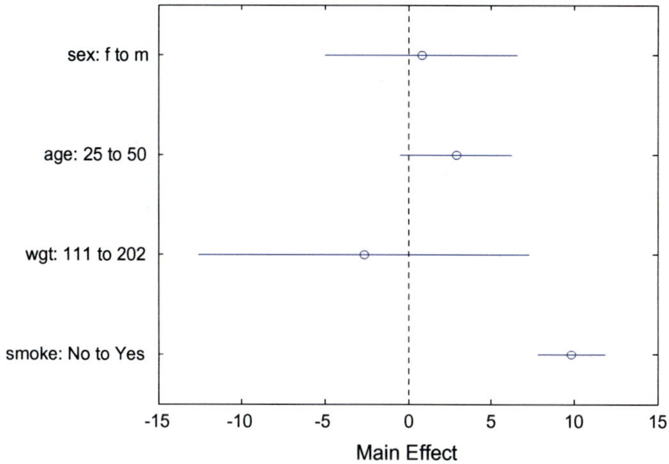

Fig. 4 Plot main effects

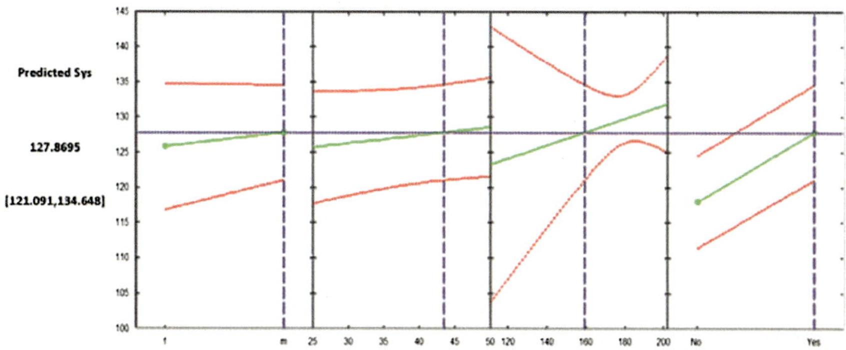

Fig. 5 Predicted systolic for parameters {sex = ‘M’, age = 43, wgt = 159, smoke = ‘y’}

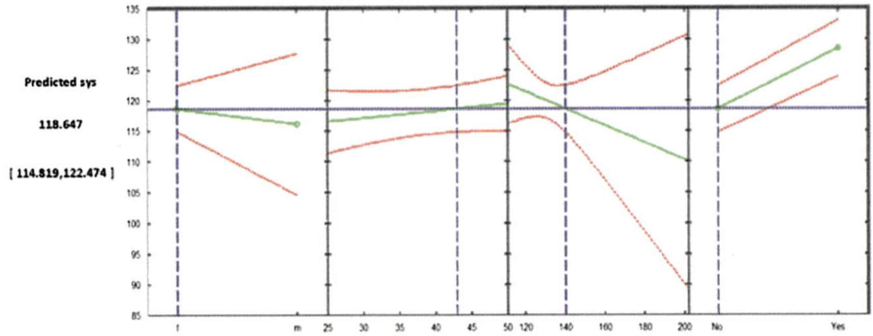

Fig. 6 Predicted systolic for parameters {sex = 'F', age = 43, wgt = 140, smoke = 'n'}

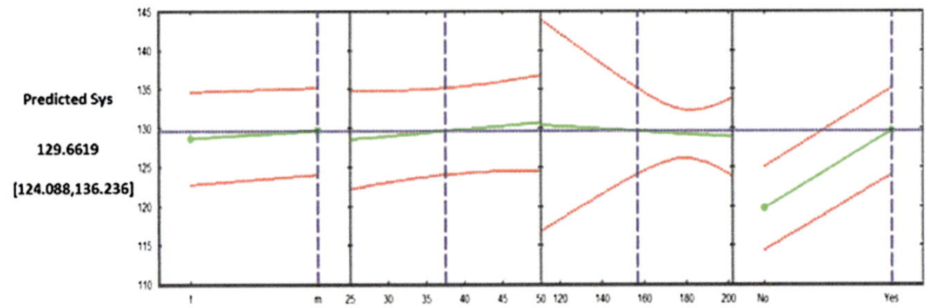

Fig. 7 Predicted systolic for Eq. 2

Sex = {f}, age = {43}, wgt = {140} and smoke value = {n}.

The results of linear Regression model values for the Eq. 2 are shown in Fig. 7. The Linear regression model applied is shown in Eq. 2:

$$\text{sys} \sim 1 + \text{sex} + \text{age} + \text{wgt} + \text{smoke} \tag{2}$$

6.2 *Unsupervised Learning Applications in Healthcare*

In Unsupervised learning approach, there are no predefined labels neither classification is done. In this approach, the algorithms extract information on the basis of similar traits and patterns. There is no prior training of input dataset is performed and the algorithms following unsupervised learning approach have to find the hidden values in the unlabeled data by themselves. There are some algorithms following unsupervised learning approaches are K-Means clustering, Hierarchical Clustering, Neural Networks, and Association Rule Mining etc. The application of unsupervised learning in Healthcare sector are:

(a) clustering of diseases based on medical symptoms.
(b) finding patients having similar traits and helps in diagnosis.
(c) identify patterns in data i.e. Medical prescriptions.
(d) medical Image Analysis.
(e) image segmentation and gene clustering.

The clustering approach is applied in WEKA (Waikato Environment for Knowledge Analysis) simulation tool on the same dataset shown in Table 5. The K-Means algorithm is used for the results in the WEKA tool. K-Means is a distance-based algorithm and uses the Euclidean distance for distance calculation in clusters. The visualization results of all the attributes in the dataset are shown in Fig. 8.

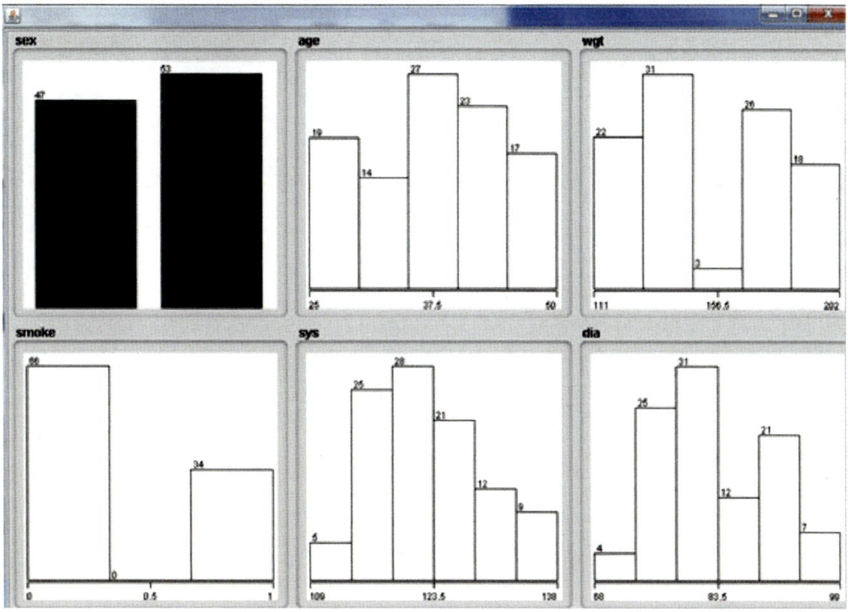

Fig. 8 Visualization of attributes

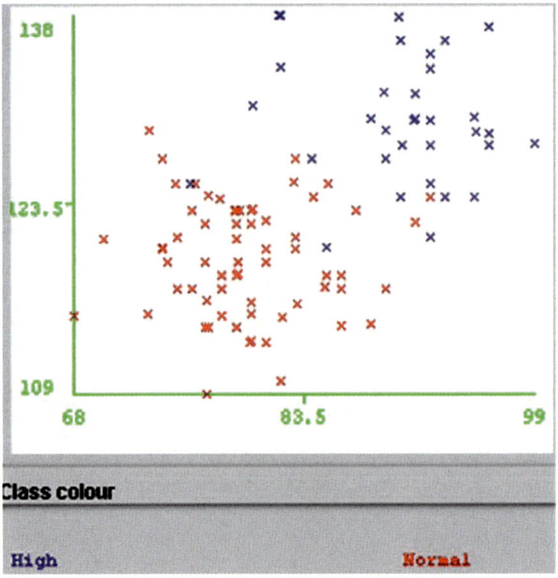

Fig. 9 Clustering of sys (y-axis) and dia (x-axis) attributes

The clustering plot of two attributes i.e. sys and dia blood pressures are shown in Fig. 9. The sys is shown on Y-Axis and dia is shown on X-Axis. The classes are categorized as High and Normal. The decision tree classification for the above dataset is performed using the J48 decision tree algorithm in WEKA. The J48 classifier model for the training dataset in Table 5 is shown in Table 6 and the decision tree is shown in Fig. 10. In the decision tree, the "sys" attribute is taken as the root node for conditional comparison.

Table 6 Classifier model

```
=== Classifier model (full training set) ===
J48 pruned tree
------------------
sys<= 127
|  dia<= 88: Normal (66.0/3.0)
|  dia> 88
|  |  age <= 39: High (5.0)
|  |  age > 39: Normal (4.0/1.0)
sys> 127: High (25.0/1.0)
Number of Leaves:      4
Size of the tree: 7
Time taken to build model: 0 seconds
=== Evaluation on training set ===
Time taken to test model on training data: 0 seconds
=== Summary ===
Correctly Classified Instances      95          95    %
Incorrectly Classified Instances     5           5    %
Kappa statistic                          0.8843
Mean absolute error                      0.0915
Root mean squared error                  0.2139
Relative absolute error                  20.6329 %
Root relative squared error              45.4805%
Total Number of Instances                100
```

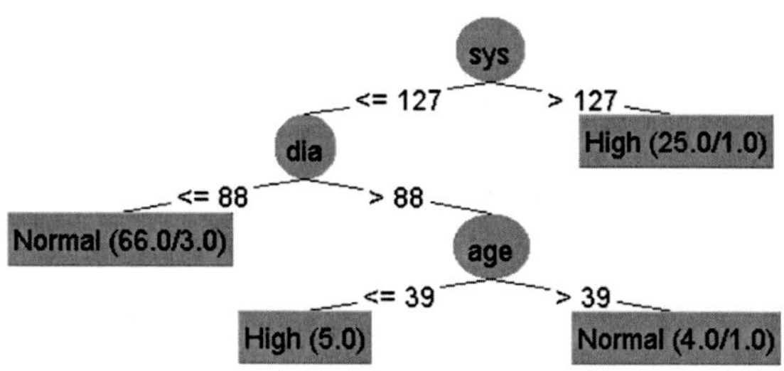

Fig. 10 Decision tree

7 Conclusion

In the present scenario, machine learning is the emerging field where a lot of research is undergoing. Machine learning in confluence with Artificial Intelligence and Deep learning is facilitating multi-disciplinary areas of society. Machine learning is representing a paradigm in healthcare sector and predicting Intelligent learning analytics in Health. The Machine learning techniques involves classification, regression, clustering algorithms for supervised and unsupervised learning approaches. With the help of data mining techniques as a part of Machine learning discovers the opportunities for innovative clinical observations, electronic health records, predictive analytics in medicine etc. In this chapter, authors have discussed the Machine Learning analytics with healthcare perspective and proposed the patient centric models using Machine Learning.

In Machine Learning, the applications of Machine learning can be explored using emerging programming languages like python. In the proposed chapter, the practical applications of Machine learning are performed using python and MATLAB considering patient led data learning. The knowledge extracted using Machine learning techniques are helpful in (a) handling future healthcare challenges, (b) developing machine learning models to discover new healthcare insights, (c) predictive learning based on patient symptoms and diagnosis.

References

1. M. Chen, Y. Hao, K. Hwang, L. Wang, L. Wang, Disease prediction by machine learning over big data from healthcare communities. IEEE Access **5**, 8869–8879 (2017)
2. M. Mozaffari-Kermani, S. Sur-Kolay, A. Raghunathan, N.K. Jha, Systematic poisoning attacks on and defenses for machine learning in healthcare. IEEE J. Biomed. Health Inform. **19**(6), 1893–1905 (2014)
3. J. Wiens, E.S. Shenoy, Machine learning for healthcare: on the verge of a major shift in healthcare epidemiology. Clin. Infect. Dis. **66**(1), 149–153 (2017)
4. G. Manogaran, D. Lopez, A survey of big data architectures and machine learning algorithms in healthcare. Int. J. Biomed. Eng. Technol. **25**(2–4), 182–211 (2017)
5. J.T. Pollettini, S.R. Panico, J.C. Daneluzzi, R. Tinós, J.A. Baranauskas, A.A. Macedo, Using machine learning classifiers to assist healthcare-related decisions: classification of electronic patient records. J. Med. Syst. **36**(6), 3861–3874 (2012)
6. F. Jiang, Y. Jiang, H. Zhi, Y. Dong, H. Li, S. Ma, Y. Wang, Q. Dong, H. Shen, Y. Wang, Artificial intelligence in healthcare: past, present and future. Stroke Vasc. Neurol. **2**(4), 230–234 (2012)
7. V. Palanisamy, R. Thirunavukarasu, Implications of big data analytics in developing healthcare frameworks—a review. J King Saud Univ. Comput. Inf. Sci. (2017)
8. A. Madabhushi, G. Lee, Image analysis and machine learning in digital pathology: challenges and opportunities (2016)
9. R. Miotto, F. Wang, S. Wang, X. Jiang, J.T. Dudley, Deep learning for healthcare: review, opportunities and challenges. Brief. Bioinform. **19**(6), 1236–1246 (2017). https://doi.org/10.1093/bib/bbx044

Pratiyush Guleria has completed his Ph.D. in Computer Science from Himachal Pradesh University Shimla, India. He has done M.Tech. in Computer Science with a Gold Medal from Himachal Pradesh University, Shimla, India. He has done B.Tech. in Information Technology from Himachal Pradesh University. He has more than 11 Years of Experience in IT Industry and Academics. He has published research papers in peer reviewed International Journals, Conferences and as Book Chapters. His research interests include Data Mining, Machine Learning, and Web Technologies.

Prof. Manu Sood is currently working as a Professor in the Department of Computer Science at Himachal Pradesh University Shimla, India. He has completed his Ph.D. in Computer Engineering under the Faculty of Technology from University of Delhi, Delhi, India. He completed his M.Tech. in Information Systems with a Gold Medal from Netaji Subhash Institute of Technology, Delhi, India. He has received his B.E. degree in Electronics and Telecommunication from Government Engineering College, Jabalpur, Madhya Pradesh, India. He has a consistent track record in academics throughout his career. Prof. Sood has over 30 years of extensive experience in IT Industry and Academics in India at various positions. He has been the Technical Program Committee Member for a number of International Conferences. He has, to his credit, approximately 35 research papers in peer reviewed International Journals, Conferences and as Book Chapters. He is actively involved in a number of research projects sponsored by various government agencies. His research interests include Software Engineering, Model Driven Software Development, Model Driven Architecture, Aspect Oriented Software Development, E-learning, Service Oriented Architecture, MANETs and VANETs.

Unsupervised Learning on Healthcare Survey Data with Particle Swarm Optimization

Hina Firdaus and Syed Imtiyaz Hassan

Abstract The behavior of a machine learning algorithm very much depends upon the structure of entities. There a pool of big data, which have lots of ambiguity, noises, granularity, unlabeled behavior that make it tough to find pattern and visualize an outcome. To solve this issue machine learning proposes an algorithm like clustering, which is mainly part of unsupervised learning. In the proposed model we used the unsupervised learning specifically the clustering algorithms like Expectation Maximization (EM), Make Density Based Analysis (MDBC) wrapped with the clustering algorithm like Kmeans, EM, GenClust++. The dataset has been chosen is a survey data on healthcare issues like chronic diseases, regular treatment etc. across various states and district of India. We have used Weka tool for formulation and training our dataset. This dataset is obtained from the Open Data Government Platform India Portal under the Government of Open Data License India, from the Department of Health and Family Welfare. The dataset contains ordinal type of values. These kinds of dataset are excellent examples of exploratory data analysis. The particle swarm optimization is applied on dataset for optimizing the attributes later trained with clustering algorithms which gave better log likelihood value comparable to the results on simple clustering algorithms. We even compare the performance speed among the algorithms to check how quickly they are able to perform on the variety of attributes.

Keywords Unsupervised learning · Particle swarm optimization · Kmeans · Genclust++ · UN SDG · Healthcare analytics · Machine learning · Big data

1 Introduction

While doing the study and designing the model of our healthcare survey dataset, we found the unprocessed data with no labels and improper distribution of values. As

H. Firdaus (✉) · S. I. Hassan
School of Engineering Sciences and Technology, Jamia Hamdard, New Delhi 110062, India
e-mail: hinafirdaus95@gmail.com

S. I. Hassan
e-mail: s.imtiyaz@gmail.com

V. Jain and J. M. Chatterjee (eds.), *Machine Learning with Health Care Perspective*, Learning and Analytics in Intelligent Systems 13,
https://doi.org/10.1007/978-3-030-40850-3_4

we have lots of datasets which are full of unfiltered and ordinal datatypes or even variable style. In that case unsupervised algorithms which mainly work on clustering techniques and making associations among the values of attributes come as a great rescuer. Which we even proved through over experiments.

The unsupervised learning techniques are based on various factors such as standard variance, similarity, dissimilarity, correlation etc. henceforth it is quite intriguing to train a dataset in this kind of arrangements. Using the supervised and some semi-supervised learning algorithm we may get good accuracy but it is not a definite way to define the correct classification of the data. Later in this chapter we will learn in what way unsupervised applications can help to train a dataset in less time.

Nature inspired algorithms are an interesting sub branch of artificial intelligence which takes inspiration from the birds, fish, bugs etc. around us in this environment. Similarly, one of the popular swarm intelligence algorithms is particle swarm optimization (PSO).

In this coming sub-section, we will learn more about machine learning algorithms, data science in healthcare, unsupervised learning, particle swarm optimization and also about the sustainable development in healthcare using data science.

1.1 Machine Learning

A buzzword in past decade still creating chaos and excitement in tech world is machine learning. Popularly machine learning in layman terms is a technology which *will take all the human jobs*. Automation in every field is increasing exponentially. From home appliances, to transportation, entertainment, food, lifestyle, study, hospitals, politics it is everywhere. The Gen-Z (newer generation after millennials) has smart-phones from the age of 2–3 year or even lesser than that in developing and developed countries. We can encounter seeing millennials and Gen-Z using only a 4–5 in. of smart phones for "finding restaurants nearby to eat", "what to do when I am getting bore?", "how to study?", "how to sleep?", "how to get rid of acne?" these kind of queries are not asked by friends, family or even tried by themselves it is all looked up online. We can really understand from this societal behavior that the flow of data from user to the database of the website can be dangerous and even helpful, mainly it is the way we use it.

Machine learning provides a great support system to model these kinds of data to understand their behavior in various scenarios and giving a prediction about the implication of that data. This is a very multidisciplinary field with huge applications in physical science, bioinformatics, health, rocket launching, elections in democracy, education, logistics etc.

Let us learn about some of the basics of machine learning.

1.1.1 Definition of Machine Learning

Earlier we saw the effect of machine learning in our day today life. Let's now learn who coined the termed machine learning and why?

In the early 1950s Samuel [1] coined the termed machine learning in one his paper where he solved the checker game problem teaching a computer to play better than its human creator in merely 8–10 h of training. In that same period parallel work of Alan Turing was getting published and came into light when he proposed an analogy "Can machine thinks?". That was the time of world war II, the field of information and technology was taking a huge turn. Fast transportation, rockets and missiles, nuclear energy and a in midst of all these a person who was very underrated in his time thinking and proposing theorems to proof the possible ways a machine can do as humans do naturally [2].

Fast forwarding it to the late 1990s era another academician and researcher Tom M. Mitchell gave a formal definition to machine learning. This is defined as [3] "*A computer program is said to learn from experience E with respect to some class of tasks T and performance measure P if its performance at tasks in T, as measured by P, improves with experience E*".

As of now we have learned about the formal definition and a brief timeline of machine learning in the next sub section we will even see the types of machine learning.

1.1.2 Types of Machine Learning

Machine learning is majorly of three types. Such as refer Fig. 1:

(a) Supervised learning
(b) Unsupervised learning
(c) Reinforcement learning

(a) *Supervised learning*

This technique is very well known for predictive analysis when our datasets are labeled. Labeling the dataset is manual work mostly done during preprocessing phase

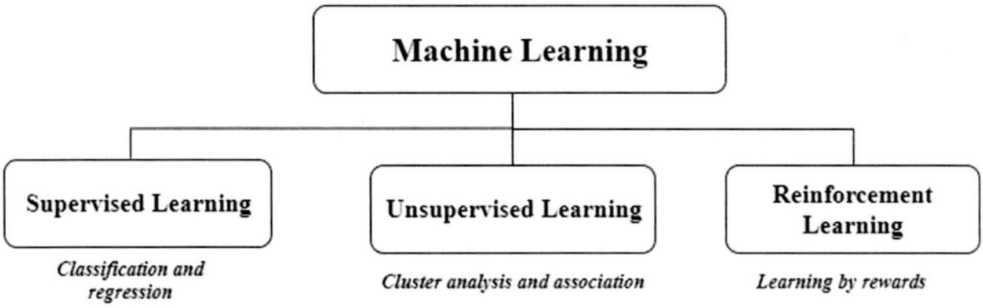

Fig. 1 A tree diagram for types of machine learning

or even while teaching the algorithm about prediction. To define the labeled dataset, we can take a simple example of Google photos. If you are using Google photos to store your pictures. It will identify the faces and try to recognizes you. If not able to recognize it will ask whether it is you or someone else. In this way it learns about you and various things, even when you upload a group image it can figure out your face and tag you in that photo. So, this is an example of classification algorithms. Where the dataset is using attributes of your images as input and labeling it with various category that you are introducing like humans, faces, cats, dogs, trees etc. In this way various algorithms like Naïve Bayes, Support vector machine, DBSCAN, linear regression, neural networks etc. are the part of supervised learning.

(b) *Unsupervised learning*

Learning in this technique is very much computer generated. The algorithms will learn by its own about the data by using correlation, association, maximum likelihood etc. like parameters to give a proper justification on the behavior of the dataset. A real-world example is like if a restaurant owner wants to open a restaurant in unknown place which area will he choose? So, the answer is simple he has to collect the data on the parameter of population, weather and climate, access to local market, water etc. Now using any of the unsupervised technique like K-means he can cluster the parameters based on high correlation. Whichever area is showing good stats or likelihood value he chooses that and start his restaurant. We can learn that clustering is major learning technique of this.

Mostly used unsupervised learning algorithms are K-Means, Expectation maximization, Canopy, association rules, density-based algorithm etc.

(c) *Reinforcement learning*

This algorithm is like a human baby learning to walk and talk. Likewise, this technique has an agent which mainly learn from the environment. The learning is done by simulating positive and negative rewards. If the actions are as expected performed by the agent it is considered as positive reward otherwise a negative reward. A real-life example is training a robot to pick-up the blocks and stacking in a place. So, environment is grid, blocks and bucket to place. Agent is hand of robot. Now the robot is programmed to pick-up the top block, move and stack it in a bucket. If the robot drops the block in mid-way after first complete simulation means it is not doing the work properly this will send a false alarm or signal. A repetitive action of picking and dropping will be performed until unless robot hand learns to do the action without any negative rewards. Major popular reinforcement learning algorithms are Q-learning, Temporal difference, deep adversarial networks.

1.2 Data Science in Healthcare

1.2.1 What Is Data Science?

Exponentially generated data every day has now got some meaning of its existence. This means when we are generating a data log in our every move from googling to checking in Facebook, watching movies online, shopping etc., has created opportunities to a section in the market which wants to analyze and process the data with precision and accuracy. So, the processing and analysis of the data in the benefit of any organization is known as data science.

Let's us take an example of a company who handle many things together. In that company if a data scientist is there it has many huge datasets to handle with very out of context attributes to find any kind of connection which can answer the query imposed to them by the boss or client or even for them to give any new angle to the existing dataset. This is the work of a data scientist in business. Well, they handle many complex problems making the analysis more automated using the help of machine learning algorithms.

Data science follow a five-step process namely:

a. Data extraction
b. Cleansing
c. Visualization
d. Insights generation

So, data science and machine learning co-exist. Because data science applies the knowledge of machine learning, AI, deep learning, data mining etc. to make a business run and giving them insights about the underlining activities going inside the dataset.

1.2.2 Applications of Data Science

The field of data science is for curious people who can't stop thinking about a certain problem until unless they exhaust all of its possibility to exist. There are many popular applications of data sciences available such as Recommendation system, Internet searching, Healthcare, Specific Advertisement flashing either to individual or group of crowds, Speech recognition, Image recognition, Gaming, Websites for comparing prices (tickets, route, housing etc.), Travel planner, Fraud and risk detection, Logistics and delivery. These applications are the most popular in industry among many other. From this list of application our focus is only on healthcare in this chapter.

1.2.3 Use of Data Sciences in Healthcare

Healthcare is the one of the most popular and targeted industry for AI, robotics, machine learning, deep learning you name it this industry has applications in all the technical field. After finance the most grossing industry is healthcare, and it is even needed the most.

Now a days, a person from remote village of Assam to city dweller of New York in USA first Google its symptoms. "My nasal is irritating, what to do?", "I am having redness in my skin" and what not. Why is that so? Why google is so reliable but not a physician? Why we are googling so many symptoms and even following the recommendation given by some websites? This is such a problem that even doctors have to put a board outside their cabin please keep your Google knowledge out of this door.

It is because of great recommendation system used by Google. Text analysis search engine which can even detect emotions perfectly and help and user to give answer. Even when you searched about your flu and visited some websites searched some links in precautions and medicine, now any of the website you used has traced your patterns of searching in form of cookies and stored the data. After some time when you use your Gmail or any mailing site, social networking site, you may find similar or that website sending you ads, newsletter, subscription form, or link for similar website will pop up to show you best recommendation to get over flu. This is specific audience targeted advertisement.

Some of the machine learning algorithms are so powerful that they can process zillions of complex data in no time give accurate classification. That is why is very much believable if this technology can be used well not only to commercialize and exploit the user it can be blessings to the human beings.

Some of the great applications and uses of data science in healthcare are:

Medical image analysis, Drug creations, Genomics and genetics, Virtual patient and customer assistant support, Prognosis and diagnosis of diseases virtually, Customer data management, Optimization of staffing, Learning data from wearable's and many more.

You must now have got some idea about the functioning of healthcare industry when the technology of machine learning, AI, robotics, data mining is applied. In the next section we will unfold and learn about the Unsupervised learning and after that particle swarm optimization.

1.3 Related Terms

1.3.1 Weka

This a tool developed for data mining and machine learning by the University of Waikato, New Zealand. It is a GNU general public license granted tool on JAVA SE platform written in Java. This can help in performing and classification and clustering operations with its in-build function. The dataset type is used in WEKA is ARFF (Attribute Relation File Format). There is a preprocessing section to clean the data, normalize it, fill the missing values. Along with that we can delete and append attributes in the tool itself. Visualization of the dataset is provided as in the form of ROC curve, Boundary visualizer, graph visualizer etc., Overall this tool is handy lightweight easy to use even when your dataset is novel, untouched with lots of error.

1.3.2 Types of Data

It is important to know the type of data before applying any algorithm. The type of data places a major role in pre-processing step and even during visualization. Majorly there are four types of data, i.e., Numerical type, categorical type, time series data, and text data.

Numerical data type is further classified into continuous and discrete type. To define the numerical data let us take the example number of red pens in a cartoon box when the length and width of pen is given and the cartoon box length, width and volume. In this the numerical data will be 1, 2, 100, 200, 3000 etc. these are called as discrete data types. But when we are asking of amount of ink in each pen then we can surely represent it as 10, 10.78, 20.55555 ml so a range can be given as 10–30 ml in between we can describe the value of quantity of ink present in the pen this kind of data is continuous data types.

Categorical data types represent the quality i.e. a food in restaurant is greasy, healthy, vegan, non-vegetarian, costly etc. these data types can be represented in numerical form. Ordinal data falls in the category of this data type where it classifies the data based on ranking. Like labeling the restaurant as good, better, best etc.

Time series data types mostly occur in logistics recommender system or in healthcare sector when the sequence of time is captured and stored over a period of time for analysis. Such like learning about the calorie burn using apple watch in every hour.

Text data type is mostly just text instead of above 3 data types. These are used as text or sentiment analysis.

1.3.3 Types of Dataset

The data used in machine learning can be of any type. We can use image dataset, voice, text, numeric etc. These datasets are further divided into three parts these are called as training set, validation set, and test set. These datasets are divided from the one original dataset mainly in 80:20 or 70:30 ratio for any state-of-art algorithm, although there is no restriction on division. The training set contains major portion of the dataset on which various machine learning algorithms are applied on varying parameters. Validation set and test set is applied for prediction purposes whether the applied method is correct or not.

1.3.4 Pre-processing of Data

Preprocessing is a step where the raw data is taken and verified about its condition. The raw data can be very noisy, with lots of outliers, empty value and even unidentified symbols. So this step help in rectifying and in other words beautifying the data. There are mainly three preprocessing applied in machine learning namely they are, rescaling data, binary conversion of data, standardization of data. Rescaling of data is a step when data is of variable scale. This step helps attribute to be in same scale which further used in optimization algorithm.

Binary conversion of data is transforming the data threshold in between the range of [0,1]. This step is mainly used in feature engineering when we have to select features for our analysis.

Standardization of data helps in making mean 0 and standard deviation as 1. This technique applies Gaussian distribution.

1.3.5 Exploratory Data Analysis

Extracting the hidden message when you receive a riddle puzzle for a hidden treasure. We will read and re-read the puzzle disintegrating every word, comma, full stop. This is done to reach the treasure. Similarly, a data scientist who has dataset as his riddle wants to disintegrate, extract maximum information, get deeper insight by visualization, make speculation of every moves, setting optimal parameters to nurture the initial dataset so he can get is some conclusion or head start on where to move. This step of exploration is called ad exploratory data analysis, which is major step to focus when we have all pre-processed data set to be used.

1.3.6 Log-Likelihood Estimation

Let us assume we are given a population with a probability density function $f(X\mathrm{i}|\theta)$ we can't really use the whole population at once for computation so we take a sample. The θ is a population parameter. From the population to the sample space θ is maximum likelihood estimator of the population parameter, it maximizes the value. So, the likelihood function describes the extent to which the sample provides the support for any particular parameter value. Higher supports correspond to the higher value for the likelihood. The likelihood L for the population sample of θ is (see Eq. 1):

$$L(\theta) = \prod_{i=1}^{n} f(X_i|\theta) \tag{1}$$

In any probability distributive function maximization is important and that can be achieve by concavity. The natural logarithm of likelihood function [10] make the distribution a concave function for maximization and that is called as log-likelihood function. Generally represented as l. We can take the derivative of the $L(\theta)$ and applying logarithm to it we get this equation (see Eq. 2):

$$l(\theta) = \sum_{i=1}^{n} \ln\, f(X_{\mathrm{i}}|\theta) \tag{2}$$

This equation is log-likelihood function. There is no standard value to denote the best log-likelihood value because likelihood value varies between [0,1] when we apply logarithm of zero or lower estimation the value becomes negative. In this case to use the function for estimation about the best model we need to compare with the other models $l(\theta)$ value, and the greatest among them is having better performance.

2 Unsupervised Learning

In a simple term we have already defined unsupervised learning algorithm in above Sect. 1.1.2. In this section we will study the types of unsupervised learning. We can understand one thing from the explanation in the previous section that unsupervised learning can't be simple defined just by-passing inputs and parameters dividing the set-in training and testing set and improving the accuracy by a greater number of repetitions. Here we have to visualize a lot before going into the conclusion whether our parameters are working according or against to the problem statement. While taking the dataset it is important for a researcher or scientist to know what this dataset can give us or what they can make out of it. If that is clear the algorithms can work on the basis of similarity and approximation and start combining in one group which later will be termed as clustering. By various parameters of correctly clustered instances, correlation, log likelihood value differences can help in making comparison and concluding the final outcome. In further discussion we will learn the demerits of unsupervised learning and how can we overcome it.

2.1 Types of Unsupervised Learning

In a broader perspective we can learn that unsupervised learning is divided in two main categories first is clustering, and secondly it is association [4]. In the coming section we will learn only about clustering algorithm. As this algorithm is only the main focus of our research.

2.1.1 Clustering

The definition of clustering comes from cluster, where the similar objects are grouped together. In the mid of 1850s there was cholera outbreak in London. One of the physician John Snow [5] plotted a graph denoting the location where the outbreak happens and to his surprise, he found out that the area where the outbreak occur has polluted well. So, it was the first use of clustering recorded which not only gave a stat about the status of an incident but also a feasible solution.

Likewise, in clustering we are also grouping the similar elements group in one class. It is obvious to have a doubt why ultimately, we need clustering? The idea

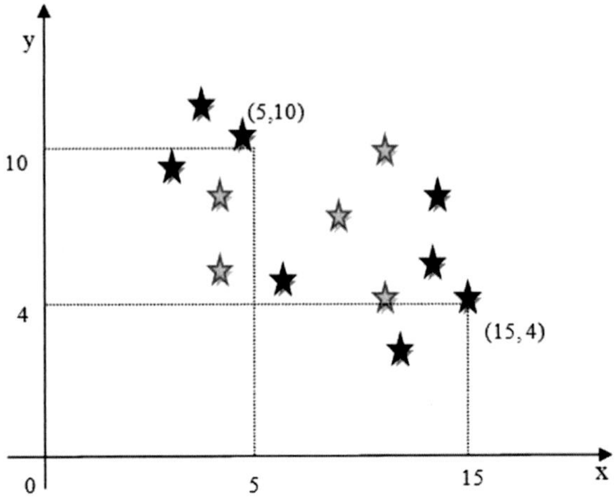

Fig. 2 The graph is plotted of housing detail where X-axis is price of houses and Y-axis is housing area code

was proposed when we are getting the data but it was not classified. This made the scientist think upon how to know what this data wants to predict or deliver. For that purpose, they introduced clustering technique like the correlated data will be clubbed together. This correlation can be obtained by matching the underlying pattern of different datasets. That was the trigger to another great use case of dimensionality reduction, which will help in reducing redundancy among the copious number of variables. Among these clustering strategies it is further distributed into hard and soft clustering. *Hard clustering* is where the observation is taken into one cluster and no probability is calculated. *Soft Clustering* is the technique where the probability and likelihood is calculated for the clusters and compared [6].

Distance Measurement in Clustering

Calculating the distribution of the objects in clustering is done using Euclidean and Manhattan distances

(a) ***Manhattan distance***

Manhattan distance is also called as city block distance. When we are having n number of coordinates in a plane. Then we need to take the sum of differences between the absolute values of various coordinates iteratively. In the given graph 2 we took a scenario of housing cheapest price in New Delhi. A graph is plotted (refer Fig. 2) where two categories of housing is selected where the x-axis is price and y-axis are area. Here the distance between each coordinate is find using the formula (see Eq. 3):

$$Manhattan\,distance\,(d(\mathrm{a_i}, \mathrm{b_i})) = \sum_{1}^{n} \left|\mathrm{a_i} - b_{\mathrm{j}}\right| \tag{3}$$

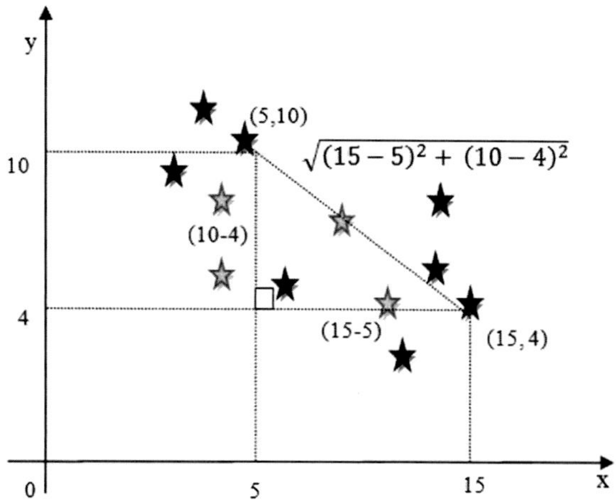

Fig. 3 X-axis is house price and Y-axis is area code. The Euclidean distance formula is applied between point (5,10) and (15,4)

We can find the distance between two points x = (a_1, b_1) = (5,10) and y = (a_2, b_2) = (15, 4), solving these two points we will get the Manhattan distance as 16.

The implementation of city block algorithm can take O (nlogn) for every input of n coordinate. The computation cost compare to Euclidean city block distance is cheaper.

(b) ***Euclidean distance***

The Euclidean distance formula is used to find distance between two points located either in a plane or 3-dimensional space. This methodology is application of Pythagorean theorem where the distance between two points are subtracted and summed then squared. Let us take the same example as above mention in part a manhattan distance. Just in this method we take the two coordinates of x and y (refer Fig. 3) and apply this formula (see Eq. 4):

$$Euclidean\ distance\ d(a_i, b_j) = \sqrt{\sum_{1}^{n} (a_i - b_j)^2}$$

$$d = \sqrt{(15-5)^2 + (10-4)^2} = 11.66 \quad (4)$$

This distance formula will produce similar result in each iteration that is why is it also called as translation invariant.

Clustering Techniques

In Fig. 4 we can see that the clustering is divided into three major categories. There are nearly hundreds of clustering algorithms available and researched upon.

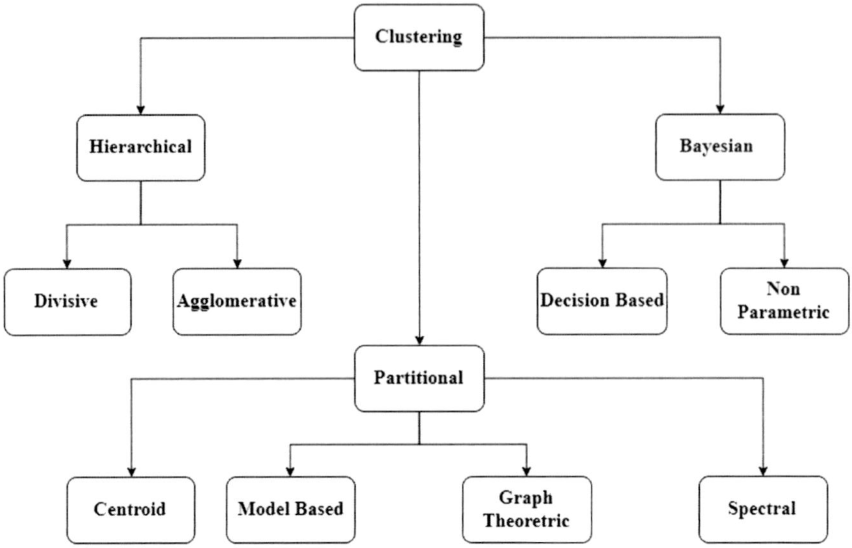

Fig. 4 A tree diagram to show the hierarchy of clustering methods

Hierarchical clustering is again divided further into two parts that is Agglomerative clustering and divisive clustering. This clustering is technique performs well when we have more normalized data where we can sight more deeper in clusters. Generally, when we are researching about the taxonomy where we need to classify the plants based on various property and of their closeness this algorithm work as charm. Agglomerative clustering on the other hand is bottom-up approach. In this technique take all the clusters as new cluster as like merge algorithm the similar clusters will be combined together until we find a common cluster for all in the end. On the other hand, divisive clustering is top-down approach. Here we take a big cluster and partition it based on common clusters together. It is done till it reaches the leaf node of cluster with unique features.

Another interesting type of technique which we will discuss in depth in our section is partition clustering. This technique is used where the observation is divided and clustered based on the similarities between various attributes. Some of the popular algorithms in this group is k-means clustering, k-medoid clustering, CLARA algorithm.

Due to the constrain of the chapter we are going to directly illustrate about the algorithms like K-Means algorithm, density-based models, expectation maximization, and combination of kmeans and genetic algorithm as GenClust++ [7].

2.2 Challenges of Unsupervised Learning

Apart from the advantages we have is it is easy to implement. In real world the data we get is mostly unlabeled and labeling those data takes really a lot a time. Also unsupervised learning is a great way to implement real-time data. Satellite imaginary dataset it is huge dataset with lots of coordinates observation and images clustering them comes very handy and help in visualization. Coming to the shortcomings, the first and foremost is about its less accuracy. As the data are very raw and unlabeled, we are really unsure what outcome is right or wrong. Which is not the case being supervised learning though. Where accuracy, correlation factor, cross fold validation etc. are various way to define the accuracy of a model. Another pitfall is that the we are really ascertaining about the results we obtained from the analysis. It always requires an expert view point before finalizing the results for unsupervised learning dataset.

In conclusion unsupervised learning comes with its own merits and demerits but has a long history of research and still an active research area. It is an application dependent and very subjective analysis field.

3 Clustering Algorithms

3.1 K-Means Algorithm

The k-means clustering [8] algorithm is a method of partition clustering. In this algorithm we are given some N number of observations in that population we need to form k-cluster based on best similar points. Generally, the number of k-cluster varies between 2 and 10. The popular method to find number of clusters to be best for this algorithm is Elbow method. Where the point like elbow is find in the graph which is forming maximum clusters.

Let us understand this algorithm. Here, we will follow a set of steps which is iterated until the best Euclidean distance is found between each of the points.
Algorithm 3.1 simple kmeans algorithm

1. Initialize *k* as initial cluster points
2. Calculate minimum distance from each point to the centroid
3. Recomputed each centroid by finding average/mean in each cluster and assign cluster to it.
4. Repeat from step 2 until the centroids doesn't change.

We have already discussed the distance formula in previous Section "Distance Measurement in Clustering". It can be calculated using Euclidean and manhattan distance. Now mathematical formulation for simple K-means we can write it as:

Let us set of data points $x = \{x_1, x_2, x_3, x_4 \ldots x_n\}$ and $v = \{v_1, v_2, v_3, \ldots, v_c\}$ be the centers (see 5).

$$f(\boldsymbol{v}) = \sum_{i=1}^{c}\sum_{j=1}^{c_i}\left(\|\boldsymbol{x}_i - \boldsymbol{v}_j\|\right)^2 \tag{5}$$

where,

$||x_i - v_j||$: Euclidean distance between x_i and v_j.
c_i: number of data points in i-th cluster.
c: number of cluster centers.

Each iteration take computation time between n objects and k clusters is O(kn). Including the computation of the cluster centers of each object get added to some clusters which will take O(n). When we combine the above steps for t iterations time will be O(tkn).

3.2 Expectation Maximization Algorithm

The expectation maximization [9] algorithm is very similar with k-means algorithm. The uniqueness of this algorithm is here we can deal with hidden data(labels). This algorithm iterates between expectation (E) step where log-likelihood parameter is getting evaluated which is discussed in Sect. 1.3.6 using the present estimated parameters, also all the missing labels get fixed with estimation. Whereas maximization (M) take the expected estimates of log-likelihood from above step and maximize it. Let us assume we are given a population of *n* data vectors i.e., $D = \{x_1, x_2, \ldots, x_n\}$ and hidden variables $H = \{h_1, h_2, \ldots, h_n\}$.

Log-likelihood for the given model can be stated as (see Eq. 6):

$$l(\theta) = \log f(D|\theta) = \log \sum_{z} f(D, H|\theta) \tag{6}$$

where θ, H are unknowns.

Now to derive the formal algorithm for EM we need Q(H) to be a distribution on hidden variables i.e. f(Q,θ).

EM algorithm steps:

Step 1: Expectation

- Maximizing hidden distribution function f with respect to the variable Q, keeping the population distribution parameter θ fixed. That can be mathematically formulated as Eq. (7):

$$Q^{k+1} = f\left(H|D, \theta^k\right) \tag{7}$$

Step 2: Maximization

- Maximize the hidden distribution variable f with respect to θ, fixing the hidden distribution variable Q. This can be represented as Eq. (8):

$$\theta^{k+1} = \arg max_{\theta} \sum_{H} f\left(H|D, \theta^{k}\right) \log f(X, H|\theta) \tag{8}$$

Maximization step in above algorithm is to maximize the lower bound of f on the likelihood of L and expectation step is to close the gap and making bound of f equals to the likelihood of L. This algorithm may converge to local optimum. In the conclusion of this algorithm we can say that this is an easy, powerful algorithm with lot more usage in unsupervised, semi-supervised and supervised learning. The popular application of EM is estimating motion models for tracking, hidden markov models, image segmentation etc.

3.3 GenClust++

Combination of the state-of-art algorithm namely k-means and genetic algorithms novel operators are generated for doing high quality clustering then that algorithm is named as GenClust++ [7]. This algorithm will take O(n) time to give the high-quality clusters. Instead of random generation of chromosomes i.e., a solution of binary string type, the best of them are chosen and stored for further generation. In this algorithm comparison is done between two consecutive chromosomes so we don't miss out the best ones. The k-means algorithm is selected using modified hill climbing algorithm [11] which is MK-means (Modified k-means algorithm). There are main eight components in this algorithm proposed mentioned as:

Component 1: normalize the dataset
Component 2: Modified k-means called MK-means
Component 3: Initial population with probabilistic solution
Component 4: crossover
Component 5: Elitism operation
Component 6: Mutation
Component 7: probabilistic cloning with MK-means
Component 8: Chromosomes selection

Let's note some of the basic notation to be used further in the explanations are, R_o is original record of numerical and categorical data types. $R^{o}_{i,j}$ is j-th normalized attribute value of the i-th record in R_o. P_s is population for the sequence of chromosomes. C = $\{c_1, c_2, \ldots, c_k\}$ are the cluster centers. A_j is the attribute of numerical domain [l,u] where l is lower limit and u is upper limit.

Component 1: normalize the dataset

Then $R^o_{i,j}$ is normalized as $\mathrm{R_{i,j}} = \frac{R^o_{i,j}-1}{u-l}$. This normalization steps make a assumption that all attributes are same and successfully avoiding the issue of scaling and mensuration.

Component 2: Modified k-means called MK-means

Distance between two records $\mathrm{R_i}$ and $\mathrm{R_l}$ is d ($\mathrm{R_i}$, $\mathrm{R_l}$). The distance is d is calculating using manhattan distance metric to reduce the total squared errors and outliers' value (see Eq. 9)

$$\mathrm{d(R_i, R_l)} = \frac{\sum_{\mathrm{j=1}}^{\mathrm{t}} \left|\mathrm{R_{i,j}} - \mathrm{R_{1,j}}\right| + \sum_{\mathrm{j=t+1}}^{\mathrm{m}} \mathrm{d}\left(\mathrm{R_{i,j}}, \mathrm{R_{1,j}}\right)}{|\mathrm{A}|} \tag{9}$$

Component 3: Initial population with probabilistic solution

The high-quality initial clusters are produced of size s with initial population of 30. The population of size s is denoted as $\mathrm{P_s}$. The probabilistic value is generated in every stage and compared. Although for the sake of GenClust++ implementation only one stage of iteration is considered generally because the rest stages may not guarantee the best chromosomes.

Component 4: crossover operation

Chromosomes are listed in list L sorted in descending order of their fitness level. $\mathrm{P_i}$ is the best fitness value and $\mathrm{P_j}$ is worst fitness value. The probability of $\mathrm{P_i}$ is found using its fitness level among of all the population of fitness in worst case. Crossover helps in eliminating duplicate chromosomes. We can write it as (see Eq. 10):

$$\mathrm{P(P_i)} = \frac{fitness(\mathrm{P_i})}{\sum_{j=1}^{|L|} fitness(\mathrm{P_i})} \tag{10}$$

Component 5: Elitism operation

We are having the under-performing clusters after iterations then we can insert the highest fitness chromosomes in them is elitism operation [11]. This operation is performed in every crossover and mutation step.

Component 6: Mutation

While finding unconventional solution mutation changes the genes of chromosomes randomly. The attributes are selected uniformly from the domain. $\mathrm{M_i}$ is mutation probability of i-th chromosomes. $\mathrm{CR_i}$ is class of clusters in record $\mathrm{R_i}$ in the population of clusters $\mathrm{P_c}$ can be calculated as (see Eq. 11):

$$M_i = \begin{cases} \frac{f_{max} - fitness(CR_i)}{2(f_{max} - \bar{f})}, & fitness(CR_i) > \bar{f} \\ \frac{1}{2}, & fitness(CR_i) \le \bar{f} \end{cases} \tag{11}$$

where f_{max} is fitness of current chromosomes and $\bar{f}$ is average fitness of the chromosomes.

Component 7: Probabilistic cloning with MK-means

At every step probabilistic cloning of chromosomes are performed. The cloning is for accelerating convergence thus it is applied over every generation. This is done to ensure the elitism and mutation operation are generating best and the fittest of all chromosomes.

Component 8: Chromosomes selection

In this step the merging of all the chromosomes are between two population of size s. This preserves the generated elite chromosomes for genetic search. In short, this algorithm ranks all the generation and chose the best amongst of all.

3.4 Density Based Clustering

The make density based is a wrapper class algorithm which enclose other clustering algorithms like k-means, canopy, EM, CLOPE etc. with their parameters and clusters are formed with a log-likelihood value. This algorithm makes it easier to make a judgment about what clustering algorithm is having good performance compares to other.

While using WEKA 3.8 the MakeDensityBased cluster technique can be used to perform the analysis easily. The clustering algorithms mostly on combining the similar property mid-points in one place or even calculating the minimum distance between the data points and cluster the less distance data points in one place. In the density based clustering the clustering is performed using for example k-means algorithm with the distance method, clustering instance, seed number etc. the algorithm performed the operation now the MakeDensityBased clusters each attribute and find the best among all the clusters by evaluating normal distribution with the reference of mean and standard deviation. This normal distribution calculates the likelihood and an estimate of likelihood is represented using log-likelihood value for the maximize value.

So, number of iterations takes place where within each cluster the sum squared error [12] is find. The sum of squared error (SSE) is the summation of squares between the differences of instance and mean. If all the cases within the clusters are identical then they will give zero value.

The mathematical representation of SSE is (see Eq. 12):

$$SSE = \sum_{i=1}^{n} \left(x_i - \bar{X}\right)^2 \tag{12}$$

where x_i is each instances of data-points and $\bar{X}$ is the mean.

The total sum of squared errors is never zero. It is represented using $SSE_{total} = SSE_1 + SSE_2 + SSE_3 + \cdots SSE_n$ for the n-th value of iteration.

4 Particle Swarm Optimization (PSO) Algorithm

This topic needs a separate section because it is very different from our above unsupervised learning algorithm. The particle swarm optimization [13] is a stochastic search nature inspired algorithm which is used in feature selection and optimization. This algorithm is inspired from the social behavior of flock of birds. As we saw in GenClust++ and various genetic algorithms that chromosomes are selected in every iteration and among them the best one is selected for further generation similarly in PSO algorithm these solutions are termed as particles. These particles are placed according to their fitness objective in best cluster in every iteration. Unlike the flock of birds and swarms of fish there is one thing in common that is best among them will lead the whole group. That one particle is best with great speed and buildup. That particles is not always same it changes based on the different position of flocks. That trait of nature is the inspiration of particle swarm optimization. The behavior of a particle totally depends on three factors that are p_i particle current position, particle current velocity s_i and particle best position b_i [14]. The particle swarms are updated in every iteration based on their p_i and s_i. We can equate this in a equation for evaluation (see Eqs. 13, 14):

$$s_{i,k}^{t+1} = \omega s_{i,k} + K_1 R_{1,k}\left(b_{i,k} - p_{i,k}\right) + K_2 R_{2,k}\left(\widehat{b_k} - p_{i,k}\right) \tag{13}$$

$$p_i^{t+1} = p_i + s_i^{t+1} \tag{14}$$

where, ω is inertia weight, K_1 and K_2 are acceleration constant. R is random constant generally value is 2 The velocity s_i calculated by contribution of these three factors [13]:

a. Selecting best velocity from previous velocity.
b. Distance between the particle from its individual personal best position is cognitive component.
c. A functional distance of particle from the swarm of best position is called as social component.

$$b_i^{t+1} = \begin{cases} b_i, \ if \ f\left(p_i^{t+1} \geq f(b_i)\right) \\ p_i, \ if \ f\left(p_i^{t+1} < f(b_i)\right) \end{cases} \tag{15}$$

In the velocity the personal best position of the particle is calculated using this formula (see Eq. 15).

The best position particle in the i-th neighbor is always attracted to its swarm. So the best particle swarm of i-th position in the neighborhood can be denoted as $K_2 R_{2,k}\left(\widehat{b_{j,k}} - p_{i,k}\right)$. This algorithm is terminated when the velocity in any of the iteration is calculated as zero.

The particle swarm optimization has various application and it is used with clustering [14], in classification [15], in our research work we tried to explore the use cases of PSO as a subset evaluator which will be surely used as feature selector. In the next section we will learn more about PSO-CFS.

4.1 PSO with Subset Evaluator

The PSO can be used as subset evaluator using cfs (correlation-based feature selection). This method helps in selecting the attributes based on PSO as subset evaluator [15] duplicity between them only the very high correlated attributes within the class are selected.

In our experiment we used Weka 3.8, CFS [15] is a filtering technique where the merit (M) heuristic is calculated for the *t* features in the dataset with this Eq. (16):

$$M_s = \frac{t * r_{cf}}{\sqrt{t + t(t-1)r_{ff}}} \tag{16}$$

where, M_s is merit of subset of attributes, r_{cf} is mean of correlation between feature class and r_{ff} average between feature of correlation. In the end of evaluation of most correlated features are collected in a subset as shown in Fig. 5.

5 Designing an Experimental Setup

In the previous section of this chapter we focused on the basics of unsupervised learning and how we are making use of this knowledge to fit our problem statement. In the current section we will learn about the problem statement where we are asking the question of why we need to make this experiment? Which will be following with a sub-section where we will learn how we can setup our experiment. We will also learn about the dataset and possible outcome obtained from the result.

5.1 Healthcare Survey Dataset with Unsupervised Learning

The health analytics datasets are very unstructured and unlabeled compare to any healthcare disease dataset. These datasets are mostly collected using manual ways

Fig. 5 PSO CfsSubsetEvaluator is applied on our healthcare analytics dataset where 25 features are selected among 91 attributes

so chances of error is very high. Which make it more challenging before applying it on any technique. The health analytics generally contains the survey of the various regions of a country and have attributes like age, sex, region, disease, government care, family income etc.

The clustering and prediction of the possibility or the higher chances of having any kind of disease in any state or district will help the government to provide any kind of necessary preparation beforehand. This analytical data can help in preventing any kind of haphazard in an area. Applying this tactic will help the government and even the citizen as they can obtain an analysis in user friendly manner like web app and mobile application so we can somewhere able to achieve the sustainable healthcare goal proposed by UN SDG. This goal is to impute any kind of chronic illness from the population and make every country a sustainable place to live.

5.2 *Procedure and Techniques*

The hybrid approach is for attribute reduction using the Particle Swarm Optimization (PSO) algorithm with CfsSubsetEval. Subsets of features that are highly correlated with the class while having low inter-correlation are preferred.

The Filtered dataset and PSO evaluated dataset both are evaluated to learn about the best model performance and pick the one model which give good evaluation. Different unsupervised learning algorithms are used like Make Density Based Cluster which is wrapper cluster model and Kmeans, EM, GenClust++ (refer Fig. 6).

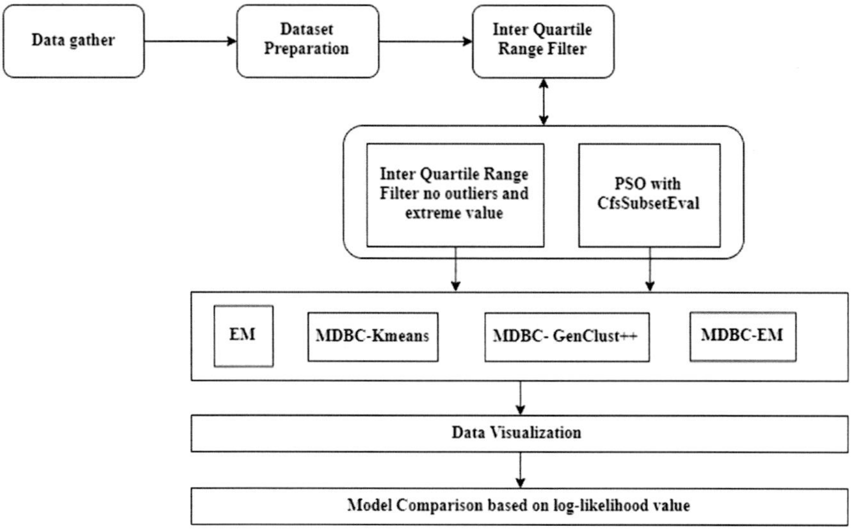

Fig. 6 The model is formulated to execute the experiment with optimized dataset for maximize result

5.3 *Various Stages of Experimentation*

Stage 1: Data gathering and preparation.

Obtain the Chronic illness survey data from the https://data.gov.in/catalog/indicators-annual-health-survey. Dataset with 92 attributes and 284 instances which is pre-processed and all the NaN and missing values where get eliminated. Visualize after the first pre-processing. The dataset is left with 91 attributes and 241 instances. These datasets are still have less outliers and noises (Fig. 7).

Stage 2: Filtering dataset using inter quartile range.

Weka 3.8.2 gives the option of filtering the dataset. As our data are unlabeled so we used Unsupervised filter techniques. It goes like Unsupervised filtering → attributes → InterQuartileRange → Set attributeIndices to first-last. The final InterQuartileRange filtered gives 91 attributes and 213 instances (refer Fig. 8).

Stage 3: Applying clustering algorithm after removing the outliers and extreme value.

EM, MakeDensityBasedCluster (MDBC) wrapped with K-means (k = 9), GenClust++, EM. The log likelihood value, clustering instance, and build time are used for evaluating the clusters. The k cluster value is 9. The dataset is divided in training set 70% and testing set of 30%.

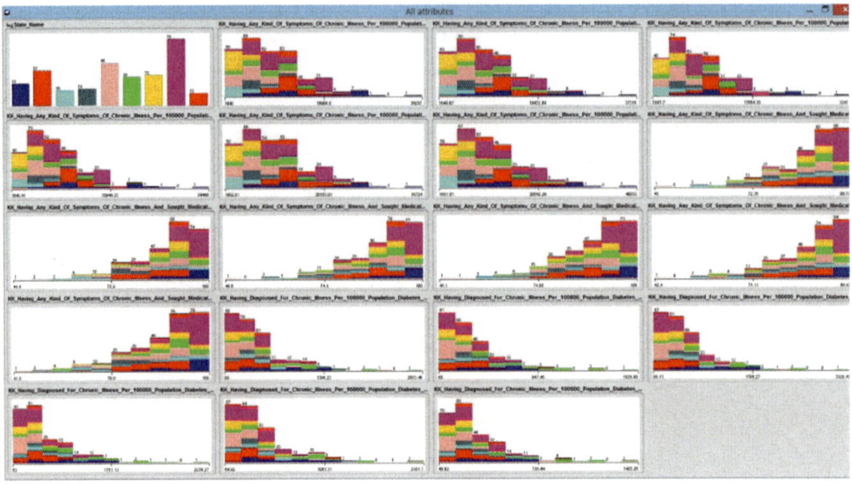

Fig. 7 The visualization of cleaned dataset using the Weka Explorer

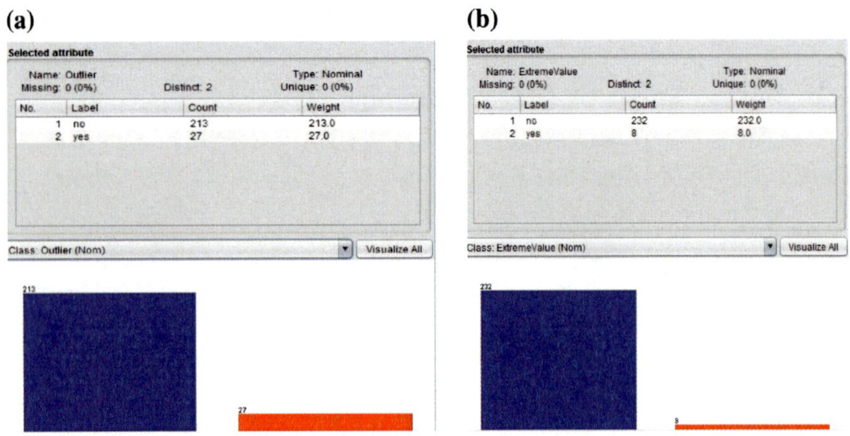

Fig. 8 **a**, **b** The visualization of the outliers and extreme value in the filtered dataset using Weka Explorer. 27 outliers and 8 extreme values are imputed from the dataset

Stage 4: The dataset is now compared in the Weka Knowledge Flow GUI with the all algorithms.

We draw all the model and visualize together in knowledgeFlow GUI for better idea about the performance of the models. The graph is plotted (refer Fig. 9) from the DataVisualizer visualization and the scatter matrix is plotted which is compared between different attributes.

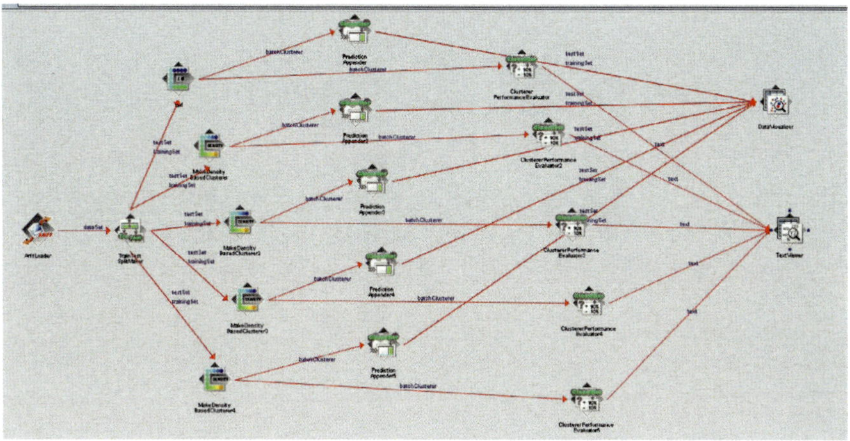

Fig. 9 Comparing all the five algorithms in filtered dataset using Weka Workbench

Stage 5: Feature selection using the PSO with CfsSubsetEval.

The Particle swarm Optimization show minimum attribute when the feature "State" is chosen and the seed value at 3 gives only 25 attributes which are highly correlated with each other.

Visualization of the reduced dataset with only 25 attributes and 213 instances (refer Fig. 10).

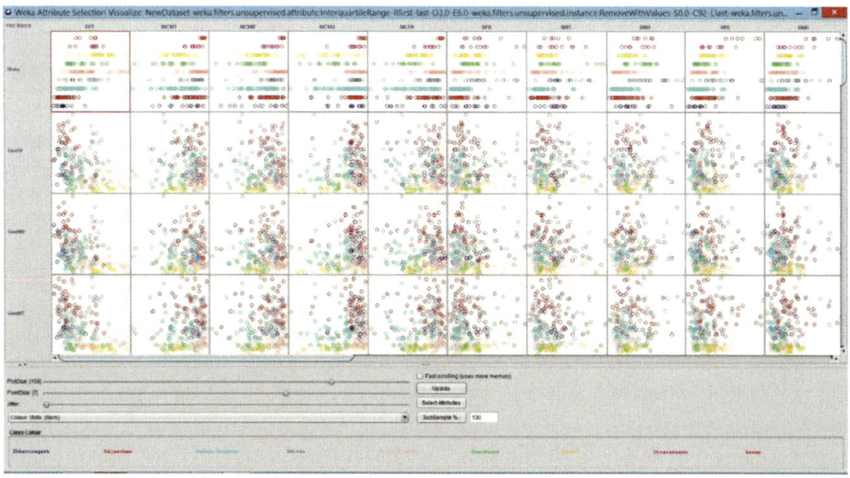

Fig. 10 Visualizing the Cluster of the PSO reduced attribute dataset using EM algorithm in Weka Explorer

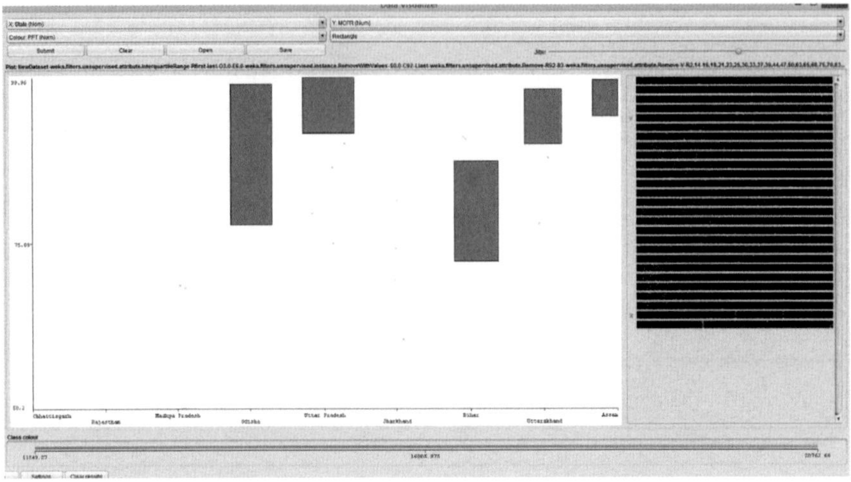

Fig. 11 Visualization of model comparison for PSO reduced attribute dataset using all five proposed algorithms in Weka Knowledge Flow DataVisualizer

Stage 6: Clustering the new obtained dataset hybrid PSO using with the same model.

Clustering the dataset using EM and cluster wrapper MakeDensityBasedCluster wrapping it with EM, Kmeans (k = 9), GenClust++. We have clustered the Total population having chronic illness and they are seeking Government treatment with Population sought medical care for the female in rural area. The cluster is highest in the end of cluster 2 which is representing the state Uttar Pradesh. Comparing all the clustering algorithms using the knowledge flow GUI in weka and plotting the dataVisualizer and textViewer for the summary.

Visualization of the whole dataset with the attribute such as state and Medical care sought by Female in rural area the cluster over the total population, we found the state of Uttar Pradesh is having high density cluster as shown in Fig. 11 where woman is having high percentage who are seeking medical care for any chronic illness.

The above six stages we can repeat stage five and six again to obtain more combination and outcome about different states.

6 Results

The model evaluation using clustering algorithm is challenging. The evaluation factor of clustering has already been discussed in Sect. 1.2.6. This model is relative clustering evaluation due to distinct pair of clusters are formed. The Weka tool have some limitations like it don't provide any relative clustering evaluation indexing.

Table 1 Comparison for IQR filtered dataset

Model	Clustered instances	Build time (s)	Log likelihood
EM	7	7.21	−583.74
MDBC-Kmeans (k = 9)	9	0.01	−586.81
MDBC-GenClust++	**10**	7.58	**−581.83**
MDBC-EM	6	6.28	−583.74

So, it is proposed in future work to make relative clustering evaluation for the same dataset.

For now, the table of outcome is constructed in different scenarios that has been evaluated under various parameters like number clustered instances, total build time, and the log likelihood value.

If the likelihood of the sample is greater under one model than another, we tend to infer that the former model is more likely than the ladder. If a likelihood is less than 1, the log-likelihood is negative, but this can be due to noises, sparse data, small sample sizes etc. We cannot objectively say anything based on a single likelihood or log-likelihood; it is strictly relative. It only compares models. This is a confirmatory, and not exploratory comparison. They do not determine the total number of clusters because you are not comparing models, which is an exploratory question. The tendency of likelihood in that case is to over fit because maximum likelihood has some high dimensional problems. Let us learn more about the result we have obtained by training our dataset under various conditions as mentioned in coming section.

6.1 Outcome of IQR Filtered Dataset

The dataset is split into training and testing dataset with 70–30%. The dataset is simply filtered so the log-likelihood value is very high. Still the MDBC GenClust++ though take more build time comparatively but it gives less log-likelihood value is less (Table 1).

6.2 Outcome of PSO Optimized Attribute Dataset

The feature Engineering is done using stochastic heuristic optimization technique PSO. Attribute is evaluated using CfsSubsetEval the PSOSearch with seed = 3 which gave 25 attributes, even the training and test dataset is divided in 70–30% ratio. Here, the build time and log-likelihood reduced exceptionally of MDBC-GenClust++. The state-of-art algorithms like EM and k-means is showing poor performance (Table 2).

To make more concrete clustering we have further divided our filtered dataset on the ten main features.

Table 2 Comparison PSO optimized attributes dataset

Model	Clustered instances	Build time (s)	Log likelihood
EM	7	4.29	−146.99
MDBC-Kmeans (k = 9)	8	0	−147.99
MDBC-GenClust++	**10**	**0.54**	**−146.86**
MDBC-EM	6	4.11	−147.00

- The first feature is *population of hundred thousand having any kind of chronic illness*. For these 213 instances with 10 attributes were selected. The "**State**" class is the only nominal attribute. Clustering is done by using 70–30% division of training and testing. Comparatively, the EM give less log-likelihood as MDBC-GenClust++. But GenClust++ is doing clustering more correctly (Table 3).
- The second feature is the *percentage of people sought medical care suffering from chronic illness*. As the dataset is getting relatively smaller so, according to the calculation EM outperform GenClust++ in small dataset (Table 4).
- The third feature is the *Diabetic patient in hundred thousand population in different states of India*. MDBC-EM give a better performance compare to another model. But if we look for the correctly clustered instances MDBC-GenClust++ is performing much better it is closely looking for chromosomes and correlation between them (Table 5).
- The fourth is the survey of *Hypertension patient in hundred thousand of population*. All the clustering has similar performance. The instances differences are less. This feature may be fully imputed from the dataset which will only create the noise. For this GenClust++ algorithm is better than another algorithm (Table 6).

Table 3 Comparison on filtered dataset for the population 100,000 having any kind of chronic illness

Model	Clustered instances	Build time (s)	Log likelihood
EM	7	1.65	**−80.11207**
MDBC-Kmeans (k = 9)	8	0.01	−82.73729
MDBC-GenClust++	**9**	**0.24**	**−81.20618**
MDBC-EM	7	1.64	−80.12846

Table 4 Comparison on filtered dataset for the percentage of people sought medical care suffering from chronic illness

Model	Clustered instances	Build time (s)	Log likelihood
EM	8	2.4	**−24.58691**
MDBC-Kmeans (k = 9)	9	0	−26.88902
MDBC-GenClust++	**12**	**0.22**	**−25.25109**
MDBC-EM	8	2.38	−24.61446

Table 5 Comparison on filtered dataset for the diabetic patient

Model	Clustered instances	Build time (s)	Log likelihood
EM	6	1.26	**−60.69583**
MDBC-Kmeans (k = 9)	9	0	−63.36254
MDBC-GenClust++	**10**	**0.22**	−61.5845
MDBC-EM	**6**	1.25	**−60.68983**

Table 6 Comparison on filtered dataset for the Hypertension patient

Model	Clustered instances	Build time (s)	Log likelihood
EM	7	1.67	−65.39522
MDBC-Kmeans (k = 9)	7	1.67	−65.39522
MDBC-GenClust++	**12**	**0.21**	**−65.2464**
MDBC-EM	7	2.41	−65.45013

Table 7 Comparison on filtered dataset for the tuberculosis patient

Model	Clustered instances	Build time (s)	Log likelihood
EM	**5**	1.79	**−51.22838**
MDBC-Kmeans (k = 9)	8	0	−53.09196
MDBC-GenClust++	12	0.32	−51.767
MDBC-EM	5	0.9	**−51.23598**

- The fifth feature is about the *hundred thousand of population suffering from tuberculosis*. As we discussed in above results the EM perform much better in small datasets. But still accuracy is maintained by GenClust++, although the state-of-art algorithm k-means is not giving any positive result in our dataset (Table 7).
- The sixth feature is about the *hundred thousand of population suffering from Asthma or any Chronic Respiratory Disease*. Although it looks that MDBC-EM is doing well but the crisp clustering is done in MDBC-GenClust++ where the correctly clustered instances are high and build time is very low (Table 8).

Table 8 Comparison on filtered dataset for the asthma or any chronic respiratory disease patient

Model	Clustered instances	Build time (s)	Log likelihood
EM	8	1.77	−60.30422
MDBC-Kmeans (k = 9)	9	0	−63.89649
MDBC-GenClust++	**12**	**0.32**	**−51.767**
MDBC-EM	**5**	**0.9**	**−51.23598**

Table 9 Comparison on filtered dataset for arthritis disease patient

Model	Clustered instances	Build time (s)	Log likelihood
EM	9	2.6	−67.92646
MDBC-Kmeans (k = 9)	8	0	−72.16498
MDBC-GenClust++	**12**	**0.32**	**−51.767**
MDBC-EM	**5**	**0.9**	**−51.23598**

Table 10 Comparison on filtered dataset for any kind of chronic illness

Model	Clustered instances	Build time (s)	Log likelihood
EM	**7**	1.63	**−79.33955**
MDBC-Kmeans (k = 9)	9	0	−81.00403
MDBC-GenClust++	12	0.42	−80.0946
MDBC-EM	7	1.98	−79.34967

- The seventh feature is about the *hundred thousand of population suffering from Arthritis Disease*. MDBC-EM and MDBC-GenClust++ closely competing with each other (Table 9).
- The eighth feature is about the *hundred thousand of population suffering from Any kind of Chronic Illness*. As the log-likelihood of MDBC-GenClust++ is approximately 0.7% high compare with EM but still the rest parameter of GenClust++ is better than EM (Table 10).
- The ninth feature is about the *population diagnosed for any kind of Chronic Illness and getting Regular Treatment (%)*.
- The tenth feature is about the *population diagnosed for any kind of Chronic Illness and getting Regular Treatment from Government Source (%)*.

In ninth and tenth feature selection dataset got really small so the EM give maximum likelihood value also the correctly clustered instances become high comparing with GenClust++. This can make us to conclude that EM performs better in small, similar distribution of value dataset. Whereas the GenClust++ is more robust and can withstand high dimension datasets because of its updated chromosomes in every iteration. We will discuss more about the obtained result in next section.

6.3 Discussion on the Result

The problem statement was to show the performance of unlabeled and unstructured dataset using the clustering algorithm. Performing our experiment using the Weka 3.8.2 tool and using the dataset form the India Government site of Analytics Health Survey of year 2012–13. This is one of the unprocessed data. We didn't labeled data due to time constraint.

The cleaning, pre-processing and visualization consumed so much of good ours, that is why only the clustering model evaluation was possible. In the given scenario where dimension reduction is applied using Particle Swarm Optimization algorithm with the seed count 3. Various features are selected and when the same models were trained under different circumstances, we have got that GenClust++ model give better clustering and log-likelihood value. Closely reading the outcome table we can conclude the EM has better performance tendency in small similar kind of dataset like when we took the percentage of population getting regular treatment from Government where all the data are between 0 to 100 (refer Tables 11 and 12) the clustering and log-likelihood stats increased compared with rest datasets when we have combined distribution of percentage and numerical value. Similarly, an interesting observation is GenClust++ algorithm doesn't work well in small dataset. Even though correctly clustered instances are high and build time were low in that situation but log-likelihood value was slightly high comparatively. We plotted a comparison graph (refer Fig. 12) where genclust++ is showed higher performance rate.

Observing the dataset under various model we can also conclude that the Indian state Uttar Pradesh has shown higher clustered density on suffering from Diabetes, Asthma, Arthritis, Hypertension, Tuberculosis, and any kind of chronic illness among hundred thousand of population in the state (refer Fig. 13). This kind of study can help the government or the healthcare industry to improve the facility and government medical care in those area.

The people suffering from chronic sickness seeking regular treatment from the government sources (%) the state of Odisha have the highest percentage 82.36% followed by Rajasthan 77.8% (refer Fig. 14). The poorest health analytics is shown

Table 11 Comparison on filtered dataset for any kind chronic illness and getting regular treatment (%)

Model	Clustered instances	Build time (s)	Log likelihood
EM	**6**	1.82	**−30.48188**
MDBC-Kmeans (k = 9)	9	0	−33.35614
MDBC-GenClust++	12	0.29	−31.55003
MDBC-EM	6	1.6	−30.49914

Table 12 Comparison on filtered dataset for any kind chronic illness and getting regular treatment from government source (%)

Model	Clustered instances	Build time (s)	Log likelihood
EM	**7**	1.78	**−29.92674**
MDBC-Kmeans (k = 9)	9	0	−32.82416
MDBC-GenClust++	6	0.24	−32.57804
MDBC-EM	7	1.82	−29.93141

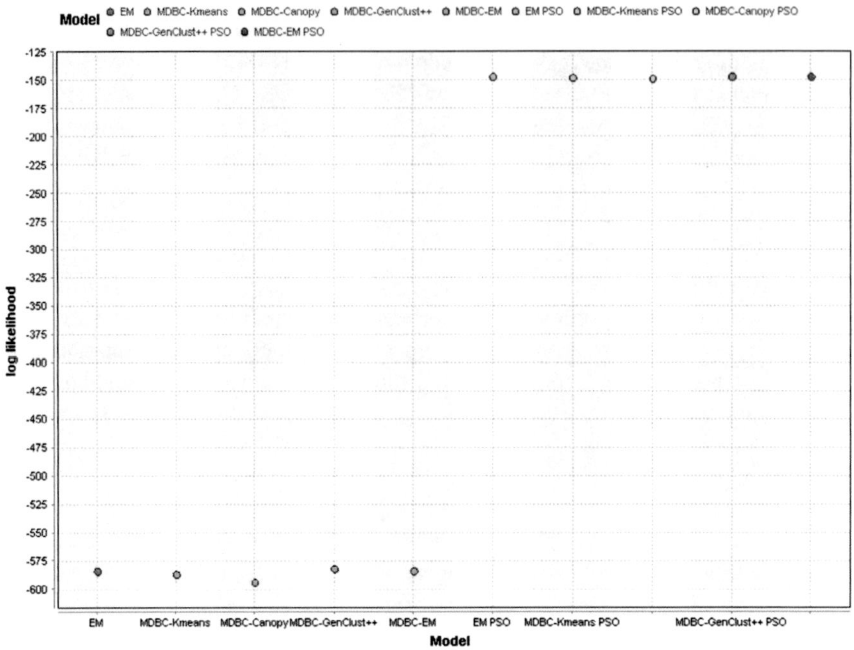

Fig. 12 Comparison in the form of scatter plot for the proposed algorithms based on factors like instances clustered, build time, and log likelihood

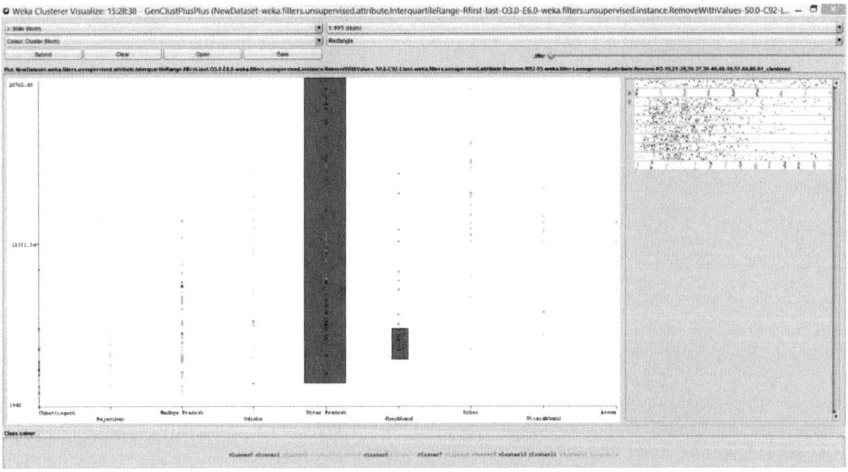

Fig. 13 Comparing Indian states where the total population likely to have more chronic illness patient compare to other states. Uttar Pradesh is highly populated with the chronic illness patient

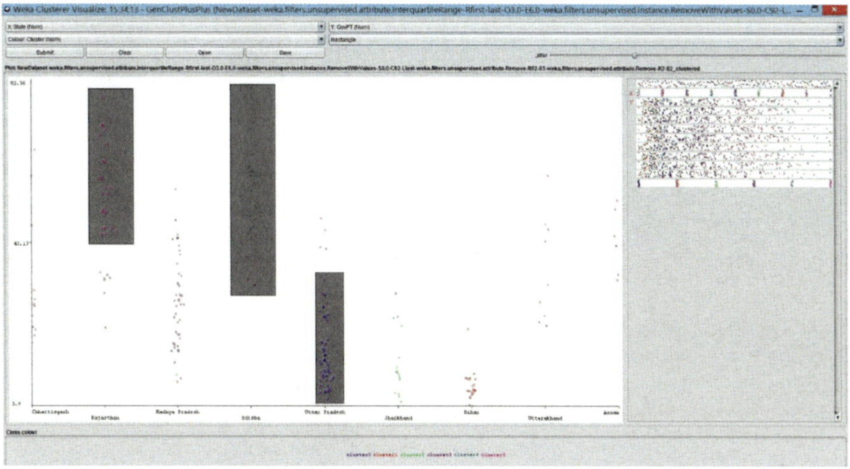

Fig. 14 Comparing Indian states where the total population likely to receive regular treatment from the Government. Uttar Pradesh performs poorly compare to Rajasthan and Odisha

in Bihar with the lowest of 3.9%. This research can help the researcher, government, doctors, and civilians to learn better about Indian states and why certain state is having that kind of health problem.

7 Conclusion

We can conclude from the above experiment that there is no thumb rule to define which clustering method is best and can be applied in all situation. As we all are less experienced in data analysis, the data every day is very different with varying style and dimension, for that I can we need to formulate a mathematical and statistical solution which can be applied in any kind of dataset to automate our work. Many papers have been written proving the state-of-art algorithms K-Means is unbeatable have extreme good results, we think this experiment doesn't prove to go with all the justification. The use of any model solely depends upon the type of data and availability of resource to proof the best model for any kind of dataset.

Also, we found that more the numerical dataset is tough for processing and labeling. The use of training and testing split in the unlabeled dataset was a challenge but, in this case, it was giving a better outcome clusters compared with the cross-validation.

We also found that the novel model GenClust++ has a better performance and clustering compared with rest algorithms, just the case of small dataset it performance was pretty low. Unexpectedly, EM shows a remarkable good performance in the small dataset.

Hence, the UN SDG 2030 goal of good health and wellbeing can also be achieved by learning and building models from the dataset available outside. This kind of experiment help in bringing stability in making decision. Like if the state government wants to improve the condition of healthcare in Bihar, they can directly look result obtained from our dataset and easily make policy and sanction money to improve the condition this way we can stop the thousands of deaths happening in our country because of lack of awareness among us.

References

1. A. Samuel, Some studies in machine learning using the game of checkers. IBM J. Res. Dev. **3**(3), 210–229 (1959). https://doi.org/10.1147/rd.33.0210
2. A. Turing, I—Computing machinery and intelligence. Mind **LIX**(236), 433–460 (1950). https://doi.org/10.1093/mind/lix.236.433
3. T.M. Mitchell, *Machine Learning*, vol. 45(37), pp. 870–877 (McGraw Hill, Burr Ridge, IL, 1997)
4. G. Hinton, T. Sejnowski, *Unsupervised Learning: Foundations of Neural Computation (Computational neuroscience)*. (MIT Press, 1999)
5. T. Koch, K. Denike, Crediting his critics' concerns: remaking John Snow's map of Broad Street cholera, 1854. Soc. Sci. Med. **69**(8), 1246–1251 (2009). https://doi.org/10.1016/j.socscimed.2009.07.046
6. Practical Guide to Clustering Algorithms & Evaluation in R Tutorials & Notes | Machine Learning | HackerEarth. (2019). Retrieved 10 October 2019, from https://www.hackerearth.com/practice/machine-learning/machine-learning-algorithms/clustering-algorithms-evaluation-r/tutorial/
7. M. Islam, V. Estivill-Castro, M. Rahman, T. Bossomaier, Combining K-Means and a genetic algorithm through a novel arrangement of genetic operators for high quality clustering. Expert Syst. Appl. **91**, 402–417 (2018). https://doi.org/10.1016/j.eswa.2017.09.005
8. J. MacQueen, Some methods for classification and analysis of multivariate observations, in *Proceedings of the Fifth Berkeley Symposium on Mathematical Statistics and Probability*, vol. 1 m No, 14 (1967, June), pp. 281–297
9. A. Dempster, N. Laird, D. Rubin, Maximum likelihood from incomplete data via the EM algorithm. J R Stat Soc Ser B (Methodological) **39**(1), 1–22 (1977). https://doi.org/10.1111/j.2517-6161.1977.tb01600.x
10. C. Stover, Log-likelihood function—from Wolfram MathWorld (2019). Retrieved 17 October 2019, from http://mathworld.wolfram.com/Log-LikelihoodFunction.html
11. M. Rahman, M. Islam, A hybrid clustering technique combining a novel genetic algorithm with K-Means. Knowl.-Based Syst. **71**, 345–365 (2014). https://doi.org/10.1016/j.knosys.2014.08.011
12. Error Sum of Squares (2019). Retrieved 17 October 2019, from https://hlab.stanford.edu/brian/error_sum_of_squares.html
13. D. van der Merwe, A. Engelbrecht, Data clustering using particle swarm optimization. In: The 2003 Congress on Evolutionary Computation, CEC '03 (2003). https://doi.org/10.1109/cec.2003.1299577
14. M. Alswaitti, M. Albughdadi, N. Isa, Density-based particle swarm optimization algorithm for data clustering. Expert Syst. Appl. **91**, 170–186 (2018). https://doi.org/10.1016/j.eswa.2017.08.050
15. S. Singh, A. Singh, Web-spam features selection using CFS-PSO. Procedia Comput. Sci. **125**, 568–575 (2018). https://doi.org/10.1016/j.procs.2017.12.07

Ms. Hina Firdaus has completed her post-graduation M.Tech. in Computer Science and Engineering, Jamia Hamdard, New Delhi (India) in July 2018. Her primary area of research is Machine learning, ICT for sustainable development, and HCI.

Dr. Syed Imtiyaz Hassan works as an Associate professor at the Department of Computer Science & Information Technology, Maulana Azad National Urdu University (Central University), Hyderabad. Assistant Professor at the Department of Computer Science and Engineering, Jamia Hamdard, New Delhi (India). His professional experience spans over more than 17 years of teaching, Research, and project supervision. He has supervised more than 80 students for interdisciplinary research and industrial projects. Over the years, he has published many research papers with national and international journals of repute. In addition to these, he is also in the Editorial Boards and Reviewers' Panels of various journals. His primary area of research is Computational Sustainability. To meet the objectives of Computational Sustainability, he explores the role of Nature Inspired Computing, Machine Learning, Data Science, Mobile Crowd Sensing, and IoT for developing Smart and Sustainable Software Systems.

Machine Learning for Healthcare Diagnostics

K. Kalaiselvi and M. Deepika

Abstract Currently healthcare domain relies more on computer technology. Medical diagnosis is an important task of intelligent systems. Machine learning systems are used to find the abnormalities at an early stage of disease diagnosis. Optimal and accurate diagnosis is a critical factor for identifying appropriate treatment. This chapter deals with the importance of Machine Learning systems in healthcare, also focuses on the types of ML systems. As the application of Machine Learning plays a vital part in public healthcare, the significance of role of ML in medical data analysis is widely discussed with few applications. Also, the available medical dataset used in ML are stated. Since the critical task in medical data analysis is the prediction of the accuracy results, the evaluation metrics used especially while dealing with healthcare data are also discussed.

Keywords Datasets · Evaluation metrics · Healthcare · Machine learning · Medical diagnosis

1 Introduction

Medical Diagnosis is an intricate and critical task that has to be performed efficiently and accurately. Mostly this is diagnosed on the basis of doctor's knowledge and experience. This may at times lead to incorrect results and expensive medical costs for treatments to patients.

Currently we require high powered computing methods that adopt Machine Learning approach which could produce intelligence for implementing computers to solve obscure perception. Rich datasets and intelligent algorithms are predominantly essential for application of machine learning in healthcare. Several algorithms help in analyzing huge datasets.

K. Kalaiselvi · M. Deepika (✉)
Department of Computer Science, VELS Institute of Science, Technology and Advanced Studies, Chennai, Tamil Nadu, India

K. Kalaiselvi
e-mail: kalairaghu.scs@velsuniv.ac.in

V. Jain and J. M. Chatterjee (eds.), *Machine Learning with Health Care Perspective*, Learning and Analytics in Intelligent Systems 13,
https://doi.org/10.1007/978-3-030-40850-3_5

2 Machine Learning

Machine Learning is an exemplar that learns from prior experience for providing improved performance in future. As a model of Artificial Intelligence, it reflects in portraying human intelligence through computer systems. It is a blend of computer science and statistics, where the former focuses on problem solving and identifying whether the problems are solvable at all stages, while the latter works for data modeling, hypothesis and measuring the reliability. The main purpose of this field is automatic learning methodologies, here learning means improvement or modification to the algorithm based on past performance.

The main aim of ML is to develop programs that use data to learn themselves without human involvement. It provides set of tools and algorithms for intelligent learning systems. These algorithms used are mainly developed to achieve speed, customizability and accuracy (Fig. 1).

Contribution of Machine Learning to healthcare targets on data analysis, disease diagnosis, therapy planning and so on. ML integrates healthcare with advanced computer systems efficient and accurate diagnosis and helps medical experts for quality treatment.

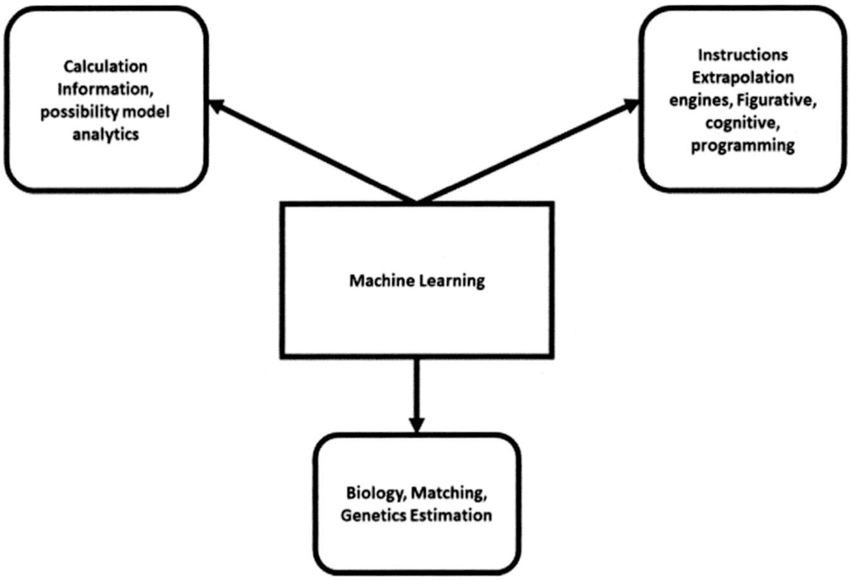

Fig. 1 Machine learning system

3 Machine Learning for Medical Diagnosis

As a discipline of Artificial Intelligence Machine learning is the simulation of human intelligence by computer systems. It combines both computer science and statistics; where computer science deals with identifying and solving problems and statistics deals with modelling data, hypothesis and reliability measure [2]. Machine learning is a paradigm that learns from past experience for the improvement of future performance. They mainly focus on the program development where data are trained to learn by themselves. Listed below are few applications of ML for healthcare:

- Identifying Diseases
- Medical Imaging Diagnosis
- Personalized Medicine
- Better Radiotherapy
- Smart Health Records
- Drug Discovery and Manufacturing
- Clinical Trial and Research
- Machine Learning-based Behavioral Modification
- Crowdsourced Data Collection
- Outbreak Prediction (Fig. 2).

Since healthcare domain deals with enormous data such as information about patients and so these massive records collected are hard to analyze and maintain by an individual. To overcome this machine learning systems paved a way by automatically discovering patterns and predicting the results on data. This enables the medical analysts to provide precision medicine by personalized care. With the help of machine learning systems, we can handle the massive amount of medical data efficiently. Machine learning techniques provide numerous algorithms for data classification and prediction, which is the primary goal of medical data analytics. Thus, it helps us to take better decisions with high rate of accuracy.

3.1 Types of Machine Learning Systems

Machine learning portrays a new vision on individuals and organizations that targets to work towards enhancing the efficiency of the systems by adopting various methods for handling data. Usually these data falls under the three types: Structured, unstructured and semi-structured data [1]. The algorithms are adopted to extract the information from data using computational methods. Many algorithms are being used and appropriate method has to be chose based on their needs. The algorithm of machine learning systems falls under these four categories such as:

- ***Supervised machine learning algorithms***: We manually act as the teacher because we train the data and feed it on the computer. The trained data holds the input i.e., the predictors and produce output, the resultant data is used to study the patterns.

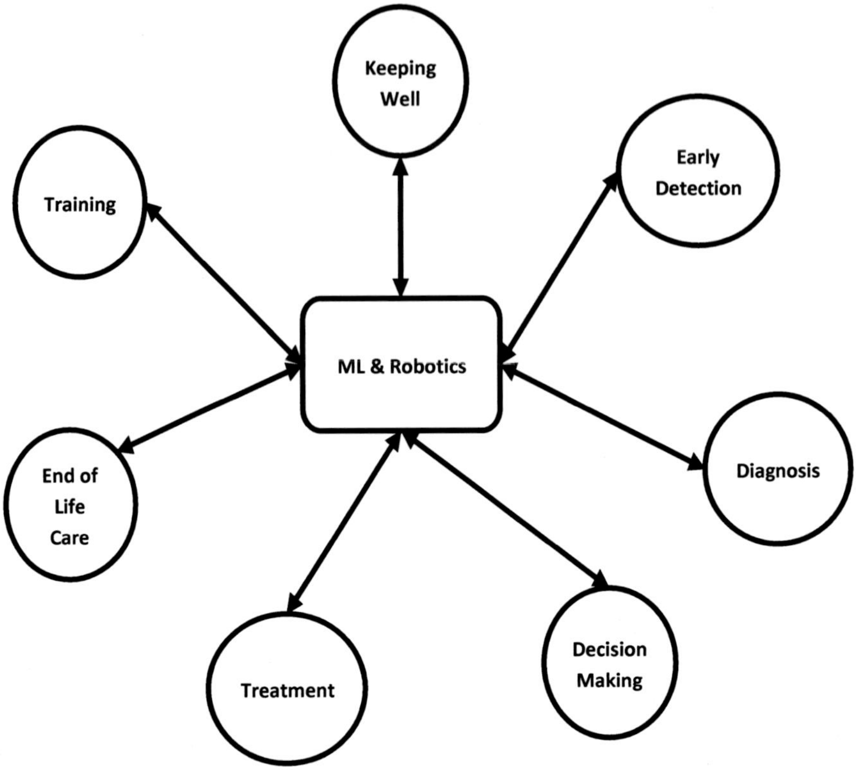

Fig. 2 Machine learning in healthcare diagnosis

They contend to build dependencies and relationships amidst the input features and targeted output values.

The primary category under supervised machine learning problems comprises classification and regression problems. Some commonly used algorithms under this type are Naïve Bayes, Nearest Neighbor, Linear Regression, Decision Trees, Neural Networks and Support Vector Machines (SVM).

- ***Unsupervised machine learning algorithms***: In this type the entity is trained with unlabeled data, unlike supervised learning which is trained in labelled data. Since no human involvement for training takes place, the system might predict new outcomes as it learns patterns on its own. These types of algorithms are specifically very useful where the experts fail to understand what to analyses from the given data.

  As we don't have targeted output categories or classified labels, we clearly use it to model relationships based on the algorithm. They are mainly used in descriptive modelling and pattern detection. The primary category of unsupervised learning algorithms includes association rule learning algorithms and clustering algorithms. Some of the commonly used algorithms are association rules and k- means clustering.

- ***Semi-supervised machine learning algorithms***: Under supervised learning we have labels for all the observation in dataset and in unsupervised learning we don't have labels for them. Semi-supervised learning comes among these two types. In current scenario the cost of labelling becomes high as it involves human expert's knowledge.

  So, some models were developed which has labels present in for fewer observations and unlabeled data in many observations. These types of semi-supervised models are best factors for model building. Inspite of unlabeled data they explore the pattern by holding the data information of group parameters (Fig. 3).

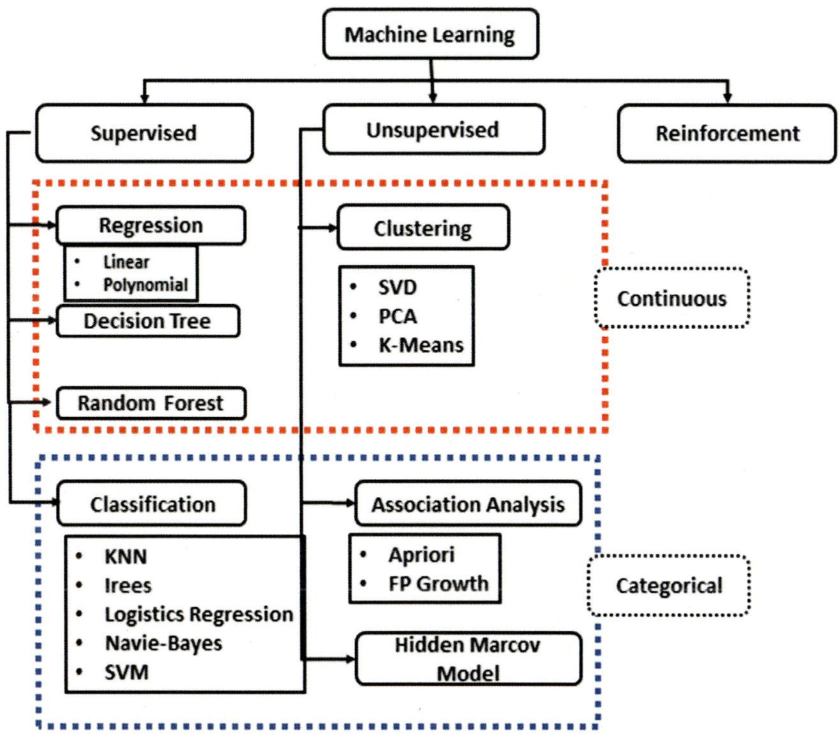

Fig. 3 Types of ML systems

- ***Reinforcement machine learning algorithms***: This type targets at using the data got by observing the communication within the environment and act to minimize the critical risk and maximize the efficiency. Reinforcement learning algorithms otherwise called agents learns from environment iteratively. Here the software agents automatically analyze the behavior and find pattern to improve system performance. Some of the commonly used algorithms are Q-learning, Deep Adversarial Networks and Temporal difference.

4 Role of Machine Learning in Healthcare

Even though enormous tasks had been performed by researchers in the ML domain, it doesn't imply that all manpower will be replaced by machines. Primarily ML is like a tool which can only be working as efficient as a user. Since 1970s ML has been an element of healthcare research; where it was initially applied to attune antibiotic dosages for infected patients. As there is a gradual raise in the count of electronic health records and outbreak inn sequencing genetic data, ML techniques is now on demand in healthcare industry.

Most of the present ML diagnostic applications comprise of the following types namely Chatbots, Oncology, Pathology and Rare disease. Chatbots determine the patterns in the symptoms from patients to develop a probable examination thus prevent disease and/or recommend an appropriate course of action. The researchers under oncology use to train algorithms to identify cancerous tissue at a level comparable to trained physicians [4].

Pathology is the process of diagnosing disease with reference to the laboratory results of bodily fluids like urine, blood and tissues. Machine vision and other machine learning technologies can augment the classical methods with microscopes. Clinicians diagnose Rare Diseases through machine learning systems by combining algorithms with facial recognition software. Using facial analysis and machine learning techniques the patient images are interpreted to determine the phenotypes that associate with rare genetic diseases (Fig. 4).

4.1 Hospital Management and Patient Care

Healthcare zone and hospitals provide vast range of critical services and so it faces many challenges delivering patient care. In the long-term effectiveness of the treatment, affording huge cost becomes difficult process as clinics and hospitals are bonded with resource management. The main aim in hospital management is to assure proper treatment through appropriate medical staff and efficiently scheduling the usage of diagnostic equipment.

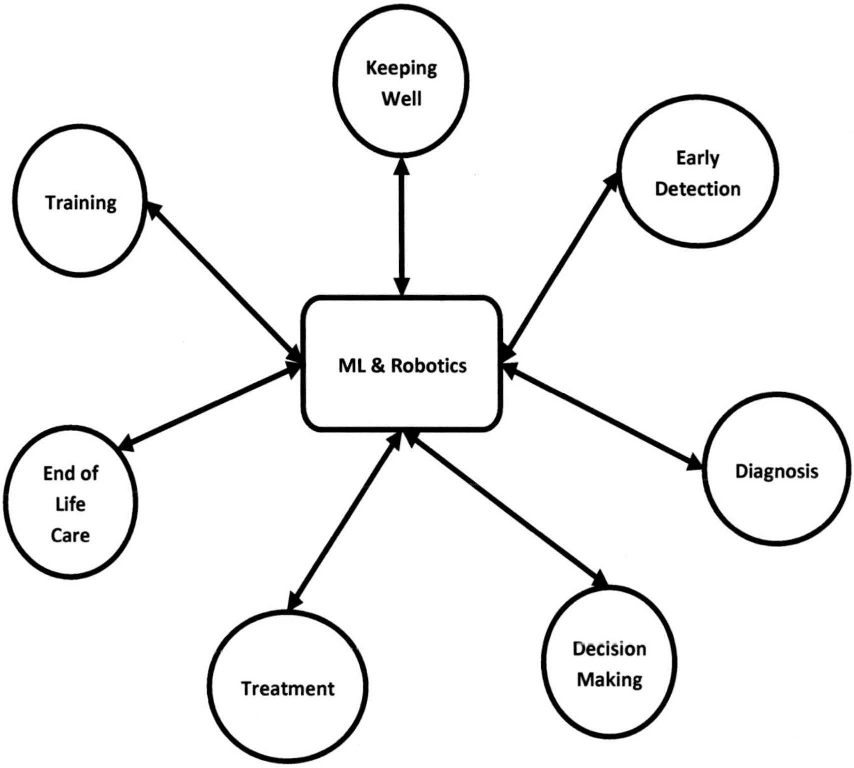

Fig. 4 ML in medical analysis

Machine learning plays an important part in all these areas right from managing inventory to alarming emergency department for patient care. The waiting time of the patient can be predicted using the staffing levels and other factors. Virtual nursing assistants can monitor patient remotely and assist them over phone. Thus saving 20% of their time also patient could avoid unnecessary hospital visits (Fig. 5).

5 Data Analysis in Public Health

Over many years data analysis has imprinted vital role in public health. Machine learning techniques are incorporated in finding insights with high degree of objectivity and develop models using larger and complex datasets. We require huge data for analyzing regardless of the types of machine learning methods used. We need the data that can be placed in repository where it is sharable and distributed.

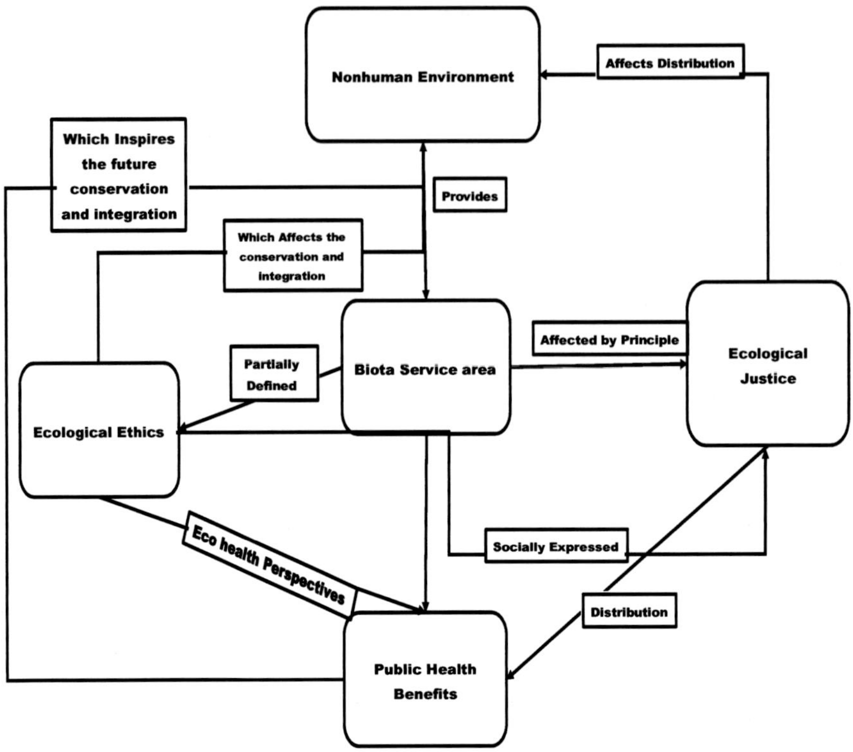

Fig. 5 Machine learning in public health

5.1 Pattern Matching in Genetic Research

One of the most prominent medical research activities for past twenty years is studying patterns of gene expression. Currently over 30,000 coding regions in human genome are there, these take care of synthesis of many proteins. So ML based solutions are important in genomic analysis as it perfectly handles the complex and high dimensional problems.

The two primary kinds of genomic analysis implemented using ML are: Gene Expression and Gene sequencing. Also, Gene Editing by using CRISPR technology in altering DNA sequencing also uses machine learning. To obtain better deterministic gene alteration we need to collect more data for developing effective machine learning models.

5.2 *Disease Diagnostics*

Medical diagnostics are set of tests performed to diagnose disease, certain medical conditions and infections. Accurate diagnosis without much time delay and effective treatment is the important factor for betterment of patients. Some diagnostic errors have accounted in almost 10% of human deaths and it attributed to around 6–17% of all hospital complications. ML provides solutions to the issues on difficulties in diagnosing especially when dealing with pathology and oncology. Also, while examining Electronic Health Records (EHR) ML provides better diagnostic prediction (Fig. 6).

In healthcare research ML is mainly focused to work on Image based diagnosis and Image analysis. Since healthcare holds enormous datasets, machine learning based imaging fits perfectly as it requires huge datasets that are accurately classified. Variations between imaging systems on medical imaging has an issue. A particular model that is trained under the imaging system from developer may fail to work from images taken from other developer.

6 ML Applied in Medical Diagnostics

Cancer Care:

MD Anderson Oncology Expert Advisor (OEA) which is a virtual expert is developed by university of Texas MD Anderson Cancer Centre (MDACC) on training IBM's Watson. This improves the aspects of caring for cancer patients without approaching

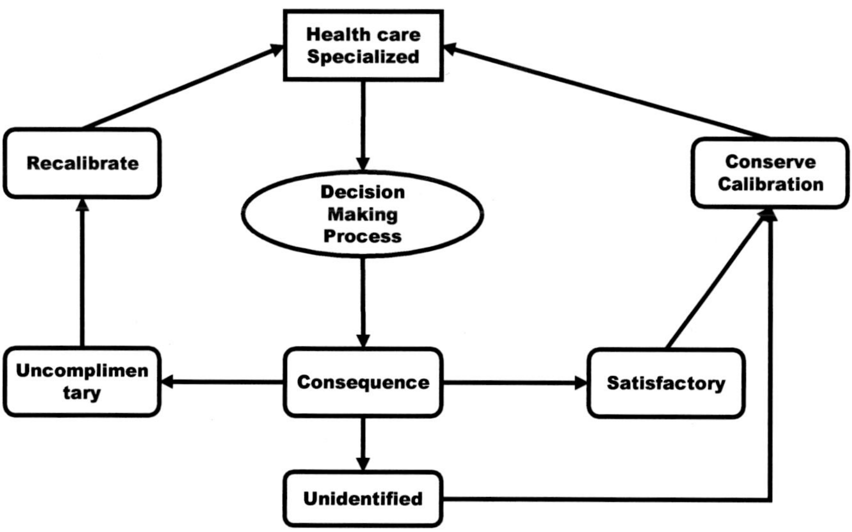

Fig. 6 Decision making process using ML in healthcare

the experts. It provides both recommendations to improve treatment and helps in therapy decisions. Hence it is designed to cancer care as well as produce clinical evidence.

Radiology and Radiotherapy:

University College London Hospital (UCLH) is working with Google's DeepMind Health in developing ML algorithms that can detect anomalies in cancerous and healthy tissues for providing effective radiation treatment. It is found that machine learning has helped in speeding up the segmentation process by ensuring that healthy and normal parts aren't damaged. Thus, it increases the accuracy and speed in radiotherapy planning.

Medical Imaging:

The project developed by Microsoft's InnerEye runs a greater research works on computer vision and machine learning to perform automatic and quantitative three-dimensional radiological images. This resulted in providing images that are very precise and are effectively turned into measuring devices. This can be used in radioactive therapy as it works better on targeted radiomics measures.

Surgery:

Smart Tissue Autonomous Robot (STAR) has surpassed the human surgeons when the same task of operating on pig's small intestines was given to them. It uses a specialized camera and high vision system to perform this task, also it created its own pattern for suturing and adjusting according to the movement of tissues during the surgery (Fig. 7).

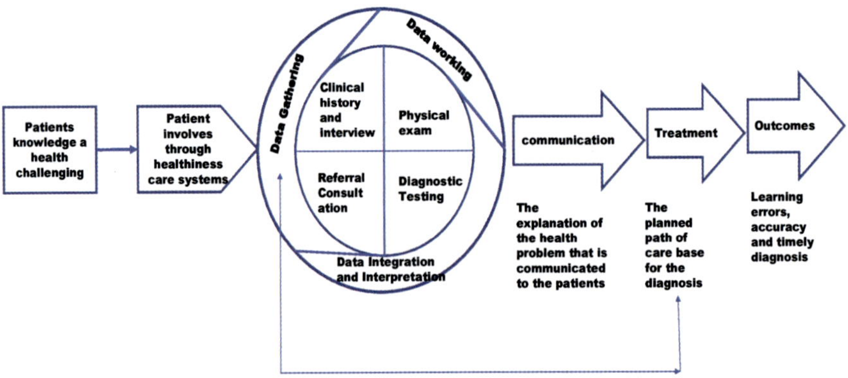

Fig. 7 Medical data analysis using ML

7 Datasets

The greatest issue that we may cross across while implementing a method for a certain problem is acquiring a proper dataset that accurately correlates to the given problem. Apart from that, the dataset needs to be pre-processed so that the proposed model will be able to perform efficiently. Ultimately this helps the model to study from that dataset. The main criteria in is that the selected data is appropriate for the models which are adopted to implement for solving a problem. However, having enormous data is trivial unless when the data is aligned or processed with respect to the developing model. We need to select data with features that can be considerable when we try to classify or predict and discard irrelevant features.

The foremost step of proper data collection plays a major role in developing a model; otherwise we should constantly loop back each time in this step. Followed by data collection it is transformed into the format that the model could understand. Generally, the input data such as images, text, videos or audio is converted into tensors and vectors on which the linear algebraic functions are applied. To increase the performance, the data has to be cleaned, standardized and normalized. The Raw data, which is unprocessed data till then has to be pre-processed. It can be a complex way to get a typical dataset to be used for various machine learning problems [3]. Few of the datasets are listed below:

- Big cities Health Inventory Data
- Healthcare Cost and Utilization Project (HCUP)
- HealthData.gov
- Surveillance, Epidemiology and End Results (SEER)—Medicare Health Outcome Survey (MHOS)
- The Human Mortality Database (HMD)
- Child Health and Developmental Studies
- Medicare Provider Utilization and Payment Data (Fig. 8).

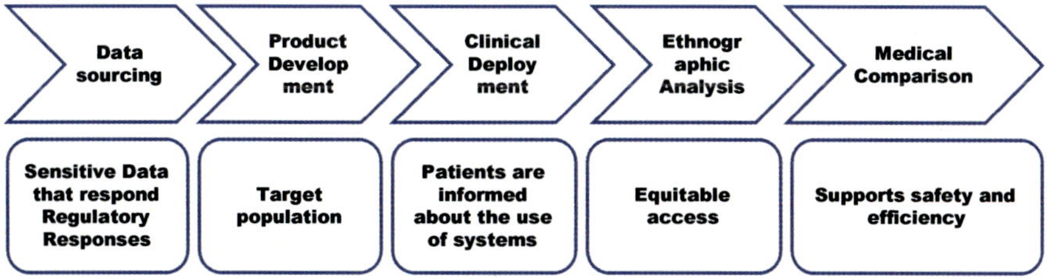

Fig. 8 Mining medical data

8 Evaluation Metrics

Analyzing and evaluating the ML algorithm is an indispensable component of a project. We may end up in producing pleasing results on considering accuracy score as evaluation metric but while evaluated when considering the logarithmic loss or any such metrics we may produce poor results. In general, we contemplate classification accuracy as the performance measure to analyze the developed model. So, the accuracy metric alone is inadequate to evaluate a model. Thus, we have varied evaluation metrics available to evaluate the performance of the developed model. Some of them are listed below [5].

8.1 *Classification Accuracy*

Classification accuracy is the proportion of number of correct predictions to the total number of input samples. The efficient results are produced if each class holds same number of samples.

8.2 *Logarithmic Loss*

Logarithmic Loss or Log Loss, evaluates by correcting the false classifications. It is important to note that while working with Log Loss, probability to each class has to be assigned by the classifier for all the samples used. Thus, they are used mostly for multi-class classification.

8.3 *Confusion Matrix*

Confusion Matrix provides the output in matrix format also it describes the overall performance of the proposed model. It is considered as the ground for the varied types of evaluation metrics

8.4 *Area Under Curve (AUC)—ROC*

This is one of the utmost extensively used evaluation metrics. AUC is mainly analyzed on issues dealt with binary classification. This classifier is evaluating the probability that the given classifier will rank a randomly chosen positive example higher than a randomly chosen negative example (Fig. 9).

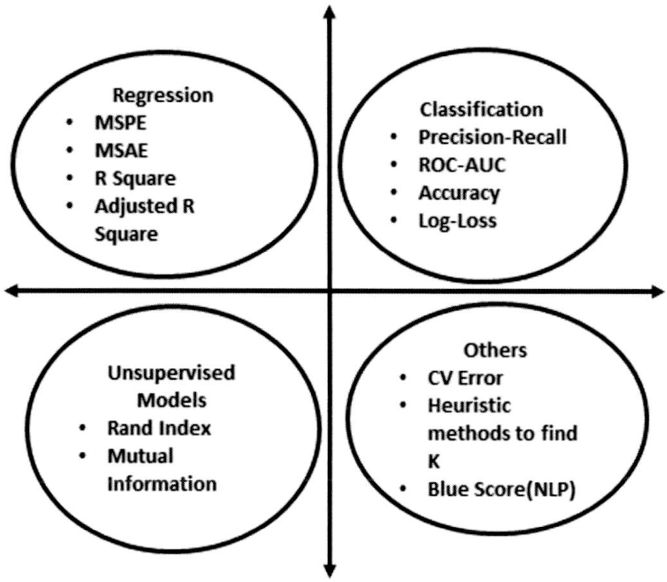

Fig. 9 Evaluation metrics adopted based on the technique

8.5 F1 Score

It is the measure of the accuracy on a specific test carried out. It can be considered as the Harmonic Mean between recall and precision. The F1 Score ranges between [0, 1]. It shows the precision of the classifier i.e., the number of instances correctly classified. It also shows the robustness i.e., it does not neglect a significant number of instances. Lowest recall rate with high precision, produces an efficient accuracy results. The main drawback here is it misses a large number of instances that are difficult to classify. Greater the F1 Score, better is the performance of the developed model.

8.6 Kolmogorov Smirnov Chart

It is specifically used in marketing campaigns and ads click predictions where we need to know the appropriate population size in order to get the maximum response rate. This chart and the statistics are widely used in credit scoring scenarios and also used in selecting the optimal population size for marketing campaigns.

8.7 Mean Absolute Error

It is the mean of difference between the predicted values and the original values. It is the measure of deviation from the predicted values to the actual output. They fail to predict the cause of error say; we have issues because of over predicting the data or under predicting the data.

8.8 Mean Squared Error (MSE)

It is similar to Mean Absolute Error; the only variation is that MSE evaluates the average of the **square** of the difference between the predicted values and the original values. Another advantage of MSE is that it's easier to compute the gradient, but on the other hand Mean Absolute Error needs complicated linear programming tools to compute the gradient. Since the square of the error is considered, the effect of major errors becomes more transparent than minor errors, thus targeting for critical errors.

9 Conclusion

There is enormous data available in the IT domain of hospitals which can be used to analyze by ML algorithms and train models on the basis of the outcomes measured. It is relatively straightforward to find patterns and predicting outcomes using machine learning. Surgical robots have the capability to bring down surgeons need and also booking the operation theatre, as well as it provides better outcomes and high success rate. Finally, Machine learning models still needs to be trained to compete with the top senior clinicians who are experts in providing better results but it can be challenged to compete with the performance of junior clinician.

References

1. https://www.edureka.co/blog/what-is-machine-learning/
2. https://www.computerworld.com/article/3479833/machine-learning-essential-to-healthcare-says-orion-health.html
3. https://towardsdatascience.com/top-sources-for-machine-learning-datasets-bb6d0dc3378b
4. https://emerj.com/ai-sector-overviews/machine-learning-medical-diagnostics-4-current-applications/
5. https://medium.com/usf-msds/choosing-the-right-metric-for-evaluating-machine-learning-models-part-2-86d5649a5428

Dr. K. Kalaiselvi has received her M.Sc., Computer Science degree from Periyar University, M.Phil. degree from Bharathidasan University, Tamil Nadu, India and Ph.D. degree in Computer Science from Anna University, Chennai, Tamil Nadu, India. She is currently working as Professor and Head in the Department of Computer Science, School of Computing Sciences, Vels Institute of Science, Technology and Advanced Studies, Chennai, India which is well known university She has more than 16 years of teaching experience in both UG and PG level. Her research interests include Knowledge Management, Data Mining, Embedded Systems, Big Data Analytics and Knowledge Mining. She has produced four M. Phil Scholars. Currently, she is guiding Ph.D. scholars and M.Phil. Scholars in VISTAs. She has published more than 32 research Papers in Various International and two papers in National Conferences. She has published a book titled "Protocol to learn C Programming Language". She has received the Best Researcher Award from DK International Research Foundation, 2018 and received Young Educator and Scholar award—6th women's day Awards 2019, from the National Foundation for Entrepreneurship Development NFED 2019. She is a professional Member of CSI. She serves as Editorial Board Member/Reviewer of various reputed Journals. She has been invited a resource person for various International National conferences and seminars organized by various Institutions. She has completed the mini project funded by VISTAS.

Ms. M. Deepika Completed her Master of Computer Science (M.Sc.), in Anna Adarsh College for Women, Chennai, Master of Computer Application (M.C.A.) from Anna University, Chennai. She is currently Pursuing her Research Work (Ph.D.) in Vels Institute of Science, Technology and Advanced Studies (VISTAS) under the guidance of Dr. K. Kalaiselvi, Head and Associate Professor in the Department of Computer Science, School of Computing Sciences, Vels Institute of Science, Technology and Advanced Studies, Chennai. She has more than three years of teaching experience in both UG and PG level. She is Interested in Machine Learning, Data Mining, Computer Networks and Data Analytics. Broad area of research is Data Mining focusing on optimal prediction of disease diagnosis. She has published a research paper in Elsevier SSRN Digital Library. She has also published three papers on various Scopus journals. She has also presented papers at International conferences.

Disease Detection System (DDS) Using Machine Learning Technique

Sumana De and Baisakhi Chakraborty

Abstract In this world, a human being suffers from many different diseases. Diseases can have a physical, but also a psychological impact on people. Mainly for four reasons, diseases are formed: (i) infection, (ii) deficiency, (iii) heredity and (iv) body organ dysfunction. In our society, doctors or medical professionals have the responsibility to detect and diagnose appropriate disease and provide medical therapies or treatments to cure or restrain the disease. Some diseases are cured after treatment, but chronic diseases are never cured despite the treatment; treatment can prevent chronic diseases to be worse over time. So, it is always important to detect and treat disease in early stage. To help doctors or medical professionals, this chapter proposes Disease Detection System (DDS) that can be used by doctors or medical professionals to detect diseases in patients using Graphical User Interface (GUI) of DDS. DDS is developed to detect some diseases such as Liver disorders, Hepatitis, Heart disease, Diabetes, and Chronic Kidney disease. Each of the diseases has different signs and symptoms among the patients. Different datasets are obtained from the Kaggle machine learning database to implement DDS. For the classification calculation, Adaboost Classifier Algorithm is used in DDS to detect diseases. This is a machine learning algorithm that results in the identification of referred diseases in DDS with 100% accuracy, precision and recall. The DDS GUI was created with the support of python as a screening tool so that doctors or medical professionals can easily detect patients with disease.

Keywords Diseases detection · Disease detection system (DDS) · Adaboost classifier algorithm · Query from user · Result from system

---

S. De (✉) · B. Chakraborty
Department of Computer Science and Engineering, National Institute of Technology, Durgapur, India
e-mail: sumanade@gmail.com

B. Chakraborty
e-mail: baisakhichak@yahoo.co.in

V. Jain and J. M. Chatterjee (eds.), *Machine Learning with Health Care Perspective*, Learning and Analytics in Intelligent Systems 13,
https://doi.org/10.1007/978-3-030-40850-3_6

1 Introduction

Human disease is an abnormal condition that affects the human organism negatively that may cause pain, uncomfortable, dysfunction, and even death. Diseases can have a physical, but also a psychological impact on people. And there are many different kinds of diseases in the world. According to [1] mainly for four reasons, diseases are formed: (i) infection, (ii) deficiency, (iii) heredity and (iv) body organ dysfunction. Infectious diseases are occurring for viruses, bacteria, fungus, worms, arthropods, etc. And it can be treated using antibiotic, antiviral, antifungal, antiprotozoal or anthelmintic medicines. Deficiency diseases are caused by long time nutrition deficiency in the human body. Hereditary diseases are caused either by genetic diseases or non-genetic diseases. Physiological diseases are caused when the proper functioning of the body malfunction because of malfunctioning of body organs. For example, asthma, glaucoma, diabetes etc.

In our society, doctors or medical professionals have the responsibility to detect and diagnose appropriate disease and provide medical therapies or treatments to cure the disease. However, chronic diseases continue to persist which may worsen over time. So, it is always important to detect and treat disease in early stage. To help doctors or medical professionals, this chapter proposes a Disease Detection System (DDS) that can be used by doctors or medical professionals to detect diseases in patients using Graphical User Interface (GUI) of DDS. DDS is developed to detect some diseases such as Liver disorders, Hepatitis, Heart disease, Diabetes, and Chronic kidney disease. Liver disorder is a damaging condition of the liver. It can be inherited or caused by other factors like viruses or excessive use of alcohol. There are many symptoms of liver disorder such as skin and eyes turning yellowish, abnormal abdomen pain, pale stool color, etc. Doctors diagnose the disorder testing CT scan, MRI reports, testing blood and analyzing tissue. Hepatitis is an inflammatory condition of the liver. There may be many different causes for hepatitis. The most common cause is a viral infection. Doctor diagnose hepatitis testing liver function and testing blood. Heart disease is a disorder of heart. There may be many types of heart disease such as diseases in blood vessels, in a coronary artery, in heart rhythm and inherited disease in the heart. A Patient can feel chest pain, chest discomfort, breathing difficulty. Doctors diagnose the disease checking physical condition, family history and testing blood, chest x-ray, ECG report, etc. Diabetes is a disorder of metabolism that increases sugar in blood. Type 1 and Type 2 can be classified by diabetes. Enough insulin is not generated in body in type 1 and for type 2, the body is able to generate sufficient insulin, but it cannot properly use the insulin. Doctors diagnose diabetes testing blood sugar level, insulin level, fasting plasma glucose, etc. Chronic kidney disease is indeed the kidney's destructive situation that can become more severe over the period. When the kidneys are very severely damaged, they will stop working. So, early diagnosis of chronic kidney disease is very significant. Doctors diagnose Chronic kidney disease testing patient's urine and blood. So, all these diseases have different symptoms and different diagnosis process. To help doctors in case of diagnosis of these diseases, Disease Detection System (DDS) has been developed as an assistant tool of a doctor. To implement DDS, different datasets for different diseases are collected from the database of machine learning in Kaggle. The Liver disorders dataset contains 583 instances of observations that have total 11 attributes, the dataset of hepatitis contains 155 instances of observations that have total 20 attributes, the dataset of Heart disease contains 303 instances of observations that have total 14 attributes, The dataset of Diabetes contains 768 instances of observations that have total 9 attributes, and the dataset of chronic kidney disease contains 400 instances of observations that have total 25 attributes. The machine learning algorithm, AdaBoost Classifier Algorithm is used in DDS to detect diseases. In machine learning, boosting is a method that merges comparatively weak and incorrect laws to create a law of prediction that is highly accurate. AdaBoost Classifier Algorithm provides the results of detection of these diseases with 100 percent accuracy, precision and recall. The DDS Graphical User Interface (GUI) was created with the support of python as a screening tool so that doctors or medical professionals can easily detect patients with any disease. The chapter's remaining details are as described: Sect. 2 concentrates on earlier relevant works, Sect. 3 on the methodology and implementation of the system, Sect. 4 deals with the proposed DDS model, Sect. 5 provides a correlation of accuracy between previous works related to DDS, Sect. 6 describes the outcome of the simulation, and then, Sect. 7 comes to a conclusion.

2 Previous Related Works

In today's world disease diagnosis through the computer becomes a very popular topic in the research area. In this research case, machine learning algorithms have a very important role. Because machine learning algorithms are fast and accurate to detect any diseases. Many machine learning diagnostic applications of AI have been successfully made in the present world for the diagnosis of diseases. Much research was done in the field of medicine in the earlier years where researchers utilized machine learning algorithms to identify diseases. This section of the chapter is going to discuss about some recent previous works in which algorithms have been utilized for machine learning to identify diseases. The journal paper [2] provides a survey that analyses different machine learning algorithms to diagnose different diseases like a disease in heart, diabetes, disease in the liver, dengue and hepatitis. This paper focused on the different machine learning algorithms and the most utilized tools to analyze different diseases.

To help doctors and patients to detect disease in early stage, a project based on machine learning algorithms has been discussed in [3]. The dataset has been used in this project is purely text based. This paper has designed a system where a doctor can enter the symptoms of patients and diagnoses common diseases. The paper [4] shows a review on disease diagnosis using machine learning techniques. This paper shows that machine learning is used for the high dimensional and the multi-dimensional data and concludes some limitations of machine learning algorithms. Another study on machine learning algorithms for medical diagnosis is done in paper [5]. This paper focused on the use of different machine learning algorithms for accurate medical diagnosis. A deep study of machine learning algorithms for disease diagnosis is done in paper [6]. This paper concentrates on latest developments in machine learning that have significantly affected the detection and diagnosis of different diseases. Many researches are done to predict heart diseases such as in [7], the research paper used various machine learning algorithms such as SVM, RF, KNN and ANN classification algorithms were used to identify early-stage heart disease. A diagnostic system was developed in the research paper [8] using NN, capable of predicting reliably the level of risk of heart problem. In this world, congestive heart failure has become one of the primary causes of death. So, to prevent congestive heart failure, the research paper [9] has designed system using algorithms like Boosted Decision Tree, CNN. The research [10] aimed to establish a rapid and accurate automatic detection of ischemic heart disease. The machine learning algorithm, XGBoost classifier is used here to establish the task. Liver disease is one of the serious diseases and it should be diagnosed in the early stage so many researches are done to predict liver diseases such as in [11], liver disease is detected using machine learning algorithms where five different phases are used and it is seen that for feature selection, J48 algorithm works better.

The paper [12] proposes an electromagnetic system that includes an antenna as a data capture tool and a supervised Machine Learning (ML) system to learn directly from gathered data an inferring model for FLD.

This paper aims to get a better diagnosis of liver diseases the paper [13] proposes 2 methods of identification, one is patient parameters and second is genome expression. The machine learning technique is used here for the diagnosis.

Four classification models are developed in [14] to diagnose fatty liver disease accurately. Among the four models this paper shows that the random forest model performs better.

The project [15] aims at improving the diagnosis of liver disease using approaches to machine learning. This research's main goal is to use classification algorithms to classify healthy individuals' liver patients.

Hepatitis disease has been predicted in [16] using machine learning models. Two ML algorithms are developed and compared in this paper; to predict the development of cirrhosis in a large cohort infected with CHC.

To predict Hepatitis B, the research article [17] has developed four machine learning models based on four different algorithms, that includes the decision tree (DCT), extreme gradient boosting (XGBoost), logistic regression (LR) and random forest (RF).

One of the popular techniques of machine learning is the Neural Network (NN). The paper [18] shows diagnosis of hepatitis disease using different NN architectures.

In paper [19] logistic regression, random forest, decision tree, C4.5 and multilayer perceptron classifier algorithms are used to predict Hepatitis—Infectious Disease. In this analysis random forest classifier algorithm predicts accurately in a minimum time.

In the paper [20], for the smart prediction of chronic kidney disease (CKD), DFS with D-ACO algorithm has been used to develop a smart health monitoring system.

Data mining algorithms are used in [21] to predict kidney disease stages. Among those PNN algorithm provides better performance in classification and prediction.

To predict chronic kidney disease, two different machine learning algorithms like DT and SVM are used in [22], among which SVM performs better.

For predicting the chronic kidney disease in [23] four machine learning methods are used, out of which SVM classifier gives highest accuracy.

A model based on XGBoost is developed with better accuracy for the prediction of Chronic Kidney Disease in [24].

Type-2 Diabetes is predicted using a machine learning algorithm on paper [25]. Support vector machine is used to implement the prediction model.

The experiment results in the research paper [26] show that the random forest algorithm is very efficient to develop a powerful machine learning model to predict diabetes.

To predict diabetic mellitus among the adult population, the paper [27] used four different machine learning algorithms. Among these algorithms only C4.5 decision tree provides higher accuracy.

Different tree classifiers for machine learning are evaluated based on their True Positive Rate (TPR) and accuracy in [28]. Higher precision was achieved in this study using Logistic Model Tree (LMT) for predicting diabetes mellitus.

3 System Implementation and Disease Detection Methodology

Figure 1 shows the steps to implement DDS and to diagnose diseases. Detail about the steps is discussed in following:

1. Datasets collection: Different datasets are obtained from the machine learning database of Kaggle to implement DDS. The dataset of liver disorders contains

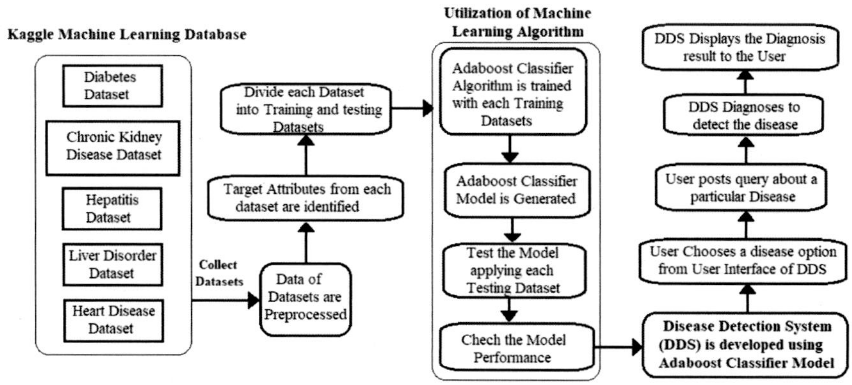

Fig. 1 DDS implementation and detection of diseases

Table 1 Attributes-values in liver disease dataset

Input attributes	*Values*
Age(in years)	4–90
Gender	Male, Female
Total_Bilirubin	0.4–75
Direct_bilirubin	0.1–19.7
Alkaline_Phosphate	63–2110
Alamine_Aminotransferase	10–2000
Aspartate_Aminotransferase	10–4929
Total_Protiens	2.7–9.6
Albumin	0.9–5.5
Albumin_and_Globulin_Ratio	0.3–2.8
Output attribute	*Values*
Dataset	1 = liver disease; 2 = no liver disease

583 instances of observations that have total 11 attributes, the dataset of hepatitis contains 155 instances of observations that have total 20 attributes, the dataset of heart disease contains 303 instances of observations that have total 14 attributes, the dataset of diabetes contains 768 instances of observations that have total 9 attributes, and the dataset of chronic kidney disease contains 400 instances of observations that have total 25 attributes. The following tables show the datasets detail (Tables 1, 2, 3, 4 and 5).

2. Data of datasets are preprocessed: Pre-processing data strategy is used to turn imperfect actual data into a valuable and usable form. For example, convert all nominal data to numerical data; the missing attribute values are substituted by the attribute's calculated average value.
3. Target attributes from each datasets are identified: Target attribute is identified from each dataset, such as liver disorder has "Dataset" attribute as the target attribute that contains two classes: "liver disease yes" and "liver disease no", hepatitis has "Class" attribute as the target attribute that contains two classes: "Die" and "alive", heart disease has "num" attribute as the target attribute that contains two classes: "heart disease yes" and "heart disease no", diabetes dataset contain 'outcome' attribute as the target attribute, that contains two classes: "Diabetic" and "Not-Diabetic", and Chronic kidney disease has "Classification" attribute as the target attribute that contains two classes: "Ckd" and "Not-ckd".
4. Divide each dataset into training and testing datasets: Excluding the column of the target attribute, the datasets are categorized into 2 sets with 7:3 proportions, 70% of it is utilized to train the model of machine learning, while 30% is utilized to test the model's precision, accuracy and recall tests.
5. Adaboost classifier algorithm is trained with each training datasets: AdaBoost classifier Algorithm is used to detect diseases. In machine learning boosting is a method that merges comparatively weak and incorrect laws to create a law of

Table 2 Attributes-values in hepatitis disease dataset

Input attributes	*Values*
Age (in years)	10–80
Sex	1 = male, 2 = female
Steroid	1 = no, 2 = yes
Antivirals	1 = no, 2 = yes
Fatigue	1 = no, 2 = yes
Malaise	1 = no, 2 = yes
Anorexia	1 = no, 2 = yes
Liver_big	1 = no, 2 = yes
Liver_firm	1 = no, 2 = yes
Spleen_palable	1 = no, 2 = yes
Spiders	1 = no, 2 = yes
Ascites	1 = no, 2 = yes
Varices	1 = no, 2 = yes
Bilirubin	0.39–4.0
Alk_phosphate	33–250
Sgot	13–500
Albumin	2.1–6.0
Protime	10–90
Histology	1 = no, 2 = yes
Output attribute	*Values*
Class	1 = die, 2 = live

prediction that is highly accurate. The classification of adaBoost is an iterative assembly process. After merging many poorly performing classifiers it creates a strong classifier to improve classifier accuracy. The main idea of adaboost is just to define the classifier weights and train the data set for each iteration to assure that uncommon events are predicted accurately. The Training Dataset trains the adaboost classifier algorithm and the Model of Machine Learning is produced afterwards. The main idea is to be to construct a machine learning model that is capable of receiving inputs and using statistical analysis detect an outcome. The model's performance is checked with respect to accuracy, precision and recall, where Testing Dataset is applied as the inputs to the Machine Learning Model. So, adaboost Classifier Algorithm is trained with each different disease's training dataset.

6. Adaboost classifier model is generated: Whenever Adaboost Classifier Algorithm is trained with the different training datasets, then Adaboost Classifier Model is generated for each of the disease detection. This Machine Learning Model is able to take input for different diseases and diagnose the specific disease appropriately.

Table 3 Attributes-values in heart disease dataset

Input attributes	*Values*
Age (in years)	29–77
Sex	(1 = male; 0 = female)
cp (Chest pain type)	1 = typical angina 2 = atypical angina 3 = non-anginal pain 4 = asymptomatic
Trestbps [Resting blood pressure (in mm Hg on admission to the hospital)]	94–200
Chol (Serum cholestoral in mg/dl)	126–564
Fbs (Fasting blood sugar > 120 mg/dl)	(1 = true; 0 = false)
Restecg (Resting electrocardiographic results)	0 = normal 1 = abnormal ST-T wave 2: probable or definite left ventricular hypertrophy
Thalach (Maximum heart rate achieved)	71–202
Exang (Exercise induced angina)	1 = yes; 0 = no
Oldpeak (ST depression induced by exercise relative to rest)	0–6.2
Slope (The slope of the peak exercise ST segment)	1 = not sloping 2 = flat 3 = down sloping
ca (Number of major vessels colored by flourosopy)	0–3
Thal	1 = normal; 2 = fixed defect; 3 = reversible defect
Output attribute	*Values*
Target [Diagnosis of heart disease (angiographic disease status)]	Value 0: <50% diameter narrowing Value 1: >50% diameter narrowing

7. Test the model applying each testing dataset: Once the Adaboost Classifier Model is generated. To test its performance the testing datasets for each disease are applied on the Model.
8. Check the model Performance: The quality of the model is tested for accuracy, precision, and the outcome of the recall. The adaboost classifier model delivers 100% accuracy, precision and recall outcomes in detection of each disease.

Table 4 Attributes-Values in diabetes disease dataset

Input attributes	*Values*
Pregnancies	0–17
Glucose	0–199
Blood pressure	0–122
Skin thickness	0–99
Insulin	0–846
BMI	0–67.1
Diabetes pedigree function	0.078–2.42
Age (in years)	21–81
Output attribute	*Values*
Outcome	0–1

9. Disease Detection System (DDS) is developed using adaboost classifier model: The adaboost classifier model is used to develop the Disease Detection System (DDS). The Graphical User Interface (GUI) of the DDS is designed using Jupyter Notebook in Python Environment.
10. User chooses a disease option from the User Interface (UI) of the system: The GUI of the system shows five options for disease selection. User can select any of the diseases for further diagnosis.
11. User posts query about a particular disease: After selecting the option, user can post queries about a particular disease through another form provided by the system.
12. DDS diagnoses to detect the disease: After getting input from the user, DDS diagnose the disease with the help of adaboost classifier model to detect the disease.
13. DDS displays the diagnosis result to the user: After completing the diagnosis calculation, the final result is to be displayed to the user through the GUI of the DDS.

4 Proposed DDS Model

4.1 Architecture of DDS

DDS is the request-response system in which the user needs to submit a question as a request form and the DDS offers a response with the detection result. Figure 2 shows DDS model has following architectural components.

1. Data Entry Module or User Interface (UI): It is the responsibility of the UI to connect the user to the system. A user can view the question form via the UI that

Table 5 Dataset attributes-values of chronic kidney disease

Input attributes	*Values*
Age (in years)	2–90
bp [Blood pressure (mm/Hg)]	50–100
sg (Specific gravity)	1.005–1.025
al (Albumin)	0–4
su (Sugar degree)	0–4
rbc (Red blood cells)	Normal, Abnormal
pc (Pus cell)	Normal, Abnormal
pcc (Pus cell clumps)	Present, Not present
ba (Bacteria)	Present, Not present
bgr [Blood glucose random (mgs/dl)]	22–490
bu [Blood urea (mgs/dl)]	1.5–391
sc [Serum creatinine (mgs/dl)]	0.4–7.6
pcv (Packed cell volume)	0–54
pot [Potassium (mEq/L)]	0–4.7
sod [Sodium (mEq/L)]	0–163
hemo [Hemoglobin (gms)]	0–17.8
wbcc [White blood count (cells/cumm)]	0–26,400
rbcc [Red blood cell count (millions/cmm)]	0–8
htn (Hypertension)	Yes, No
dm (Diabetes mellitus)	Yes, No
cad (Coronery artery disease)	Yes, No

(continued)

Table 5 (continued)

appet (Appetite)	Good, Poor
pe (Pedal edema)	Yes, No
ane (Anemia)	Yes, No
Output attribute	*Values*
Classification	ckd, notckd

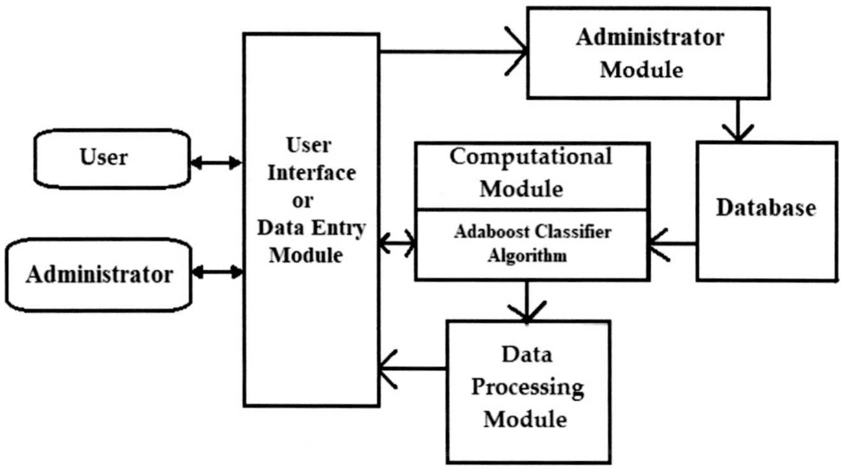

Fig. 2 Components of architecture in DDS

is provided by the system, as well as the user fills out and sends the form to the system with values.

2. User: The user is meant the doctors or medical professionals. They log into the system, provide data on the health status of the patient to the system via the UI. The user can supply the system with any of the five disease-related data.
3. Computational Module: The responsibility of Computational Module is to classify whether the recently posted data indicates disease or not. For example, if the user provides data for checking hepatitis, then The Computational module is responsible for classifying whether or not the posted input data belong to the class of hepatitis. In this way, similarly, other diseases may be detected. Computational Module uses adaboost Classifier model to calculate the classification. And it estimates the accuracy, precision, and recall outcome of the system as well.
4. Data Processing Module: Once the Computational Module performs the classification the Data Processing Module tests the outcome of the diagnosis. When it determines that the patient has the disease, it will send the user a message that the patient is suffering from that particular disease, otherwise it will demonstrate

that the patient has no particular disease. The system accuracy is also displayed to the user by this module.

5. Administrator Module: This module supports DDS administration, administrators. Just administrators can insert, remove, upgrade, and alter the database dataset information.
6. Administrator: doctors or medical professionals should be the Administrator, who will have adequate knowledge of the diseases. With valid data, they can upgrade the datasets or erase unwanted records from Datasets.
7. Database: The database stores all of the Datasets of diseases (Liver disorders, Hepatitis, Heart disease, Diabetes, and Chronic Kidney disease).

4.2 Use Case Diagram of DDS

Multiple consecutive actions performed by the DDS are shown in Fig. 3 use case diagram. In the use case diagram, the actor is DDS. The detail about the actions, performed by DDS is discussed in below:

Check User Input: User can choose a disease option from the user interface of DDS. Whenever the user selects the disease, to diagnose the disease, the system provides a form and asks users to fill up the form with required data. Then through the user interface of the system user submits the required data and fills up the form.

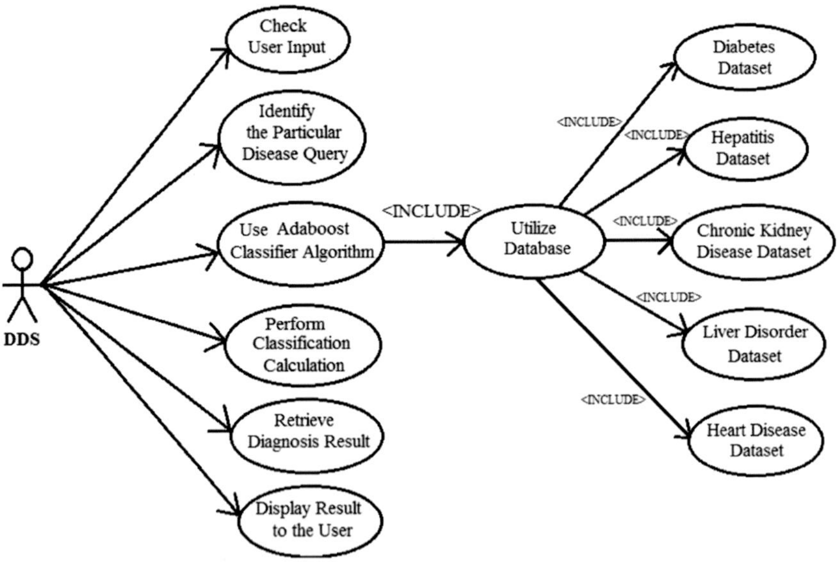

Fig. 3 DDS use case diagram

- Identify the Particular Disease Query: Whenever a user submits the form, the system receives the set of data and can identify which particular disease it should diagnose.
- Use Adaboost Classifier Algorithm: DDS utilizes the system database that includes the different datasets. For a particular disease diagnosis, 70% of that particular dataset is used to train the adaboost Classifier algorithm. Then adaboost classifier model is generated that is able to classify the new input.
- Perform Classification Calculation: Now, DDS uses the adaboost classifier model to classify that according to the user data for particular disease the patient has the disease or not. For example, if the user wants to diagnose heart disease and he submits data according to system demands, then system utilize 70% of heart disease dataset to train the adaboost classifier algorithm then an adaboost classifier model is generated which can classify that according to the user inputted data the patient has the heart disease or not.
- Retrieve Diagnosis Result: After classification calculation is done, the system gets the diagnosis result.
- Display Result to the User: The diagnostic result will then be displayed to the user through the system's UI module.

4.3 Context Diagram of the DDS

Figure 4 depicts the DDS context diagram. The user enters into the Disease Detection System and the system displays a form that includes five diseases options such as Liver disorders, Hepatitis, Heart disease, Diabetes, and Chronic Kidney disease options. The user must select one of the diseases from the options that he/she wants to diagnose. Then the system checks the selected option submitted by the user and provides another form according to that particular selected disease. If the user does not select any option and submits the form, then system finds an error and asks the user to select an option again. When the system finds what disease, the user has selected, then it displays a form that asks some questions about that selected disease. The query form asks the user to provide some values of the disease input attributes. Whenever the user completes the form and sends it to the system, the system must approve the form otherwise the system will show the user's error message and the user will have to resubmit the form. After accepting the form, the computation module of the system uses the adaboost classifier algorithm to calculate the classification. The computational module generates adaboost classifier model and then feed the new inputted set of data to the model. The model will classify that according to the inputted data the patient has the disease or not. For example, if the user has provided Hepatitis data, then the computational model uses Hepatitis dataset from the database to train the adaboost classifier algorithm and the adaboost classifier model will classify that according to that inputted hepatitis data the patient has hepatitis or not. After classification is done the Data Processing Module will obtain

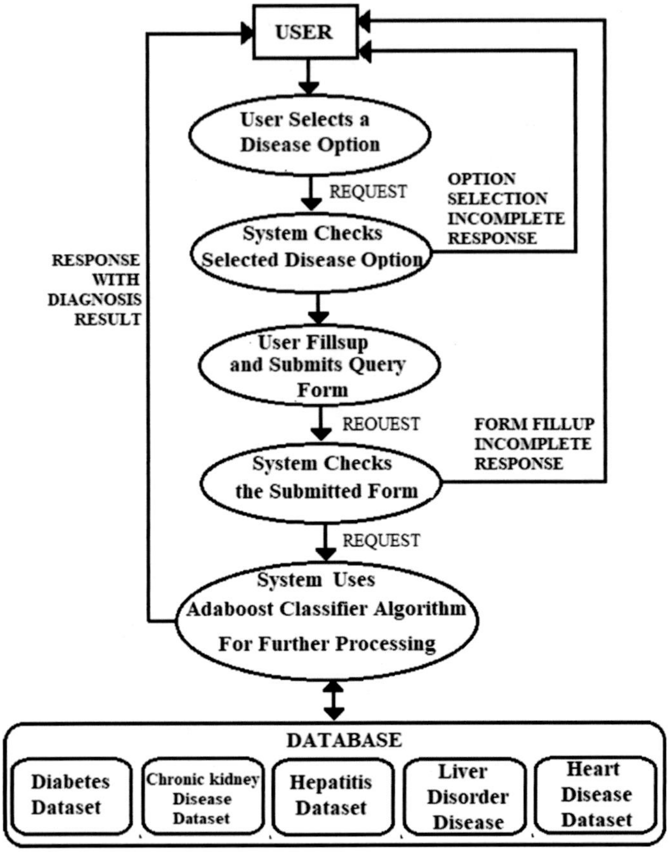

Fig. 4 DDS context diagram

the outcome of the diagnosis and show it to the user. If the diagnosis result is '1' then the Data Processing Module shows the user a message that 'The patient is suffering from Hepatitis' otherwise, for the diagnosis result '0', the user gets the message 'The patient is not suffering from Hepatitis'.

5 Accuracy Comparison

In the past many researches were done in the medical field where researchers used different machine learning algorithms to detect different diseases. This segment of the chapter shows the best accuracy of machine learning algorithms from various previous works. And compares their performances with our proposed machine learning model's performance.

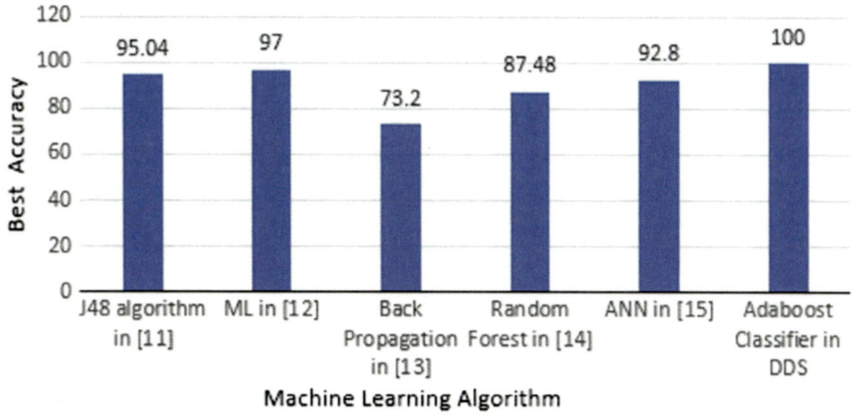

Fig. 5 Accuracy comparison for liver disease detection

5.1 Accuracy Comparison of DDS with Previous Works

5.1.1 Liver Disease Detection

From the different previous works, mentioned in Sect. 2 in this chapter, to diagnose liver disease, different machine learning algorithms are used. And different machine learning algorithms provide different best accuracies to detect the disease. Figure 5 shows the comparison graph where the accuracy of the adaboost classifier algorithm in DDS is compared with other best accuracies from different previous works those diagnose liver Disease.

5.1.2 Hepatitis Disease Detection

From the different previous works, mentioned in Sect. 2 in this chapter, to diagnose Hepatitis disease, different machine learning algorithms are used. And different machine learning algorithms provide different best accuracies to detect the disease. Figure 6 shows the comparison graph where the accuracy of the adaboost classifier algorithm in DDS is compared with other best accuracies from different previous works those diagnose Hepatitis Disease.

5.1.3 Heart Disease Detection

From the different previous works, mentioned in Sect. 2 in this chapter, to diagnose heart disease, different machine learning algorithms are used. And different machine learning algorithms provide different best accuracies to detect the disease. Figure 7 shows the comparison graph where the accuracy of the adaboost classifier algorithm

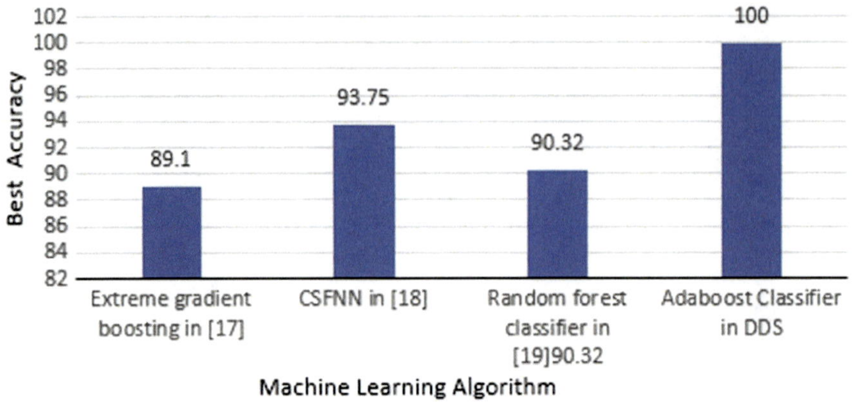

Fig. 6 Accuracy comparison for hepatitis disease detection

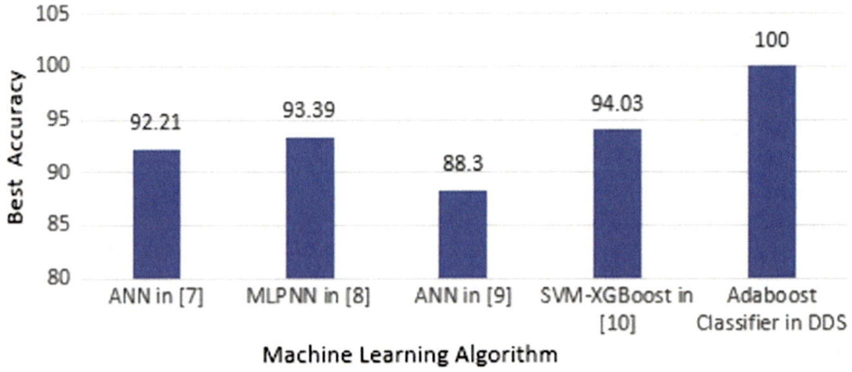

Fig. 7 Accuracy comparison for heart disease detection

in DDS is compared with other best accuracies from different previous works those diagnose Heart Disease.

5.1.4 Diabetes Disease Detection

From the different previous works, mentioned in Sect. 2 in this chapter, to diagnose Diabetes disease, different machine learning algorithms are used. And different machine learning algorithms provide different best accuracies to detect the disease. Figure 8 shows the comparison graph where the accuracy of the adaboost classifier algorithm in DDS is compared with other best accuracies from different previous works those diagnose Diabetes Disease.

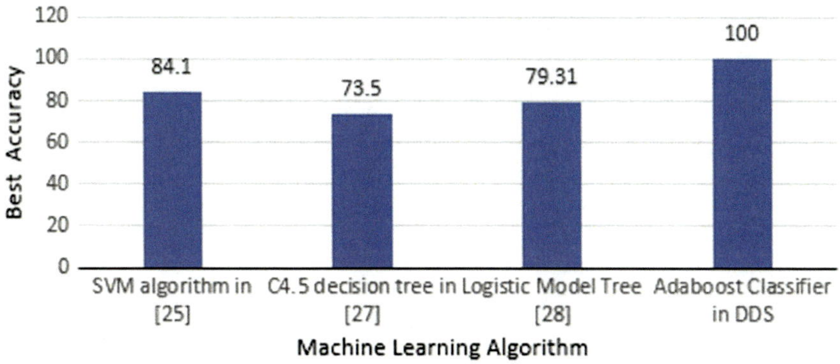

Fig. 8 Accuracy comparison for diabetes disease detection

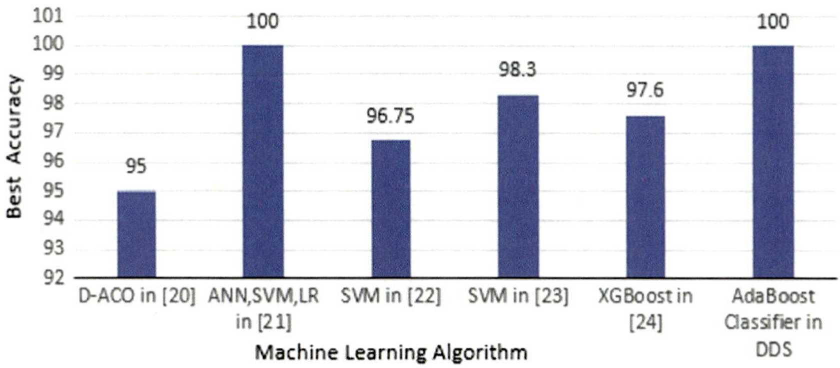

Fig. 9 Accuracy comparison for chronic kidney disease detection

5.1.5 Chronic Kidney Disease Detection

From the different previous works, mentioned in Sect. 2 in this chapter, to diagnose Chronic Kidney disease, different machine learning algorithms are used. And different machine learning algorithms provide different best accuracies to detect the disease. Figure 9 shows the comparison graph where the accuracy of the adaboost classifier algorithm in DDS is compared with other best accuracies from different previous works those diagnose Chronic Kidney Disease.

6 Simulation for Result

In this chapter, to design the Graphical User Interface (GUI) of the DDS, Jupyter Notebook is used as the simulation tool in Python Environment. As the python library Scikit Learn module has been used that supports different classifications, regression and clustering algorithms and includes Python numerical and scientific libraries NumPy and SciPy respectively. Figure 10 shows the first display form of the DDS that is used by the doctors as the gateway to detect different diseases.

Whenever the doctor wants to detect a disease, then he has to click the appropriate button for diagnosis the disease and then the Following forms will appear. Figures 11 and 12 show the liver disease diagnosis results. Figures 13 and 14 show the diagnosis results for Hepatitis Disease. Figures 15 and 16 show the diagnosis results for Heart Disease. Figures 17 and 18 show the diagnosis results for Diabetes Disease. Figures 19 and 20 show the diagnosis results for Chronic Kidney Disease.

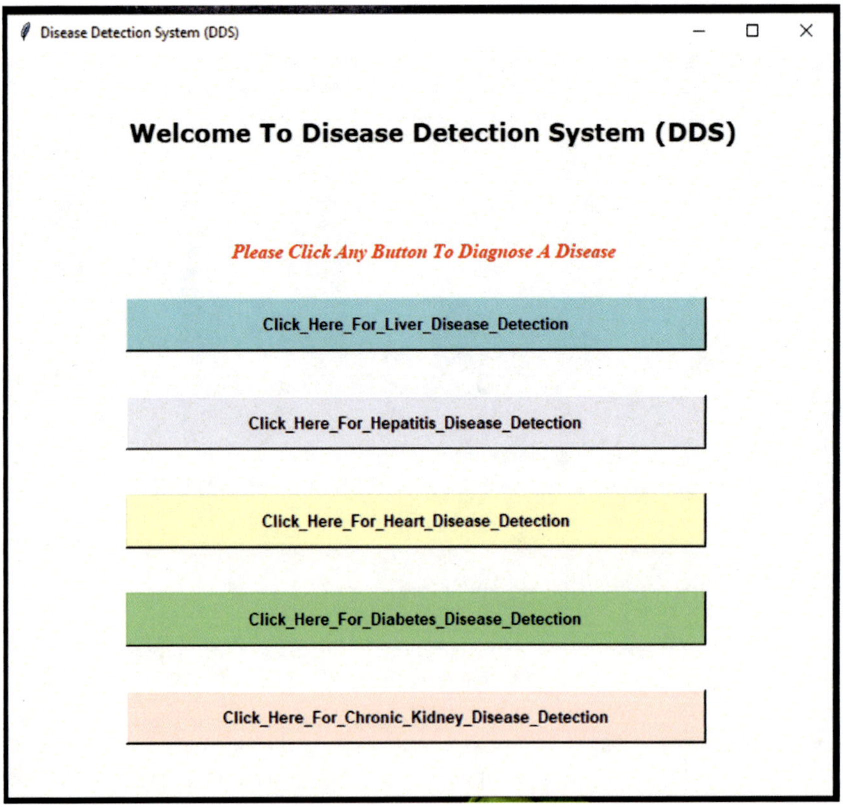

Fig. 10 First display form of DDS

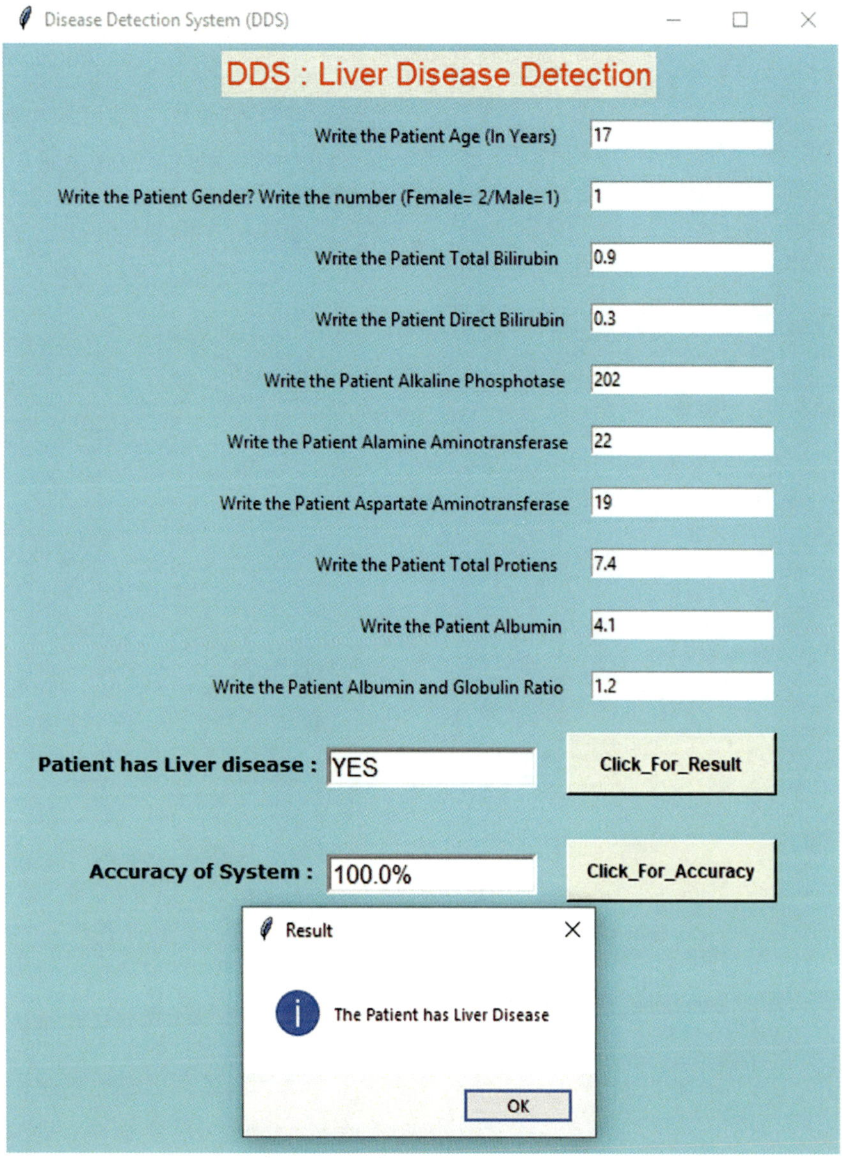

Fig. 11 Diagnosis result is positive for liver disease

7 Conclusion

This chapter proposes a Disease Detection System (DDS) that uses Adaboost Classifier Algorithm as a machine learning algorithm. DDS is designed to help doctors or medical professionals to detect five diseases such as Liver disorders, Hepatitis,

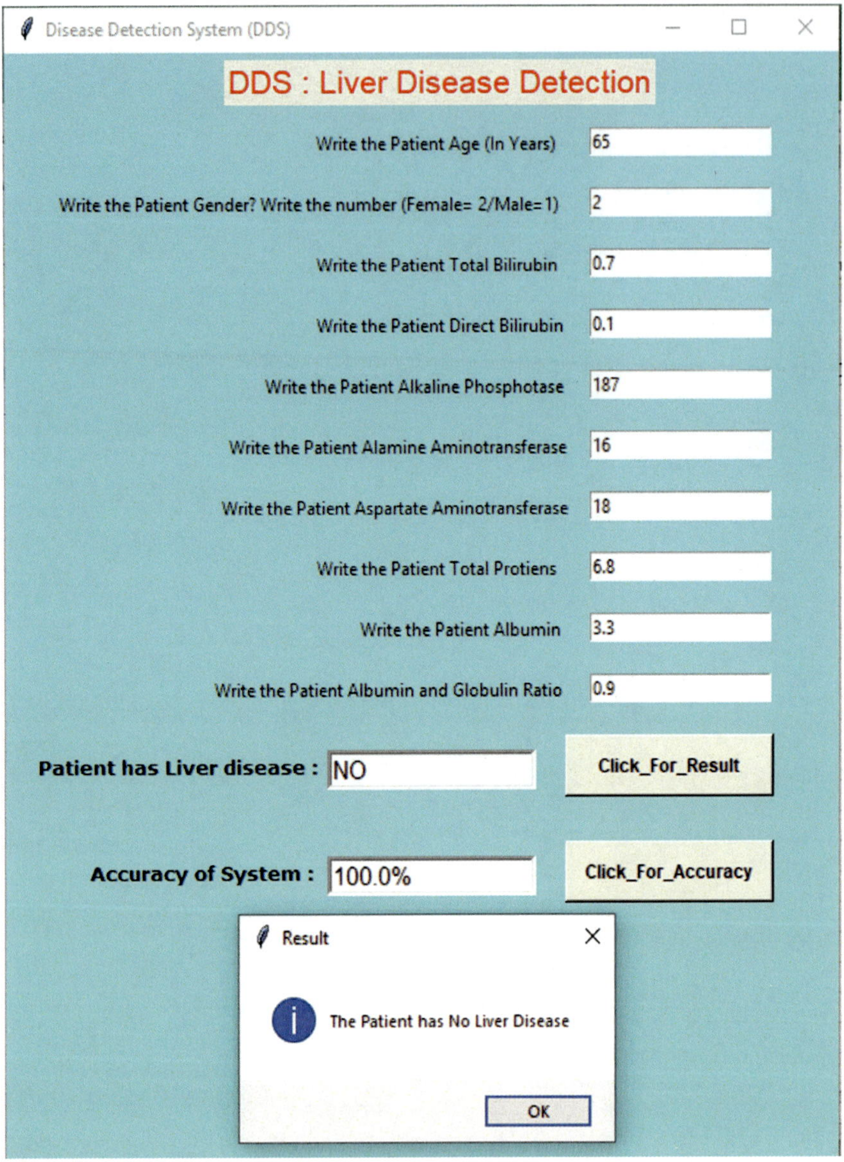

Fig. 12 Diagnosis result is negative for liver disease

Heart disease, Diabetes, and Chronic Kidney disease. For the implementation of the DDS and detect diseases this chapter follows some steps such as: from the machine learning database of Kaggle, different datasets for different diseases are obtained, pre-processing data strategy is used to turn imperfect actual data into a valuable and usable form, target attribute is identified from each dataset, excluding the column

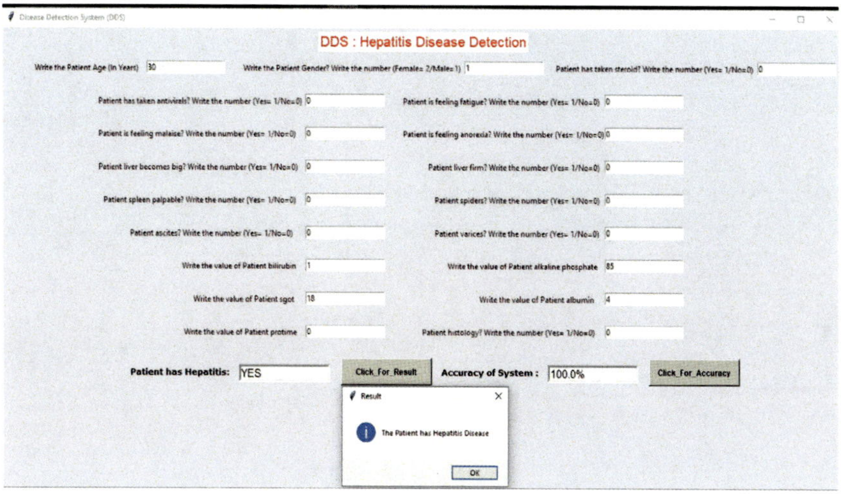

Fig. 13 Diagnosis result is positive for hepatitis disease

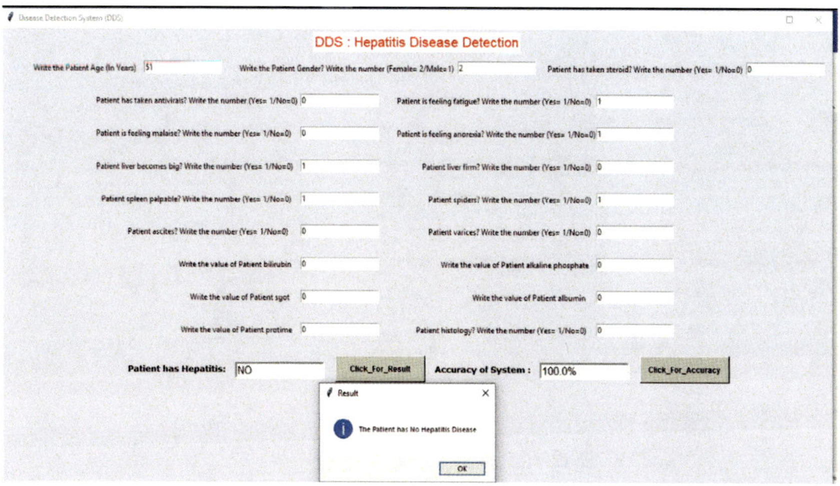

Fig. 14 Diagnosis result is negative for hepatitis disease

of the target attribute, the datasets are categorized into 2 sets with 7:3 proportions, adaboost classifier algorithm is trained with different disease's training dataset and adaBoost classifier model is generated for each of the disease detection that is able to detect different diseases. To test the Performance of adaboost classifier model, different disease's testing dataset are applied to their adaboost classifier model. It is found that adaboost classifier model provides 100% accuracy, precision and recall results in detection of each disease in the DDS, finally Disease Detection System (DDS) is implemented using the adaboost classifier model. In Python Environment,

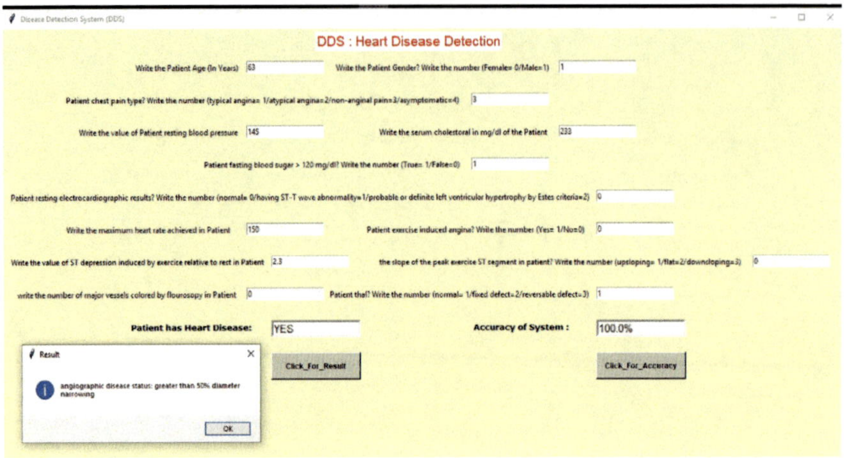

Fig. 15 Diagnosis result is positive for heart disease

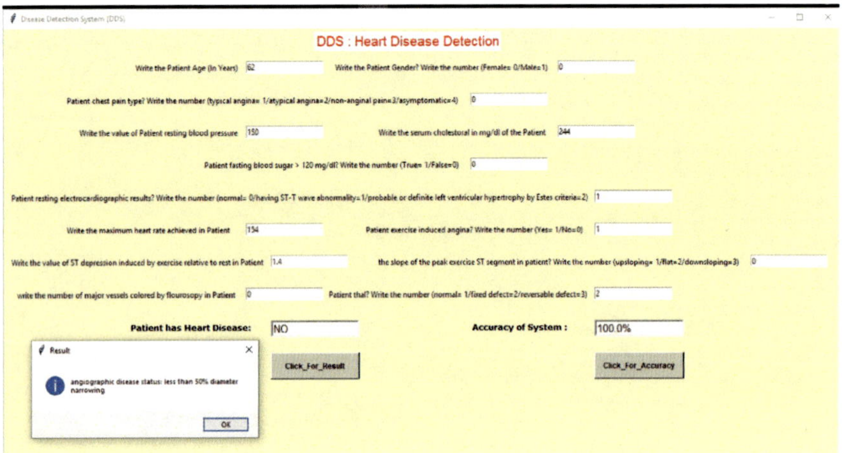

Fig. 16 Diagnosis result is negative for heart disease

DDS Jupyter Notebook's Graphical User Interface (GUI) is used as a simulation tool so that doctors or professionals in medicine can easily diagnose any disease among patients. Once DDS is developed, doctors or medical professionals select one disease from the five disease options in the GUI of the DDS for detection. Then through the GUI of the system he/she submits the required patient's data and fills up the form. Lastly system uses the adaboost Classifier model to detect whether the patient has the specified disease or not and displays the result to the doctor or medical professional. Accuracies of various machine learning algorithms from various previous related works were compared in this chapter and it is shown that DDS works better.

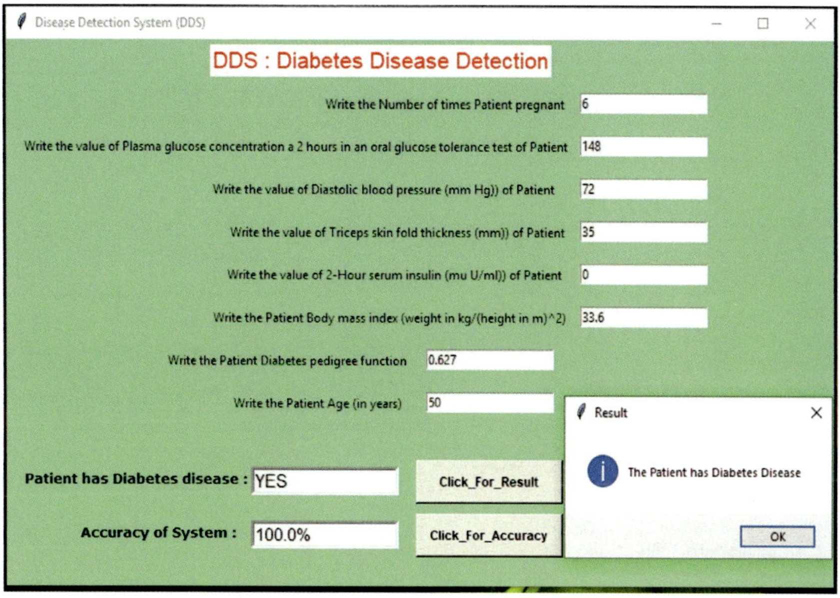

Fig. 17 Diagnosis result is positive for diabetes disease

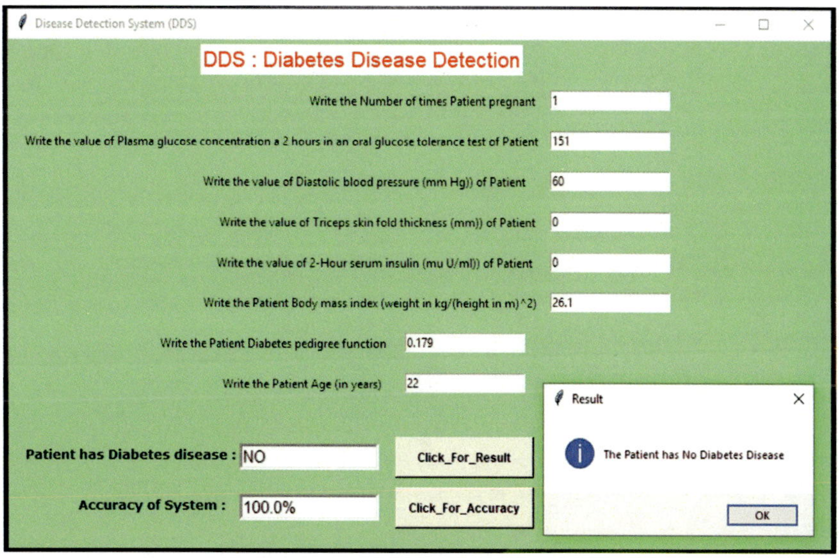

Fig. 18 Diagnosis result is negative for diabetes disease

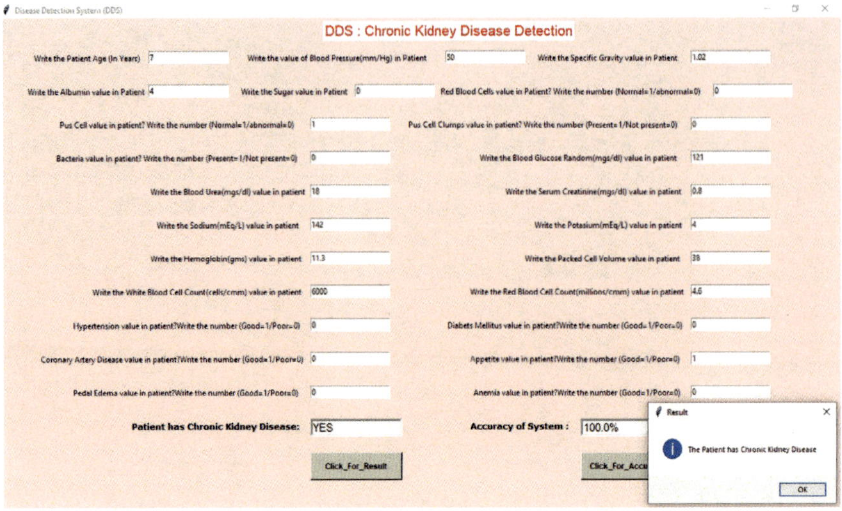

Fig. 19 Diagnosis result is positive for chronic kidney disease

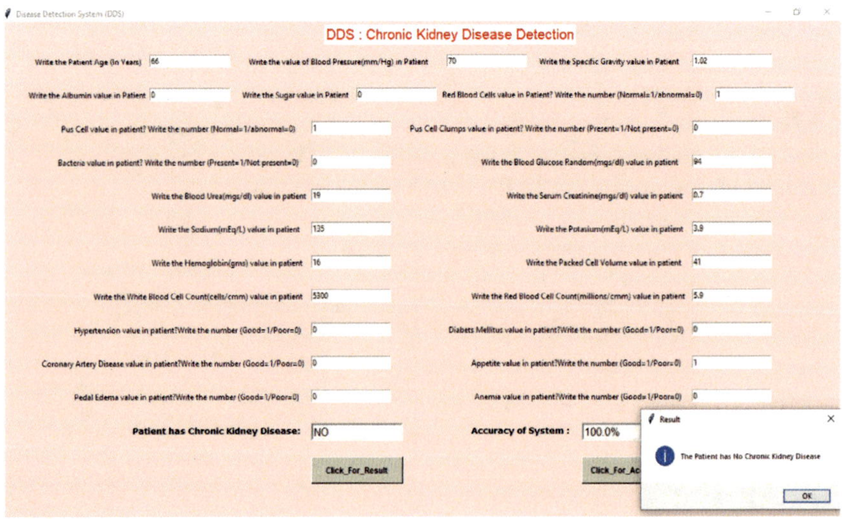

Fig. 20 Diagnosis result is negative for chronic kidney disease

References

1. Disease. Available: https://en.wikipedia.org/wiki/Disease. Last accessed 2019
2. M. Fatima, M. Pasha, Survey of machine learning algorithms for disease diagnostic. J. Intell. Learn. Syst. Appl. **9**, 1–16 (2017)

3. A.D. Sunny, S. Kulshreshtha, S. Singh, Ba M. Srinabh, H. Sarojadevi, Disease diagnosis system by exploring machine learning algorithms. Int. J. Innov. Eng. Technol. (IJIET) **10**(2), 14–21 (2018)
4. S. Razia, P. Swathi Prathyusha, N. Vamsi Krishnan, S. Sumana, A review on disease diagnosis using machine learning techniques. Int. J. Pure Appl. Math. **117**(16), 79–85 (2017)
5. D. Pavithra, A.N. Jayanthi, A study on machine learning algorithm in medical diagnosis. Int. J. Adv. Res. Comput. Sci. **9**(4), 42–46 (2018)
6. A.J. Dinu, R. Ganesan, F. Joseph, V. Balaji, A study on deep machine learning algorithms for diagnosis of diseases. Int. J. Appl. Eng. Res. **12**(17), 6338–6346 (2017)
7. P. Mamatha Alex, S. P Shaji, Prediction and diagnosis of heart disease patients using data mining technique, in *International Conference on Communication and Signal Processing* (2019), pp. 0848–0852
8. K. Subhadra, B. Vikas, Neural network based intelligent system for predicting heart disease. Int. J. Innov. Technol. Exploring Eng. (IJITEE) **8**(5), 484–487 (2019)
9. S. Khade, A. Subhedar, K. Choudhary, T. Deshpande, U. Kulkarni, A system to detect heart failure using deep learning techniques. Int. Res. J. Eng. Technol. (IRJET) **6**(6), 384–387 (2019)
10. R. Tao, S. Zhang, X. Huang, M. Tao, J. Ma, S. Ma, C. Zhang, T. Zhang, F. Tang, J. Lu, C. Shen, X. Xie, Magnetocardiography based ischemic heart disease detection and localization using machine learning methods. IEEE Trans. Biomed. Eng. **66**(6), 1658–1667 (2019)
11. V. Durai, S. Ramesh, D. Kalthireddy, Liver disease prediction using machine learning. Int. J. Adv. Res. Ideas Innov. Technol. **5**(2), 1584–1588 (2019)
12. A. Brankovic, A. Zamani, A. Abbosh, Electromagnetic based fatty liver detection using machine learning, in *13th European Conference on Antennas and Propagation (EuCAP 2019)* (2019), pp. 1–3
13. S. Sontakke, J. Lohokare, R. Dani, Diagnosis of liver diseases using machine learning, in *2017 International Conference on Emerging Trends & Innovation in ICT (ICEI)* (2017), pp. 129–133
14. C.-C. Wu, W.-C. Yeh, W.-D. Hsu, MdM Islam, P.A. Nguyen, T.N. Poly, Y.-C. Wang, H.-C. Yang, Y.-C. Li, Prediction of fatty liver disease using machine learning algorithms. Comput. Methods Programs Biomed **170**, 23–29 (2019)
15. J. Jacob, J.C. Mathew, J. Mathew, E. Issac, Diagnosis of liver disease using machine learning techniques. Int. Res. J. Eng. Technol. (IRJET) **5**(4), 4011–4014 (2018)
16. M.A. Konerman, L.A. Beste, T. Van, B. Liu, X. Zhang, J. Zhu, S.D. Saini, G.L. Su, B.K. Nallamothu, G.N. Ioannou, A.K. Waljee, Machine learning models to predict disease progression among veterans with hepatitis C virus. PLoS ONE **14**(1), 1–14 (2019)
17. X. Tian, Y. Chong, Y. Huang, P. Guo, M. Li, W. Zhang, Z. Du, X. Li, Y. Hao, Using machine learning algorithms to predict hepatitis B surface antigen seroclearance. Comput. Math. Methods Med. **2019**, 1–7 (2019)
18. L. Ozyilmaz, T. Yildirim, Artificial neural networks for diagnosis of hepatitis disease, in *Proceedings of the International Joint Conference on Neural Networks*, vol 1 (2003), pp. 586–589
19. K.K. Napa, V. Dhamodaran, Hepatitis-infectious disease prediction using classification algorithms. Res. J. Pharm. Technol. **12**(8), 2019 (2003)
20. M. Elhoseny, K. Shankar, J. Uthayakumar, Intelligent diagnostic prediction and classification system for chronic kidney disease. Sci. Rep **9**(1), 1–4 (2019)
21. E.-H.A. Rady, A.S. Anwar, Prediction of kidney disease stages using data mining algorithms. Inf. Med. Unlocked **15**, 100178 (2019)
22. S. Tekale, P. Shingavi, S. Wandhekar, A. Chatorikar, Prediction of chronic kidney disease using machine learning algorithm. Int. J. Adv. Res. Comput. Commun. Eng. **7**(10), 92–96 (2018)
23. A. Charleonnan, T. Fufaung, T. Niyomwong, W. Chokchueypattanakit, S. Suwannawach, N. Ninchawee, Predictive analytics for chronic kidney disease using machine learning techniques, in *2016 Management and Innovation Technology International Conference (MITicon)* (2016), pp. 80–83
24. A. Ogunleye, Q. Wang, Enhanced XGBoost-based automatic diagnosis system for chronic kidney disease, in *2018 IEEE 14th International Conference on Control and Automation (ICCA)* (2018), pp. 805–810

25. H. Abbas, L. Alic, M. Rios, M. Abdul-Ghani, K. Qaraqe, Predicting diabetes in healthy population through machine learning, in *2019 IEEE 32nd International Symposium on Computer-Based Medical Systems (CBMS)* (2019), pp. 567–570
26. S. Benbelkacem, B. Atmani, Random forests for diabetes diagnosis, in *2019 International Conference on Computer and Information Sciences (ICCIS)* (2019), pp. 1–4
27. M.F. Faruque, Asaduzzaman, I.H. Sarker, Performance analysis of machine learning techniques to predict diabetes mellitus, in *2019 International Conference on Electrical, Computer and Communication Engineering (ECCE)* (2019), pp. 1–4
28. D. Vigneswari, N.K. Kumar, V. Ganesh Raj, A. Gugan, S.R. Vikash, Machine learning tree classifiers in predicting diabetes mellitus, in *2019 5th International Conference on Advanced Computing & Communication Systems (ICACCS)* (2019), pp. 84–87

Sumana De received the B.Tech. and M.Tech. degree in Computer Science and Engineering from West Bengal University of Technology (WBUT), Kolkata, West Bengal India, in 2009 and 2012 respectively. She is at present a part-time Ph.D. research scholar in Computer Science and Engineering in the National Institute of Technology (NIT), Durgapur, West Bengal, India. Her research area includes Development of Knowledge Management Systems (KMS): System for Problem detection and diagnosis.

Baisakhi Chakraborty received the Ph.D. degree in 2011 from National Institute of Technology, Durgapur, India in Computer Science and Engineering. Her research interest includes knowledge systems, knowledge engineering and management, database systems, data mining, natural language processing applications and software engineering. She has several research scholars under her guidance. She has more than 30 international publications. She has a decade of industrial and 14 years of academic experience.

Knowledge Discovery (Feature Identification) from Teeth, Wrist and Femur Images to Determine Human Age and Gender

K. C. Santosh and N. Pradeep

Abstract Age estimation and gender identification process can be generally assessed from analysis of bone maturation in the human skeleton and dentition status. Age and gender determination of an unknown becomes a principal element in identification process and it plays an important part in forensic investigation. Teeth plays vital role in forensic since they can be preserved better than other parts. Bones such as femur and wrist bones also plays vital role in age and gender identification since they are very unique in nature. Long bone helps in forensic investigation for a better age and gender identification. Femur bone is the only long bone in human body and its features like femur head diameter, shaft length, femoral neck angle, condyles width can be considered as best features in age estimation and gender identification. Wrist bone has more unique features than any other bones in human body like carpal bones, fusion of radius and ulna. The current study aspires to accomplish the purpose in developing a computer aided system that identifies features from teeth, wrist and femur bone that will provide accurate gender identification and age estimation in less time. There is more scope for researchers to contribute towards age estimation and gender determination from the analysis of medical images.

Keywords Femur bone · Femur head diameter · Shaft length · Femoral neck angle · Condyles width · Ulna · Radius

1 Introduction

Gender identification and age estimation can be done through many means. Identifying all possible gender and age related attributes and features are challenging task and they play vital role in personal identification process. Some of the features utilized

K. C. Santosh (✉) · N. Pradeep
CSE Department, Bapuji Institute of Engineering and Technology, Davangere, Karnataka 577004, India
e-mail: kcsantoo@googlemail.com

N. Pradeep
e-mail: nmnpradeep@gmail.com

V. Jain and J. M. Chatterjee (eds.), *Machine Learning with Health Care Perspective*, Learning and Analytics in Intelligent Systems 13,
https://doi.org/10.1007/978-3-030-40850-3_7

for gender and age determination are: General physical examination (e.g.—Height, Weight etc.), Dental eruption, Radiology examination of epiphyseal union etc. Age estimation and gender identification is a vast researching field and new techniques by using the dentine features and other skeletal parts yet to be investigated under study [1]. Age assessment is a major step in identifying from post-mortem remains. Identification of age and gender of unknown human bodies will be the fundamental task of forensic investigation. This identification process also plays a vital role in context with crime inspection or a mass disaster like earth quake, tsunami etc. since the age and gender at the time of death can provide useful clue to investigators in order to achieve accuracy in identification of human. Age determination by Teeth and bones is one of the main means of determining gender and age estimation.

Teeth play a vital role in personal identification process since teeth are better preserved for long time and it's one of the strongest body parts in human body. Mainly teeth are helpful in identification of gender using many oral techniques. Size of crown, root length, measuring the mesiodistal and buccolingual dimensions, permanent canine teeth and their intercanine distance, mandibular canine index are some of the essential key features in teeth.

Wrist bones are made up of carpal bones and phalanges that have more unique features than any other bone features of human. Ossification of all Carpal bones, lower ends of radius and Ulna bones are the key features in Wrist that will helps us for estimation of age and gender identification. Radiography of the wrist is the most commonly used to calculate bone age.

Long bones have many distinct features like the diaphysis, or shaft; the metaphysis, or the flared end of the shaft; and the epiphysis, or end cap of the bone. Hence, long bones plays major role for the assessment of age and gender. Since femur is the long bone in human body and it is often easy to assess their gender and age, the length of femur bone in male tends to be heavier, longer, and massive compared to female bone [2]. The femur bones total length, the head diameter and the condyles width of femur are few features used in identification of gender and age.

Data Acquisition: Dataset collection (acquisition) is a major step in our research work, the sample size of data collected should be sufficiently enough to validate the results, since digital images of teeth, wrist and femur bones are not publically available in the Internet and hence digital X-ray images were collected from various hospitals in Davangere. Dataset sample collected from different hospitals were listed in Table 1.

Table 1 Sample size of digital images of teeth, wrist and femur collected from different hospitals

S. No.	Hospital	Sample size		
		Teeth	Wrist	Femur
1.	Bapuji Dental College and Hospital, Davanagere	147	–	–
2.	College of Dental Sciences, Davanagere	995	–	–
3.	Jagadguru Jayadeva Murugarajendra Medical College, Davangere	–	190	40

2 Features of Teeth to Identify Age and Gender

From infant to age of 21, teeth are the important field for true age calculator. Human beings have two sets of teeth—one is primary teeth also called as baby teeth and other set known as permanent teeth also known as secondary teeth. Primary teeth start to come out at about six months of child age, this tooth eruption starts in the lower jaw with the central incisors. Figure 1 depicts the set of toddler teeth for different age group. Second set of teeth will eventually take place of first set of teeth when they fall out.

The primary set of teeth as depicted in Fig. 1 has a primary set of 20 teeth, in which every child will have complete set of these teeth by the age of 3 years. The primary teeth are made up of: two set of incisor teeth in both upper and lower jaw, two set of central canines and four molars namely first and second molar in each jaw. Incisors teeth used to bite pieces of food, canine teeth helps to hold and tear food and molars are used for grinding.

Every tooth type—incisors, canines, premolars, molars—erupts on a predictable schedule. Figure 2 depicts the complete set of adult human teeth, which are the most accurate age indicators. Several contributions have been contributed in identifying age and gender using teeth features by many experts and few of them have been explained below.

2.1 Related Works on Teeth

Nagi et al. [3] in 2018 developed a method for age estimation using digital panoramic radiographs of human teeth, their study is to estimate human age through tooth

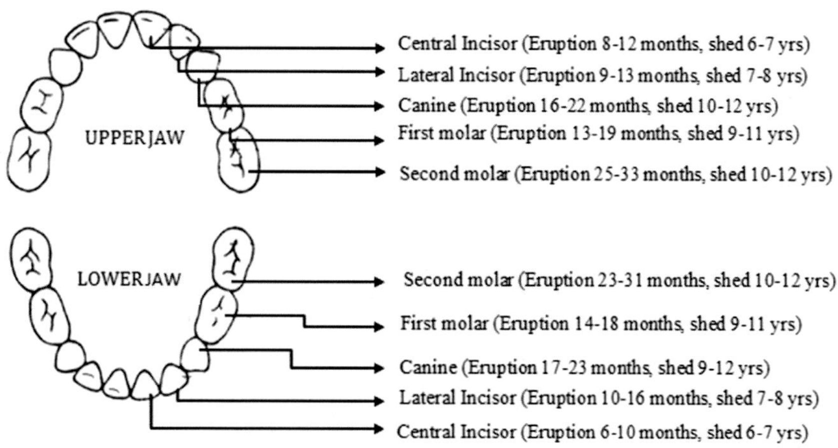

Fig. 1 Growth of children's teeth

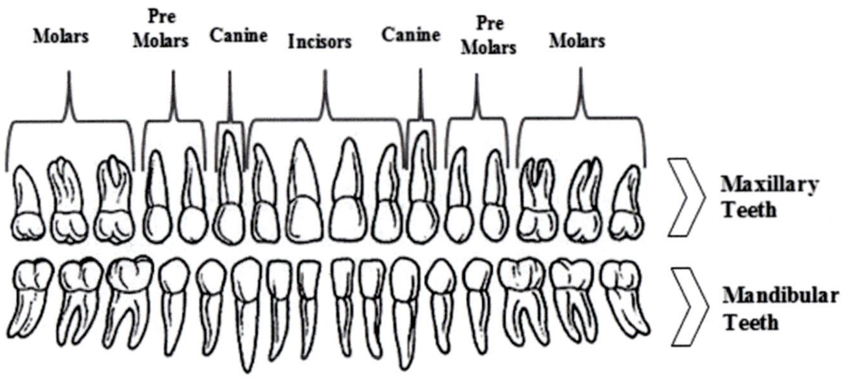

Fig. 2 Complete human teeth

coronal index (TCI) method. Some of the features extracted by authors to estimate age of individual are: (i) height of the crown, i.e., coronal height (ii) height of the coronal pulp cavity, second and first premolars in mandibular was measured in millimeter (mm) and (iii) each tooth's TCI was calculated.

Measurement of right mandibular first molar teeth in panoramic view is shown in Fig. 3. TCI was calculated by measuring CH through a cervical line to the peak of cusp. CPCH was calculated (yellow line) using distance between cervical lines to the peak of the pulp horn (red line). The values provided from the TCI of tooth, it was then calculated as shown in Eq. (1)

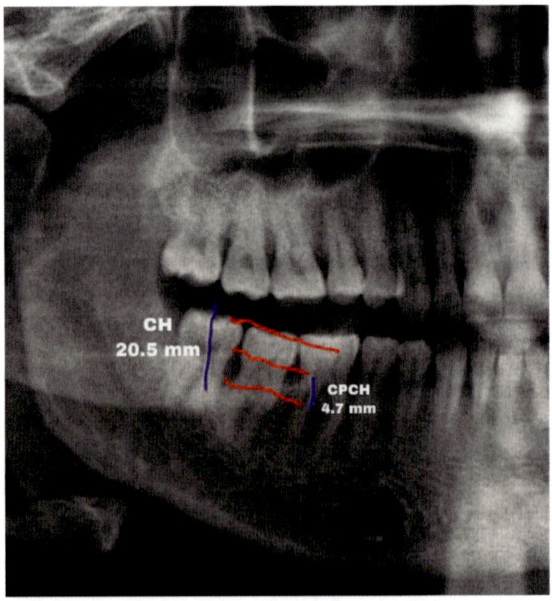

Fig. 3 Rt. mandibular first molar measurement depicted in panoramic image (from JIAOMR 2018, vol. 30 p. 64–67)

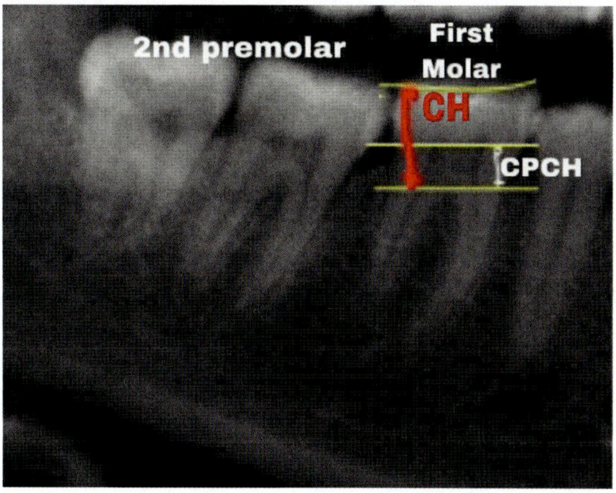

Fig. 4 First and second premolar in right mandibular

$$TCI = CPCH \times 100/CH \quad (1)$$

Figure 4 is a digital panoramic image that represents the distance between first and second premolar present in right mandibular jaw.

Pagare et al. [4] in 2017 proposed a study based on odontometric unique features in gender identification since teeth is the hardest and stable tissue in human being. Their study revealed that the arch length in male tends to be higher than the female arch length. Figure 5 showing the graph of arch length of male and female and also length of arch from maxillary jaw is more than length of mandibular.

The authors also found that the Inter canine distance was more in Male maxilla and mandible than Female maxilla and mandible as depicted in Fig. 6.

VikramSimha Reddy et al. [5] in 2017 conducted a study on Gender identification using Barr bodies from teeth and they observed barr bodies presence to a maximum of 400 °C from a female teeth and also they identified the absence of barr bodies in male teeth.

Larissa Chaves et al. [6] in 2016 determined human gender by odontometric analysis of molars to evaluate the presence of sexual dimorphism between the first and second permanent molar. Authors used 50 pairs (25 male, 25 female) of dental casts for their study.

The measurements performed by authors using digital calliper and the following variables were recorded. Human teeth model set purchased from dental accessories shop and measurement with respect to the upper and lower jaw of teeth are depicted in Fig. 7: Mesiodistal width is the greatest distance between the proximal surfaces of the molar tooth is shown in Fig. 7a, Buccolingual/palatal width is the distance between the outermost points of the molar crowns is shown in Fig. 7b and the distance between the lingual cusps of the corresponding molar teeth in opposite quadrants is depicted in Fig. 7c.

The mean and standard deviation of the measurements according to tooth and gender of human are depicted in Table 2. There is a major difference in the Mesiodistal

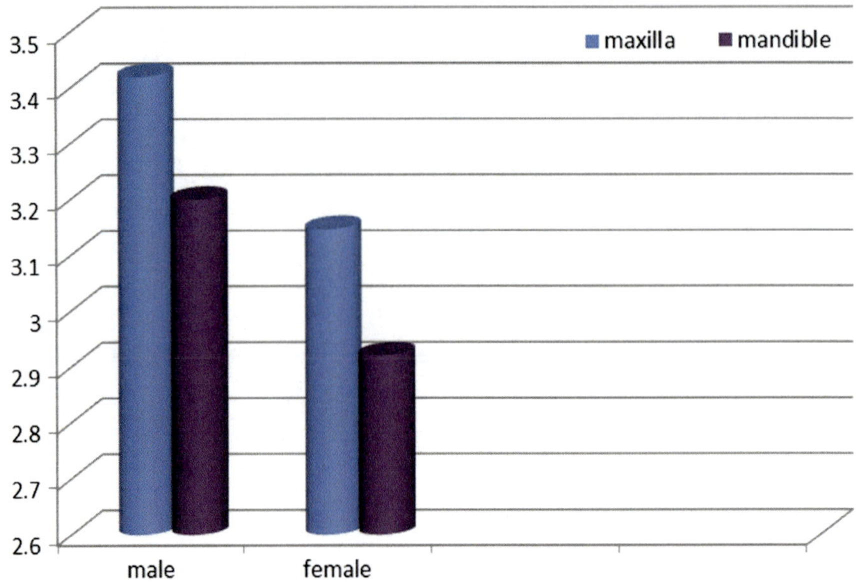

Fig. 5 Bar chart of arch length (from AJFSC 2017, vol. 4, p. 1060)

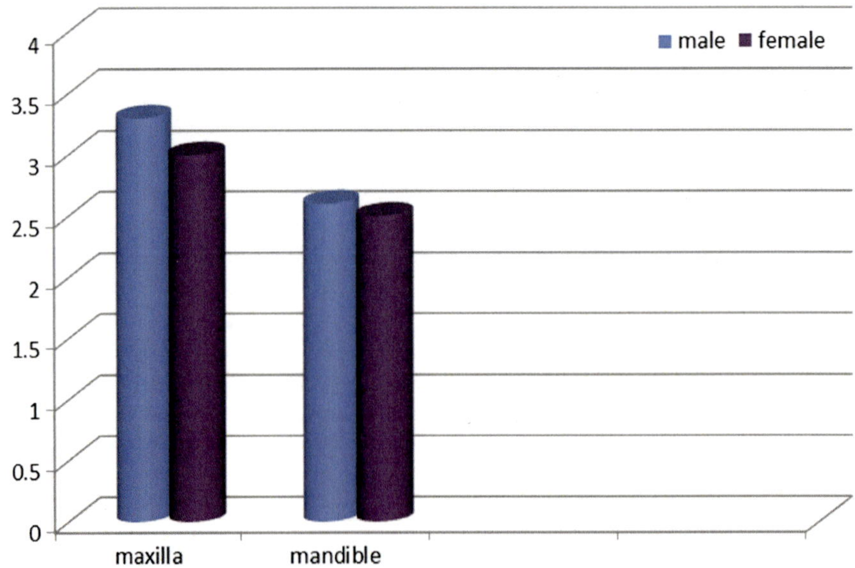

Fig. 6 Bar chart of inter canine distance

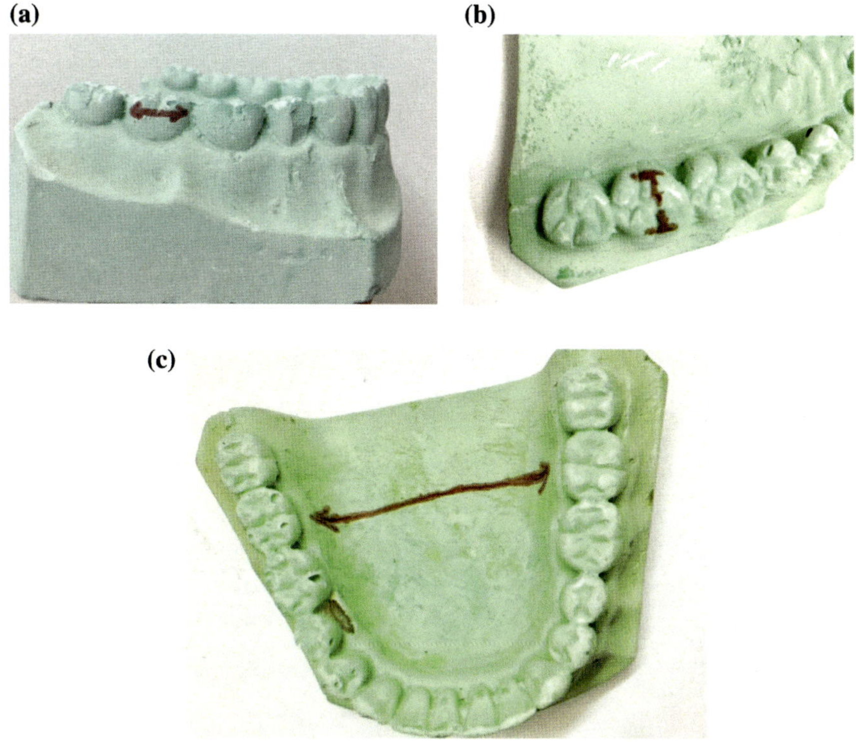

Fig. 7 **a** Mesiodistal distance. **b** Buccolingual distance. **c** Distance between cusps of the molar (from BJOS 2016, vol. 15, p. 35-38)

distance (MD) and Buccolingual distance (BL) widths in first molars of left upper jaw and right upper jaw as well left lower jaw and right lower jaw and male teeth were found to be larger than the female ones. Also for second molars in the mandibular right and maxillary right (tooth no. 47 and tooth no. 17, respectively) showed a statistically significant difference concerning their MD width.

Kewal Krishan et al. [7] in 2015 stated that Radiographic methods are used in different stages of mineralization and this will help further for a better accuracy in age assessment. The mineralization of teeth is a better estimation of chronological age than the mineralization of bone.

Acharya [8] in 2014 proposed a method estimation of age using digital dental features. Few researcher examined dentine translucency with the help of callipers and its length was measured for sectional and unsectioned teeth. Dentine translucency can be examined using digital technique. To achieve more accuracy in age estimation, digital methods were used in the place of callipers. Lorentsen and Solheim [9] found alternate method for age estimation measuring dentine translucency area that was better correlation than translucency length and contributed more in multiple

Table 2 Mean and standard deviation of the measurements according to tooth and gender

Measures	Sex				
	Tooth #	Male		Female	
		Mean	Std. deviation	Mean	Std. deviation
Mesiodistal	16	10.32	0.57	9.84	0.43
	26	10.26	0.57	9.84	0.50
	36	10.77	0.67	10.25	0.57
	46	10.66	0.61	10.27	0.55
Buccolingual	16	10.48	0.65	10.05	0.46
	26	10.49	0.55	10.09	0.52
	36	10.14	0.57	9.72	0.46
	46	10.08	0.63	9.64	0.53
Mesiodistal	17	9.70	0.61	9.34	0.55
	27	9.59	0.62	9.28	0.57
	37	10.34	0.81	9.99	0.54
	47	10.26	0.77	9.85	0.65
Buccolingual	17	10.40	0.77	10.07	0.50
	27	10.43	0.78	10.08	0.46
	37	9.96	0.59	9.51	0.55
	47	9.82	0.66	9.36	0.51
Distances	16-26	37.81	2.51	36.13	2.42
	36-46	34.30	2.40	33.04	2.99
Distances	17-27	43.29	2.52	40.72	2.58
	37-47	40.68	2.62	39.02	2.33

regression method. Hence researchers recommended to measure translucency area for better estimation of age.

In case of mass disaster, flood, building collapse, crime investigation, chemical and nuclear explosions gender identification comes to be first preference in order to identify an unknown body. Several methods have been used for the gender identification. Gender is determined from Morphological analysis as well as from molecular analysis. Morphological analysis is performed by odontometric methods and orthometric methods. Few dentine methods like (a) mesiodistal (MD) dimensions and buccolingual (BL) dimension of teeth and (b) Mean Canine Index (MCI) (dental index) are included.

Khangura et al. [10] in 2011 analyses that in male teeth dimension of mesiodistal are more compared that with female teeth. This difference may be because of the two reasons (1) due to presence of amelogenesis over a long period, the thickness of enamel in males is more compared to females and (2) Y chromosomes production rate is slower in male maturation stage.

Table 3 Difference in mesiodistal and buccolingual measurement of tooth among male and female

Tooth No.	Mesiodistal (MD)		Buccolingual (BL)	
	Male	Female	Male	Female
L1	8.9	8.5	7.1	7.0
L2	7.0	6.65	6.5	6.2
L3	8.3	7.6	8.4	7.9
L4	6.9	6.8	9.3	8.9
L5	6.7	6.65	9.8	9.3
L6	11.0	10.6	11.0	10.9
L7	10.4	9.9	11.0	10.7
U1	5.5	5.3	6.2	6.1
U2	6.1	5.9	6.5	6.5
U3	7.2	6.6	7.55	7.4
U4	7.1	7.0	7.9	7.6
U5	7.4	6.9	8.6	8.2
U6	11.1	10.8	10.4	10.2
U7	10.5	10.2	10.3	9.9

Garn et al. [11] recommends that in the identification of gender of an unknown body, buccolingual measurement is a reliable source than other variables. Garn et al. stated that buccolingual length measures larger in male dentition than female.

A prior study based on teeth dentition recommended that mesiodistal measurement is best suitable for gender determination than buccolingual measurement. The arch size of both upper jaw and lower jaw impacts the size of tooth, i.e. largest size in male bodies may be associated to largest mesiodistal measurement of tooth compared with female. Though various investigation on mesiodistal measurement have been concluded that mesiodistal dimension is a better indicator of gender than buccolingual dimension, but still some miscalculation may exist in scaling the mesiodistal dimension because of close proximal contacts. Hence, both mesiodistal and buccolingual measurements are considered as a reliable parameter in determination of gender (Table 3).

2.2 *Gender Assessment Using Canine Index Method*

Gender identification based on features of teeth is mainly focused on comparing the tooth dimensions in males and females. Canines were found to exhibit the greatest genderual dimorphism among all teeth because of the following reasons: (a) Canines are less exposed to plaque and calculus. (b) Lesser pathological migration of canines than other teeth. (c) Canines are usually survived than any other teeth in many conditions. Mesiodistal size in the canines and inter-canine measurement in the mandible gives better estimation in identifying gender. Teeth dimensions are widely used to establish the gender of the individual [12]. At the age between 15 to 23 years, all canine teeth were erupted and there would be minimum attrition at this level and also the inter-canine distance is fixed by the age of 13 years. Canine index method consisted of measuring the upper jaw and lower jaw canine widths and their inter-canine measurements of respective jaws.

Largest MD width of canine was measured at a point with adjacent teeth using a divider, by placing the dividers two end point over the digital callipers and this distance was recorded. The obtained readings are used for further analysis and to calculate Canineindex and Standardcanineindex for all canines as shown in Eqs. (2) and (3) formulated by Muller et al. [9]

$$Canineindex = \frac{Mesio{-}distal\_crown\_width}{Intercanine\_distance} \tag{2}$$

$$Standardcanineindex = \frac{(mean\_maleCI - SD) + (mean\_femaleCI + SD)}{2} \tag{3}$$

Right mandibular and left mandibular canines from male is measured and compared with female and it was found that mesiodistal crown width in male is more as shown in Table 4. Similarly right maxillary and left maxillary canines was measured for both gender and it was found that mesiodistal crown width is more in male compared with female.

The mandibular inter-canine and maxillary inter-canine distances and gender-wise distribution of inter-canine distances are shown in Table 5.

Table 4 Gender-wise distribution of canines mesiodistal crown width

Position	Gender	Mandibular canine		Maxillary canine	
		Mean	±S.D.	Mean	±S.D.
Right	Male	7.20	0.44	7.80	0.45
	Female	6.89	0.41	7.55	0.45
Left	Male	7.26	0.44	7.85	0.45
	Female	6.94	0.41	7.60	0.44

Table 5 Inter-canine distances (ICD) measurement for gender identification

	ICD (mm)	Range	Mean	±S.D.
Mandible	Male	21.26–27.78	25.25	1.22
	Female	21.49–27.90	24.75	1.08
Maxilla	Male	30.02–39.51	34.17	1.68
	Female	29.20–37.47	33.47	1.32

Table 6 Features of teeth with ground truth values for age and gender identification

S. No.	Feature name	Ground truth value			
		Male (mm)	Female (mm)	Age (years)	
1.	Mesiodistal distance in first molars of maxillary	10.32	9.84	–	
2.	Buccolingual distance in first molars of maxillary	10.48	10.05	–	
3.	Distance between the lingual cusps in molar	37.81	36.13	–	
4.	Maxillary arch length	Higher	Inferior	–	
5.	Mandibular arch length	Higher	Inferior	–	
6.	Size of pulp chamber	11.3	10.44	–	
7.	Mandible Inter canine distance	25.45	24.75	–	
8.	Maxillary Inter canine distance	34.17	33.47	–	
9.	Translucency Measurements	51.5	48.5	Conventional 41	Digital 40

Various features from teeth are identified for age estimation and gender identification, their ground truth values for gender and age estimation are depicted in Table 6.

3 Features of Wrist Bone to Identify Age and Gender

The wrist joint and hand bones contains the radius, ulna and 19 other short bones like 5 metacarpals bones and 14 phalanges, also it consists of 8 carpals. Figure 8 shows the carpal bones of wrist and other parts of human hand. The sequential orders of the individual carpal bones are: (i) capitate, (ii) hamate, (iii) triquetral, (iv) lunate, (v) trapezium, (vi) trapezoid, (vii) scaphoid or navicular and (viii) pisiform. The carpal bone set in wrist joint, plays an important role which differs from individual with respect to the age, mainly from the young stages of childhood to adult. The distal epiphysis of the radius ossifies before the triquetum and that of the ulna before the pisiform.

The maturation rate of the carpal bones differs from person to person. The carpal bone maturation completes earlier than that of long and short bones. Usually the right hand tends to be used more frequently for day to day work and this reflects on the normal growth of bones, and hence left hand digital x-ray images are extensively preferred and used than right hand for age estimation.

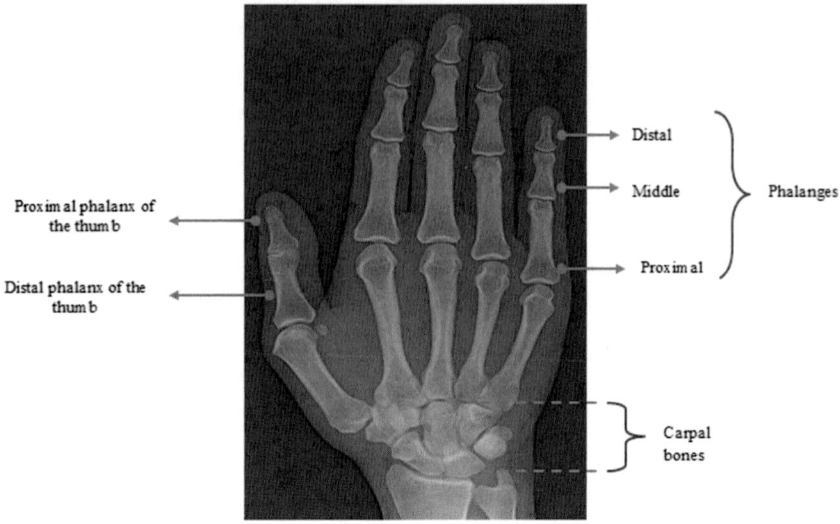

Fig. 8 Bones of a human hand and wrist

3.1 Related Works on Wrist

Greulich and Pyle [13] in 1959 developed an atlas that includes complete development stages of hand bone. They started their research on hand bone assessment on gender and age estimation. The related works in identifying age and to determine gender using wrist features have been explained below:

Pietka et al. [14] in 1991 developed a method for personal identification based on computer aided system based on phalangeal bones. This identification process was advanced by integrating the system with a Computer aided PACS lab (Picture Archiving and Communication systems). Researchers saved individual database of bones in the laboratory. These digital images were classified into uniform and non-uniform digital image pixels to eliminate background noise. After classification stage, outer lines were obtained by identifying the edge from image and further these edges may be included or excluded based on error-correction technique. After finding edges from image, edges can be rotated accordingly. Sobel edge detector is used to identify the region of interest from digital images of phalangeal bones.

Fishman [15] in 1982 proposed an evaluation method of skeleton maturation based on hand wrist images. Fishman's 11 stages of grading system is depicted in Table 7.

Eleven discrete adolescent skeletal maturational indicators covering the entire period of adolescent development is found on these six sites as depicted in Fig. 9.

A Skeletal maturation Indicator as depicted in Fig. 10 was used to facilitate SMI evaluation. Based on this approach, key stages were checked first.

Hand-wrist radiographs are helpful in identifying the different levels in the maturation of skeleton and also it is used in predicting pubertal growth. The existence of unique and accurate bone structure with specific features and sequence of maturity

Table 7 Skeletal maturation indicator (SMI)

Stages (SMI)	Description of hand-wrist maturation stage
1	The third finger of proximal phalanx having equal width of the epiphysis and diaphysis
2	The third finger of middle phalanx having equal width of the epiphysis and diaphysis
3	The fifth finger of middle phalanx shows equal width of the epiphysis and diaphysis
4	Presence of Sesamoid in thumb
5	Capping of the epiphysis of distal phalanx on the third finger
6	Capping of the epiphysis on the middle phalanx of the third finger
7	Capping of epiphysis of middle phalanx on fifth finger
8	Union of the epiphysis and diaphysis in the distal phalanx on the third finger
9	Union of the epiphysis and diaphysis in the proximal phalanx on the third finger
10	Union of the epiphysis and diaphysis in the middle phalanx on the third finger
11	Union of the epiphysis and diaphysis appeared in the radius

has made wrist and hand digital image as a useful clinical data to determine maturity from skeleton. Digital x-ray images of metacarpal bones and wrist joint carpal bones helps in determining maturation level of bones that will be a reliable choice in prediction of individual age.

Michael and Nelson [16] in 1989 proposed very first bone age automated system called HANDX. This system is developed based on three steps, pre-processing step where digital image is enhanced used Model-Based Histogram modification algorithm, segmentation step using Gaussian distribution and measurement stage.

Roche et al. [17] in 1989 proposed a method called FELS (Fels Longitudinal Study) method. It is a computerized technique used for measuring scores and grading of bone of different age groups. More than 130 different marks are selected for analysis from every bone. It also estimates the error correction for assessment. Error rate may vary for boys of age one month to two years are 0.3–0.6 years and for girls of age 1–14 years is 0.2–0.3 years. Due to fusion of bone, this error rate may be more in youth and old age.

Da Silva et al. [18] in 2001 developed a semi-automated system of skeleton maturity using digital image processing methods. A fixed point is selected from digital image of a finger bone and contours parts are identified. A profile curve is drawn from the middle finger which was detached from the image and using this profile curve cartilage and epiphysis thickness is measured. Results are obtained from measuring the distance from tip of the profile curve and hence age estimation is done from this distance.

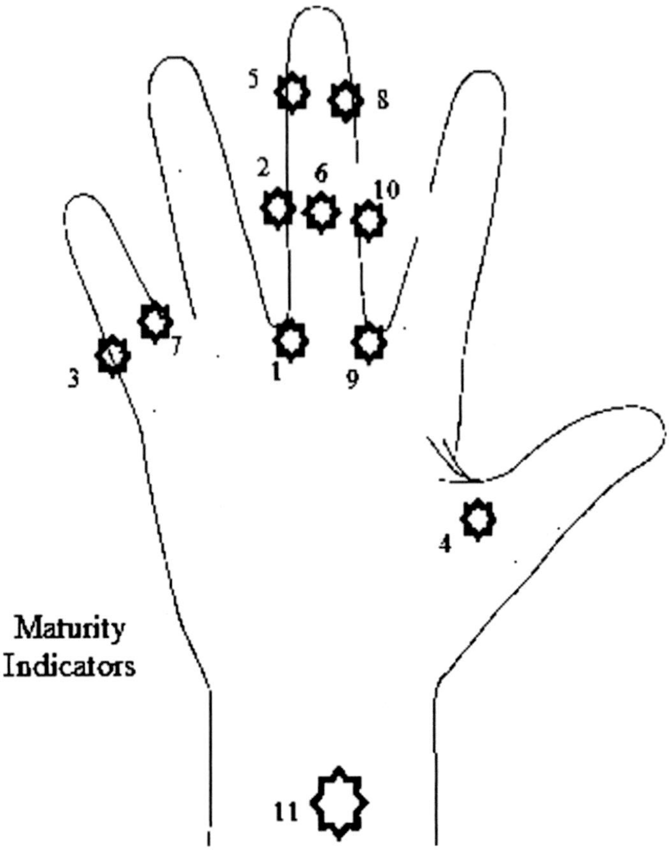

Fig. 9 Eleven skeletal maturity indicators (from NACES, Boston 1982 vol. 52, p. 90)

3.2 Tanner-Whitehouse (TW) Method

TW method depends on values based on the bones from a skeletal analysis, and from those points that assigned to the bones. Various grading scale were used in identification of gender in each stage of every bone. Figure 11 shows the hand and wrist image with the carpal, metacarpal bones, radius, ulna, short bones and bones which are subject to ossification analysis.

Gender identification can be achieved in two methods: first method is by using carpal bone score that involves the growth/appearance of carpal bones and other method is by using Radius Ulna Score, this method includes estimation of ulna and short bones.

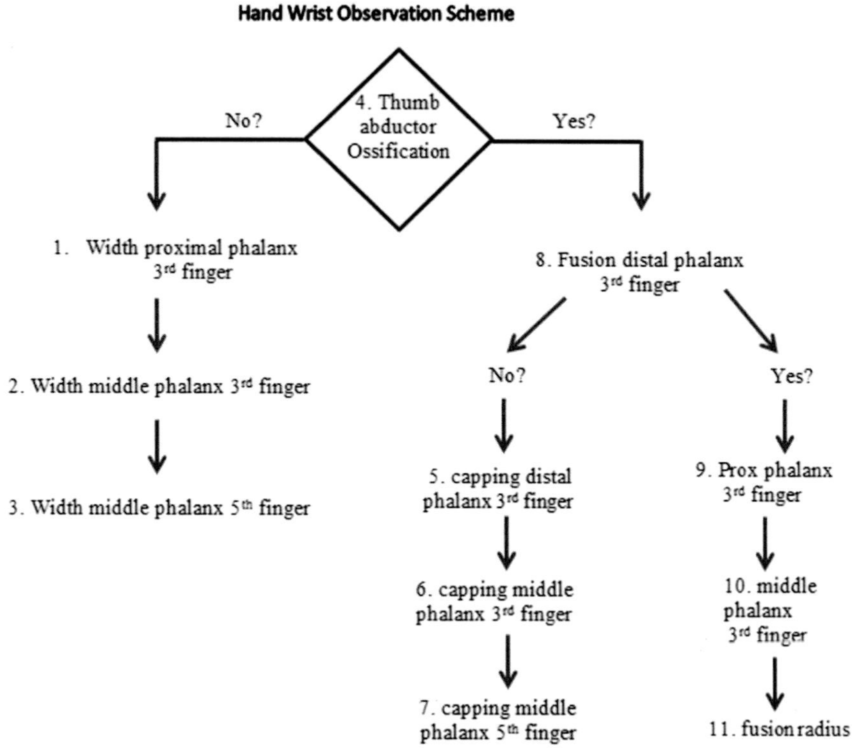

Fig. 10 SMI evaluation scheme on a wrist image

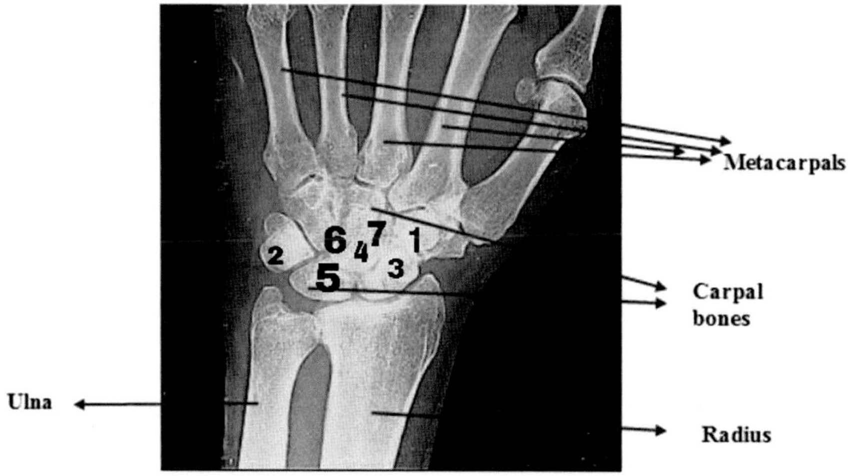

Fig. 11 Hand and wrist bones

3.3 Computer-Assisted Wrist Bone Age Assessment System

Compared to other techniques for bone age estimation and gender identification, TW method is reliable and much preferred for automation [19]. Fully automated computer system for bone age estimation are typically based on digital image techniques with reliable, fast and accurate results based ossification in carpal bone and maturity levels of metacarpal and various hand bones.

Computer assisted age identification system involves various stages as shown in Fig. 12. Input for this system is a digital image of hand and wrist, which will be preprocessed by removing the background noise and digital x-ray artifacts if present in the image. Apart from this, segmentation process and knowledge discovery (feature identification) is done. Important features are identified after analysing phalangeal bones and carpal bones. The results of SMI score and RUS score are evaluated for estimation of age.

Appearance of Carpal bones in children's from toddler to seven year old is demonstrated in Fig. 13. The numbers on the image is representation of the appropriate age of each object. Eight Carpal bones are developed in sequential order, starting with Capitate and ends with pisiform [20] as mentioned earlier.

In phalangeal analysis, growth pattern of metaphysis and epiphysis is monitored as shown in Fig. 14. Ossification of epiphyses generally takes after birth. Both Metaphyses and Epiphyses are divided at initial development stage of skeleton.

Ascending age growth pattern from initial stages in phalanges from male children is depicted in Fig. 15. Gaps in between metaphyses and epiphysis will be disappearing after some level and fusion begins, the process will be carried till fusion completes and a adult bone is found [21].

Various features from wrist are identified for age estimation and gender identification and their ground truth values for gender and age estimation are depicted in Table 8.

4 Features of Femur Bone to Identify Age and Gender

Figure 16 depicts the anterior and posterior view of Femur bone. The only one long bone in human body available is Femur bone. Long bones are very much essential for gender identification. Gender assessment is easy from femur bone since male femur bone tends to be longer and heavier compared to female femur bone, male femur bones are comprises with muscular attachment [22]. Other important criteria's in gender differentiation is the measurement of Femur bone, the head diameter, and the condyles width. Femur is the most reliable criterion to identify gender, if the thickness of femur head is over 46 mm, it can be considered as male and if thickness of head is less than 42 mm, the object can be consider as female. Several contributions have been contributed in identifying age and gender using Femur bone features by many experts and few of them have been explained below.

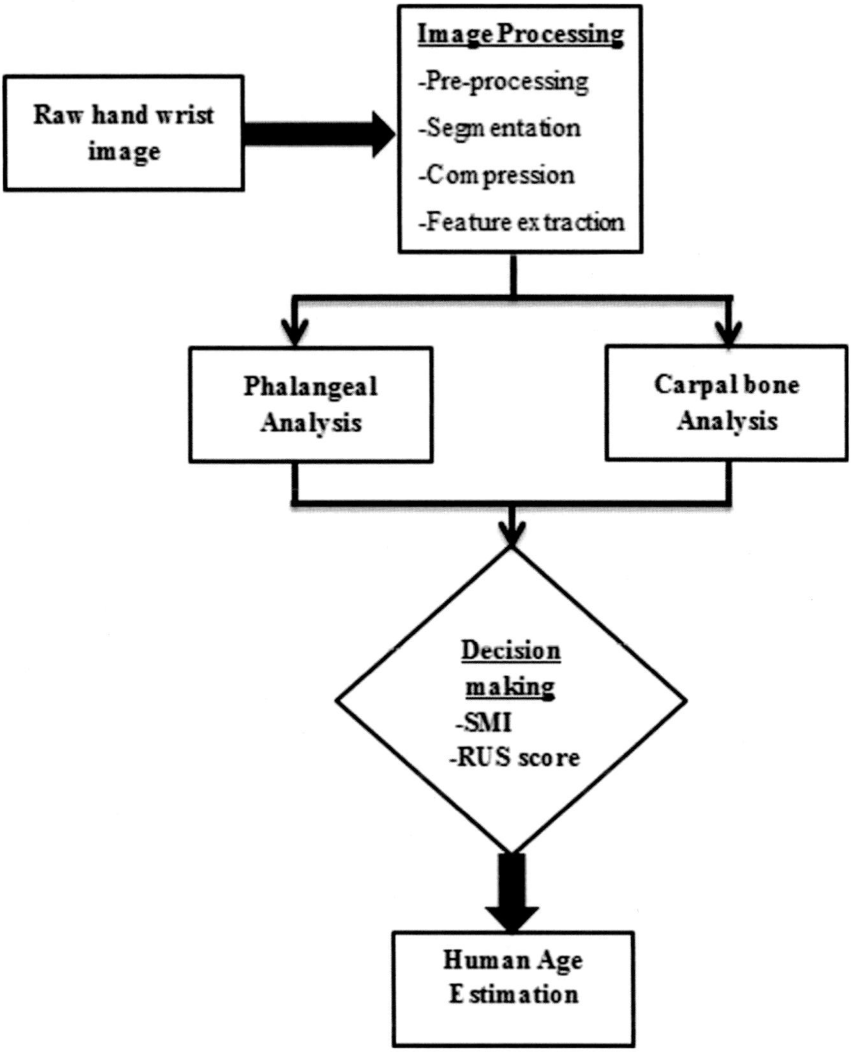

Fig. 12 Procedure for age estimation using Computer-Assisted Bone Age Assessment System

5 Related Works on Femur

Raghavendra et al. [23] in 2014 proposed a method to identify gender based on the digital radiographs of femur bones. These radiographs were focused and captured at a fixed distance of 130-cm and the zooming was maintained constantly for all the x-rays. Their experimental finding states that femoral angle is the range 136°–143° is a male, whereas femoral angle in the range 141°–146° is a female. The experimental findings of the authors have been tabulated in Table 9.

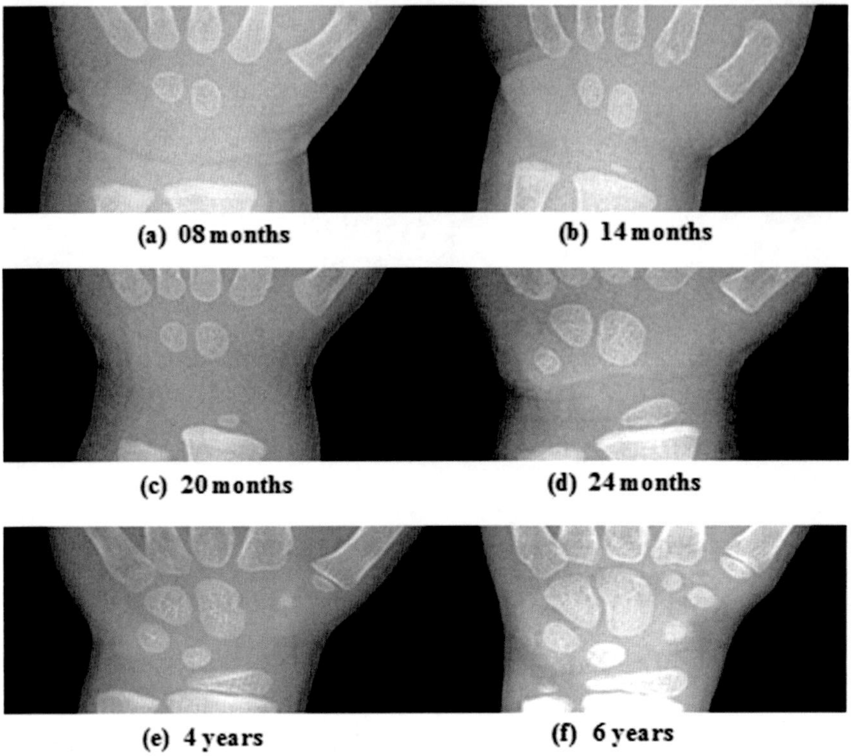

Fig. 13 Appearance of carpal bones in male children

Fig. 14 Phalange hand image parts

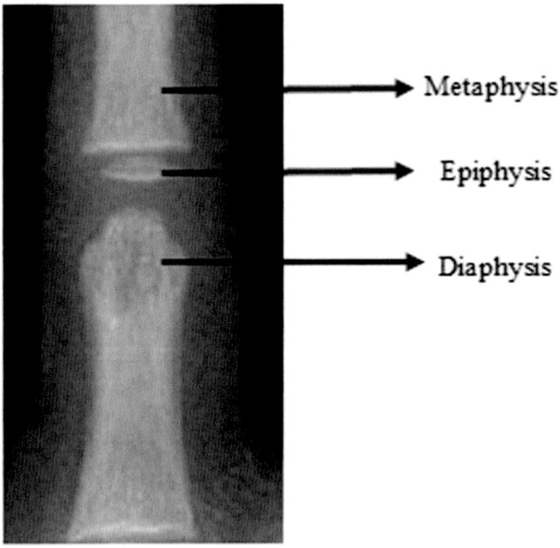

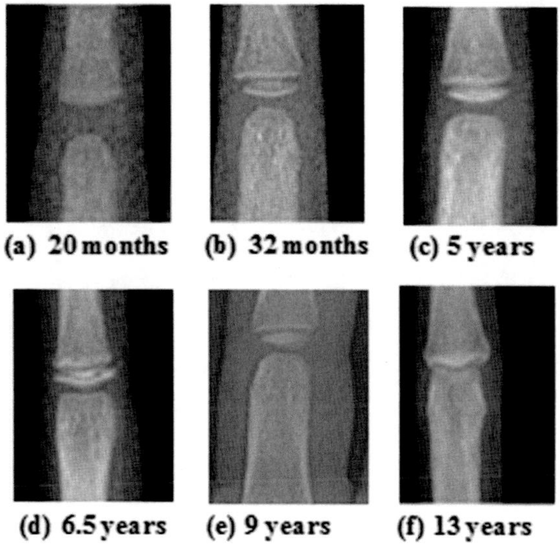

Fig. 15 Growth pattern of phalanges of male children

Table 8 Features of wrist with ground truth values for age and gender identification

S. No.	Feature name	Ground truth value			
		Male	Female	Age	
1.	Ossification of the capitate, the hamate	–	–	Male 14 months to 3 years	Female 10 months to 2 years
2.	Width of the epiphyses	–	–	Male 3–9 years	Female 2–7 years
3.	Growth of the epiphyses larger than the metaphyses	–	–	Male 9–14 years	Female 7–13 years
4.	Unification of the distal phalanges, metacarpals, and phalanges	–	–	Male 14–16 years	Female 13–15 years
5.	Appearance of sesamoid bone	–	–	Above 17 years of age in both gender	
6.	Union of lower end of ulna	33.3%	61.54%	–	
7.	Union of lower end of radius	16.67%	38.46%	–	
8.	carpal bone volumes	2.31	1.53	–	

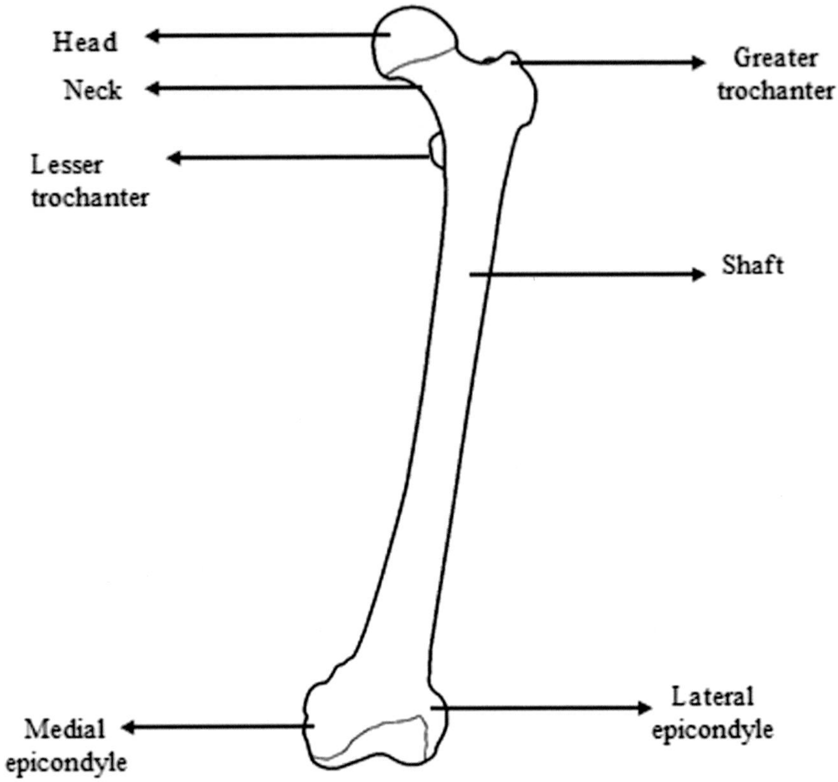

Fig. 16 Femur bone representation

Table 9 Femur bone angles for male and female

Gender	No. of training samples images	Min C_{min}	Max C_{max}	Mean $\overline{C}$
Male	34	136°	143°	138.085
Female	26	141°	146°	145.327

Femur bone in human body is identified as the longest bone and the angle in the femur bone formed at the intersection point of femoral neck and mid line of shaft is called as femoral angle [24]. Figure 17 illustrates how femoral angle C is obtained.

The proposed methodology involves the following stages:

i. Digital X-ray images are transformed into.JPEG format.
ii. Applying canny edge detector for converted image.
iii. Longest edge is considered after applying edge detector.
iv. Femoral angle is determined.

Longest edge from the femur bone is created as a series of pixels. Step by step process in obtaining the femoral angle at various stages is illustrated in Fig. 18.

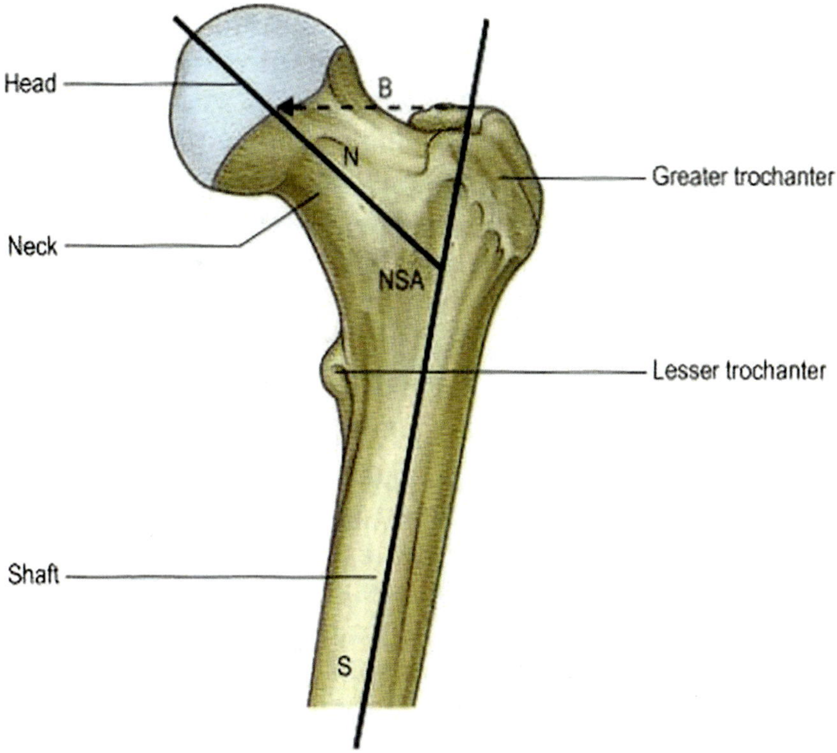

Fig. 17 The femoral angle *C* (from IJCR, vol. 9, issue 12, Dec 2017, p. 62465)

The details of the femur dataset collected by researchers and their experimental results are tabulated in the Table 10.

5.1 Gender Identification Rule Based on Femoral Angle

Gender can be classified by comparing the femoral angle *C*, which was obtained from the two line segments. Gender determination has the following rules:

Rule 1: $C_{min\text{-}male} \leq C \leq C_{max\text{-}male}$:

If the obtained femoral angle *C* is in the range between the C_{min} of male and C_{max} of male, then the gender is identified as male. i.e. femoral angle C is above 136° and below 143° indicates the gender is male.

Rule 2: $C_{min\text{-}female} \leq C \leq C_{max\text{-}female}$:

If the obtained femoral angle *C* is in the range between the C_{min} of female and C_{max} of female, then the gender is identified as female. i.e. femoral angle C is above 141°

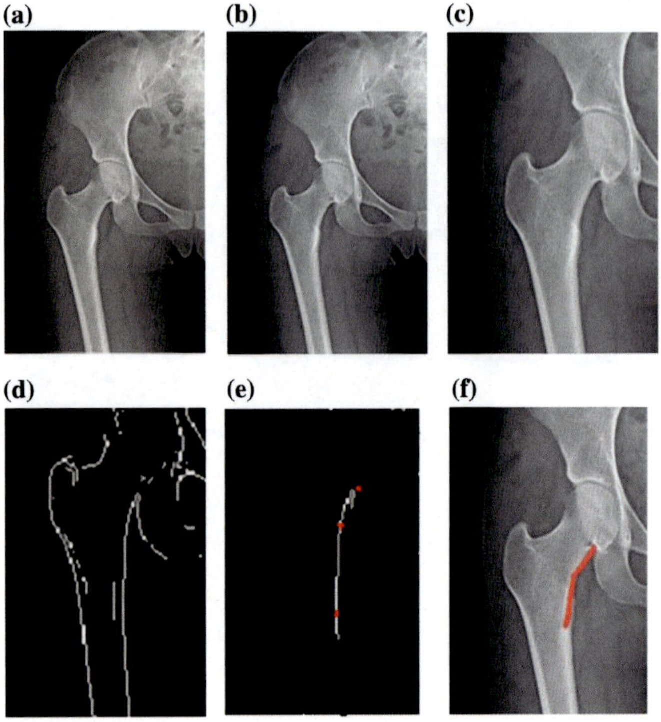

Fig. 18 Resultant images to obtain femoral angle: **a** Femur image FI, **b** FI transformed to .JPEG FI1, **c** FI1is cropped FI2, **d** Edge detected FI3, **e** Longest edge FI4, **f** Overlay of images in (**e**) and (**c**) (from IJTEL 2014, vol. 3 n0. 2, p. 422)

Table 10 Femur bone angles for male and female

Gender	No. of training samples images	Min $C_{\min}$	Max $C_{\max}$	Mean $\overline{C}$
Male	34	136°	143°	138.085
Female	26	141°	146°	145.327

and below 146° indicates the gender is female. Otherwise gender is unknown. Few samples that are identified from femoral angles of male and female are depicted in Fig. 19.

Various features from femur are identified for age estimation and gender identification and their ground truth values for gender and age estimation are depicted in Table 11.

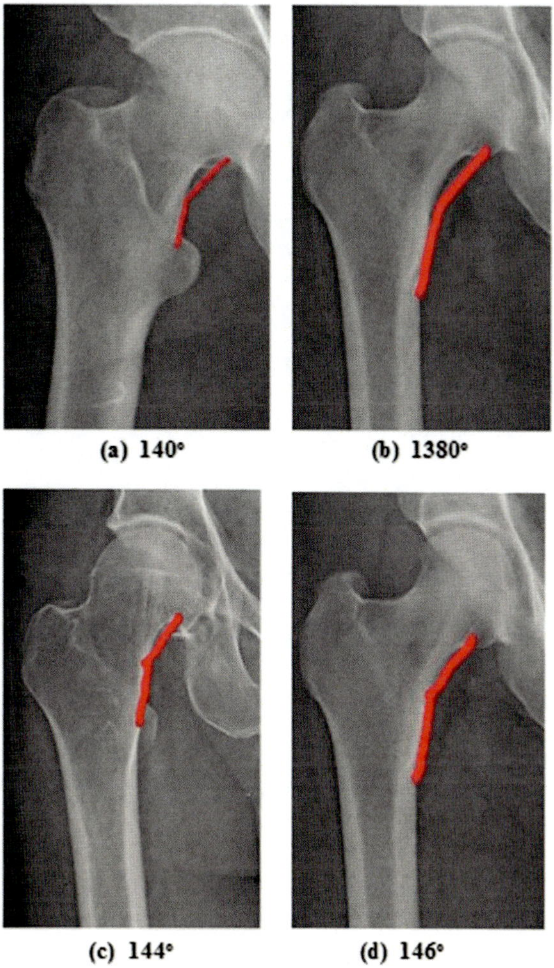

Fig. 19 Images **a**, **b** represent Male and **c**, **d** represent Female

6 Conclusion

This chapter is mainly concentrated on the latest research challenges in assessment of human age and gender identification using digital x-ray images of teeth, wrist and femur bones. Findings of the present studies are useful in medico legal cases for identification of gender and estimating age from available digital images of teeth, wrist or femur. In case of gender and age assessment, a manually performed procedure requires around 30 min to few hours of doctor's time per a patient. When the same procedure is done using software based on classical computer vision methods, it takes 2–5 min, but still requires substantial doctoral supervision and expertise.

Table 11 Features of femur with ground truth values for age and gender identification

S. No.	Feature name	Ground truth value		
		Male (mm)	Female (mm)	Age (years)
1.	Maximum femoral length	440.16	410.60	0–4 = 210.25 mm
				5–9 = 283.29 mm
				10–14 = 339.67 mm
				15–19 = 432.63 mm
				20–25 = 444.95 mm
				26–39 = 443.29 mm
				40–98 = 440.63 mm
2.	Femoral head diameter	44.45	39.89	0–4 = 19.14 mm
				5–9 = 28.59 mm
				10–14 = 35.20 mm
				15–19 = 43.45 mm
				20–25 = 44.76 mm
				26–39 = 45.29 mm
				40–98 = 45.00 mm
3.	Diaphyseal length	392.42	372.11	0–4 = 160.18 mm
				5–9 = 251.64 mm
				10–14 = 326.00 mm
				15 and above = 392.83 mm
4.	Diaphyseal length + epiphysis	419.76	402.67	0–4 = 193.40 mm
				5–9 = 272.71 mm
				10–14 = 328.31 mm
				15 and above = 418.83 mm
5.	Femoral neck shaft angle	136°–143°	141°–146°	–
6.	Bicondylar width	76–78	72–74	–
7.	Trochanteric oblique length	432.17	403.82	–
8.	Mid shaft diameter	37	24	–

7 Future Scope

Gender determination and age assessments are highly researched area, and new methods using the dentition and other skeletal parameters continue to be developed. Further the work will be expanded by identifying some more important features and knowledge discover (if available), and these feature are fused into a feature matrix and it will fed into appropriate classifier that will give a better classification accuracy.

Datasets collected from different hospitals are needed to be standardized since these images are of different resolution and different size.

Acknowledgements It is an occasion to express our deep sense of thanks and gratitude for the following persons for their sincere cooperation and valuable guidelines in preparation of this book chapter.

We would like to express our gratitude to Dr. Ashok L. Prof. and Head, Department of Oral medicine and Radiology, for there valuable suggestions and constant encouragement and we also thankful for Bapuji Dental College and Hospital, Davangere for providing us the required datasets to achieve our goals.

We are very thankful for the teaching and non-teaching staffs of College of Dental Science, Davanagere for providing us valuable information and datasets of teeth.

We are also thankful to Dr. Nagarathnamma B. Asst. Prof., Department of Anatomy, JJM Medical College, Davangere, for her endless support, assistance and suggestions.

Last but not least we would like to express our sincere gratitude for teaching and non-teaching staff of Bapuji Institute of Engineering Technology, Davangere for their helped us by providing valuable assistance and time during the preparation of this book chapter article.

Bibliography

1. C. Stavrianos, I. Stavrianos, E.M. Dietrich, P. Kafacs, Method of human identification in forensic dentistry: a review. Internet J. Forensic Sci. (2009)
2. A. Schmeling, C. Grundmann, A. Fuhrmann, Criteria forage estimation in living individuals. Int. J. Legal Med. **122**(6), 457–460 (2008)
3. R. Nagi, S. Jain, P. Agrawal1, S. Prasad, S. Tiwari, G.S. Naidu, Tooth coronal index: key for age estimation on digital panoramic radiographs. J. Indian Acad. Oral Med. Radiol. **30**, 64-7 (2018)
4. N. Pagare, S. Chourasiya, H. Dedhia, Study of odontometric parameters in gender identification. Austin J. Forensic Sci. Criminol. **4**(2), 1060 (2017)
5. A. VikramSimha Reddy, A. Ravi Prakash, K.K. Lakshmi, M. Rajinikanth, G. Sreenath, P.B. Sabiha, Gender determination using barr bodies from teeth exposed to high temperatures. J. Forensic Dent. Sci. **9**(1), 44 (2017 Jan–Apr)
6. C.F. Larissa Chaves, L.V. Carolina Vieira, A.O. de Julyana, G. Paloma Rodrigues, S. Bianca Marques, R. Patrícia Moreira, Odontometric analysis of molars for sex determination. Braz. J. Oral Sci. **15**(1), 35–38 (2016)
7. K. Krishan, T. Kanchan, A.K. Garg, Dental evidence in forensic identification—an overview, methodology and present status. Open Dent. J. **9**, 250–256 (2015)
8. A.B. Acharya, Forensic dental age estimation by ensuring root dentin translucency area using a new digital technique. J. Forensic Sci. **59**(3) (2014)
9. M. Lorentsen, T. Solheim, Age assessment based on translucent dentine. J Forensic Odontostomatol **7**, 3–9 (1989)
10. R.K. Khangura, K. Sircar, S. Singh, V. Rastogi, Gender determination using mesiodistal dimension of permanent maxillary incisors and canines. J. Forensic Dent. Sci. **3**(2), 81-5 (2011 July)
11. S. Garn, A. Lewis, D, kerewsky R (1967) Genetic control of genderual dimorphism in tooth size. J. Dental Res. **4**, 963–972
12. G.C. Townsend, T. Brown, Tooth size characteristics of Australian aborigines. Occas. Pap. Hum Biol. **1**, 17–38 (1979)
13. W.W. Greulich, S.I. Pyle, *Radiographic Atlas of Skeletal Development of Hand Wrist*, 2 ed. vol. 14 (Stanford Univ. Press, Palo Alto, CA, 1959), pp. 234–247

14. E. Pietka, M.F. McNitt-Gray, M.L. Kuo, H.K. Huang, Computer assisted phalangeal analysis in skeletal age assessment. IEEE Trans. Med. Imaging **10**(4) (1991 Dec)
15. L.S. Fishman, Radiographic evaluation of skeletal maturation. A clinically oriented method based on hand-wrist films. Angle Orthod. **52**(2), 88–112 (1982)
16. D.J. Michael, A.C. Nelson, HANDX: a model based system for automatic segmentation of bones from digital hand radiographs. IEEE Trans. Med. Imaging **8**, 1–21 (1989)
17. W.M.C. Chumlea, A.F. Rouche, D. Thissen, The FELS method of assessing the skeletal maturity of the hand-wrist. Am. J. Hum. Biol. **1**, 175–183 (1989)
18. A.M.M. Da Silva, S.D. Olabarriaga, C.A. Dietrich, C.A.A. Schmitz, On determining a signature for skeletal maturity, in *Proceedings of XIV Brazilian Symposium on Computer Graphics and Image Processing* (2001)
19. M.K. Schneider, P.W. Fieguth, W.C. Karl, A.S. Willsky, Multiscale methods for the segmentation and reconstruction of signals and images. IEEE Trans. Image Process. **9**(3), 456–468 (2000)
20. J.M. Tanner, R.H. Whitehouse, W.A. Marshall, *Assessment of Skeletal Maturity and Prediction of Adult Height* (Academic Pres, London, 1975)
21. S. Aydogdu, F. Basciftci, *Methods Used in Computer-Assisted Bone Age Assessment of Children*, vol. 2, No. 1 (2014 Mar)
22. A. Zhang, A. Gertych, B.J. Liu, Automatic bone age assessment for young children from newborn to 7-year-old using carpal bones. Comput. Med. Imaging Graph. **31**, 299–310 (2007)
23. C.S. Raghavendra, D.C. Shubhangi, Gender identification in digital X-ray images of femur bone. IJTEL **3**(2) (2014). ISSN: 2319-2135
24. B. Nagarathnamma, Measurement of neck shaft angle in cadaveric femora. Int. J. Curr. Res. **9**(12), 62462–62467 (2017)
25. N. Lynnerup, Cranial thickness in relation to age, sex and general body built in a danish forensic sample. Forensic Sci. Int. **117**, 45–51 (2001)

Santosh K. C. has 10 years of teaching experience and 4 years of research experience. Presently he is working as Assistant professor in the department of Computer Science and Engineering, Bapuji Institute of Engineering and Technology, Davangere, Karnataka, India, affiliated to Visvesvaraya Technological University, Belagavi, India. From 2010 to 2013 he was a lecturer in CS and E department, Bapuji Institute of Engineering and Technology, Davangere. From 2013 to till date he is working as Assistant Professor in CS and E department, Bapuji Institute of Engineering and Technology, Davangere. He has published 2 research papers in International Journal and International Conference. His research interests are Image Processing, Machine Learning, Pattern Recognition and Medical Image Analysis.

Dr. Pradeep N. has 17 years of teaching experience in CS and E dept., Bapuji Institute of Engineering and Technology, Davangere, Karnataka, India, affiliated to Visvesvaraya Technological University, Belagavi, Karnataka, India. His experience includes 9 years research experience. He has worked at various levels namely Lecturer, Sr. Lecturer, Assistant Professor and presently working as Associate Professor and PG coordinator, CS and E dept. and Placement officer, Bapuji Institute of Engineering and Technology, Davangere. He has published 6 research papers in various peer-reviewed International Journals and 13 research papers in various International Conferences. He has guided more than 20 students at PG level and 30 batches at UG level and guiding 2 research scholars in Visvesvaraya Technological University, Belagavi, Karnataka in the area of Pattern Recognition, Natural Language Processing and Machine Learning. He has successfully edited a book titled "Modern Techniques for Agricultural Disease Management and Crop Yield Prediction", IGI Publishers, USA. Another edited book entitled "Demystifying Big Data, Machine Learning and Deep Learning for Health Care Analytics" is in progress and to be published by Elsevier publishers. Also, he has filed Indian Patent and the patent application has been published.

Deep Learning Solutions for Skin Cancer Detection and Diagnosis

Hardik Nahata and Satya P. Singh

Abstract Skin cancer, a concerning public health predicament, with over 5,000,000 newly identified cases every year, just in the United States. Generally, skin cancer is of two types: melanoma and non-melanoma. Melanoma also called as Malignant Melanoma is the 19th most frequently occurring cancer in women and men. It is the deadliest form of skin cancer [1]. In the year 2015, the global occurrence of melanoma was approximated to be over 350,000 cases, with around 60,000 deaths. The most prevalent non-melanoma tumours are squamous cell carcinoma and basal cell carcinoma. Non-melanoma skin cancer is the 5th most frequently occurring cancer, with over 1 million diagnoses worldwide in 2018 [2]. As of 2019, greater than 1.7 Million new cases are expected to be diagnosed [3]. Even though the mortality is significantly high, but when detected early, survival rate exceeds 95%. This motivates us to come up with a solution to save millions of lives by early detection of skin cancer. Convolutional Neural Network (CNN) or ConvNet, are a class of deep neural networks, basically generalized version of multi-layer perceptrons. CNNs have given highest accuracy in visual imaging tasks [4]. This project aims to develop a skin cancer detection CNN model which can classify the skin cancer types and help in early detection [5]. The CNN classification model will be developed in Python using Keras and Tensorflow in the backend. The model is developed and tested with different network architectures by varying the type of layers used to train the network including but not limited to Convolutional layers, Dropout layers, Pooling layers and Dense layers. The model will also make use of Transfer Learning techniques for early convergence. The model will be tested and trained on the dataset collected from the International Skin Imaging Collaboration (ISIC) challenge archives.

H. Nahata (✉)
Department of Computer Science and Engineering, Institute of Aeronautical Engineering, Hyderabad, India
e-mail: hardiknahata@gmail.com

S. P. Singh
Biomedical Informatics Lab, School of Computer Science and Engineering, Nanyang Technological University, Singapore, Singapore
e-mail: satya@ntu.edu.sg

V. Jain and J. M. Chatterjee (eds.), *Machine Learning with Health Care Perspective*, Learning and Analytics in Intelligent Systems 13,
https://doi.org/10.1007/978-3-030-40850-3_8

Keywords Neural networks · Skin cancer · Deep learning · Machine learning · Cancer detection · Cancer diagnosis · Convolution neural network · CNN · Melanoma

1 Introduction

1.1 Background

High occurrence of skin cancer compared to other cancer types is a dominant factor in making it one of the most severe health issues in the world. Historically, melanoma is a rare cancer, but in the past five decades, the worldwide occurrence of melanoma has drastically risen. In fact, it is one of the prominent cancers in average years of life lost per death. Adding to the strain, the financial burden of melanoma treatment is also expensive. Out of the $8.1 Billion in all skin cancer treatment costs in the USA, $3.3 Billion are spent only on Melanoma. Squamous Cell Carcinoma and Basal Cell Carcinoma, are eminently curable if diagnosed and treated on in the early stages. The five-year survival rate of patients with early stage diagnosis of melanoma is around 99%. Therefore, timely detection of skin cancer is the key factor in reducing the mortality rate.

1.2 Motivation

Prior to the 1980s, melanoma detection was carried out by spotting the macroscopic features, as they were mostly recognized when they were considerable in size. This made the early detection unlikely and mortality rates continued to increase. In 1985, a research team at the New York University came up with the ABCD acronym which stands for Asymmetry, Border irregularity, Color variegation, Diameter (ABCD) as a simple yet effective tool to educate the general public for the early recognition of melanoma. After 1990, screening conducted through the physicians and the use of baseline full-body imaging became the common approach to detect melanoma early. Later, computer augmented digital image analysis became the new trend due to its strong sensitivity and specificity of detection compared with the manual dermoscopy.

1.3 Objective and Scope

The cardinal objective of this project is to develop state of the art Convolutional Neural Network (CNN) model to perform the classification of skin lesion images into respective cancer types. The model is trained and tested on the dataset made available by International Skin Imaging Collaboration (ISIC). The model can be used for analyzing the lesion image and find out if it's dangerous at early stage.

2 Background

2.1 Convolutional Neural Network for Image Classification

Artificial Neural Networks are made of artificial neurons inspired by biological neurons present in our brain. Convolutional Neural Network (CNN) is a modified variant of feed-forward neural network which is generally used for image classification tasks. CNNs can recognize a particular object even when it appears in different ways, as it understands translation invariance. This is a key point which makes CNN advantageous over feed-forward neural networks which cannot understand translation invariance.

In layman words, feed-forward neural networks only recognize an object when it is right in the center of the image, but fails notably when the object is slightly off position or placed elsewhere in the image. Basically, the network understands/learns only one pattern. This is precisely not convenient as the real world datasets are usually raw and unprocessed.

2.2 Working of Convolution Neural Networks

The question which arises here is how does CNN understand translation invariance? Is it the magic of Machine Learning? Yet again, it comes down to mathematics again.

The following operations are the various layers/steps of the CNN:

- Convolution
- Pooling
- Flattening
- Full Connection

We will now see what happens in each step in detail.

2.2.1 Convolution

The first operation, Convolution, extracts important features from the image. It is a mathematical operation which clearly requires two inputs, an image matrix and a filter or kernel. The filter is traversed through the image and multiplied with the pixel values to obtain feature map.

Figure 1 shows how the convolution operation happens.

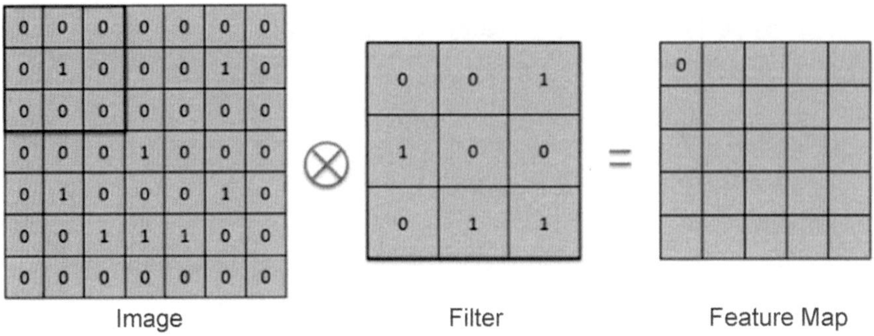

Fig. 1 Convolution operation

Convolution does lose information, but the point here is to reduce size and learn the integral information. Performing convolution with different kind of filters can assist in image sharpening, edge detection, blurring and other image processing operations.

2.2.2 Pooling

Pooling operation helps in decreasing the number of parameters when the image is very large in size. Subsampling also called Spatial Pooling curtails the dimensionality of each feature map but retains significant information.

Pooling is of basically divided into three types:

- Max Pooling (mostly used)
- Sum Pooling
- Average Pooling

Max pooling is a sample-based discretization process. It is done by applying a N × N max filter over the image, which selects the highest pixel value in each stride and builds the feature map. Similarly, in average and sum pooling, the average and sum of pixel values are taken into the feature map.

Figure 2 depicts the Max Pooling operation.

2.2.3 Flattening

To feed our feature maps in to the artificial neural network, we need a single column vector of the image pixels. As the name suggests, we flatten our feature maps into column like vector.

Figure 3 depicts the flattening process.

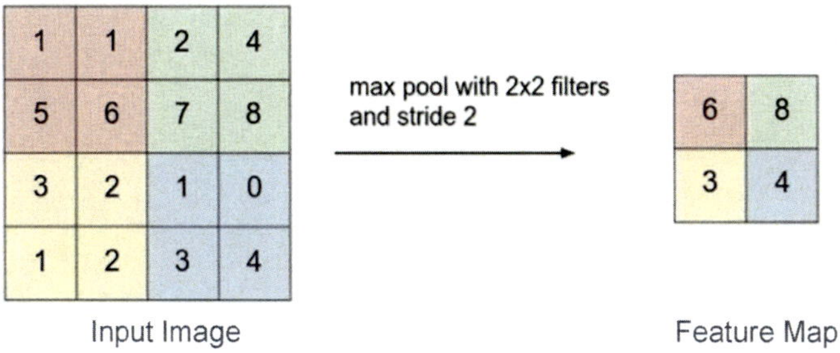

Fig. 2 Max pooling operation

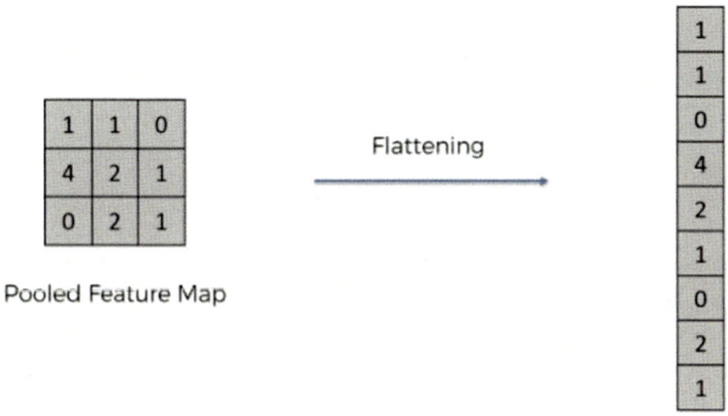

Fig. 3 Flattening operation

2.2.4 Full Connection

The fully connection layer takes the input from the preceeding convolution/pooling layer and produces an N dimensional vector where N is the number of classes to be classified. Thus, the layer determines the features most correlating to a particular class based on the probabilities of the neurons.

Figure 4 shows the fully connected layer in a neural network.

2.3 *Skin Lesion Classification Using CNN*

From past research in this field, it is evident that CNN has an extraordinary ability to perform skin lesion classification in competition with professional dermatologists. In fact, there have been instances where CNN has outperformed professional dermatologists as well [6].

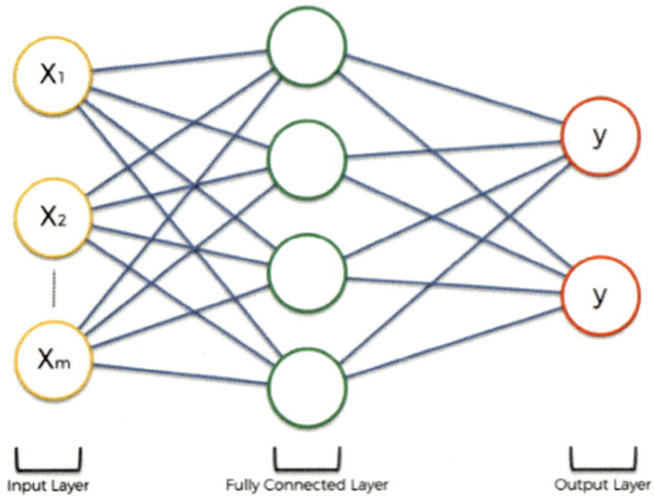

Fig. 4 Full connection layer

CNN can perform skin lesion classification in two ways. In the first case, a CNN is used as the feature extractor of the images and the classification is performed by another classifier. For the other case, CNN is used to perform end-to-end learning which can be further divided into learning from scratch or learning from pretrained model. To train the CNN from scratch, large number of images are required in order to tackle the overfitting issue. Since the number of skin lesion images to do the training is not sufficient, training CNN from scratch is less feasible. Training from a pretrained model is a better approach which is generally referred to as Transfer Learning (TL). TL helps the model to learn well even with less data, and also introduces generalization property to the trained model.

3 Methods and Dataset

3.1 Dataset

The International Skin Imaging Collaboration (ISIC): Melanoma Project is a partnership between industry and academia in order to facilitate the application of digital skin imaging to help curtail skin cancer mortality. Starting from 2015, ISIC started to organize global challenges for skin lesion analysis for melanoma diagnosis and detection.

Figure 5 shows samples of each class from the dataset.

There are two datasets used to develop this project, which were published by ISIC in 2018 and 2019 through a challenge.

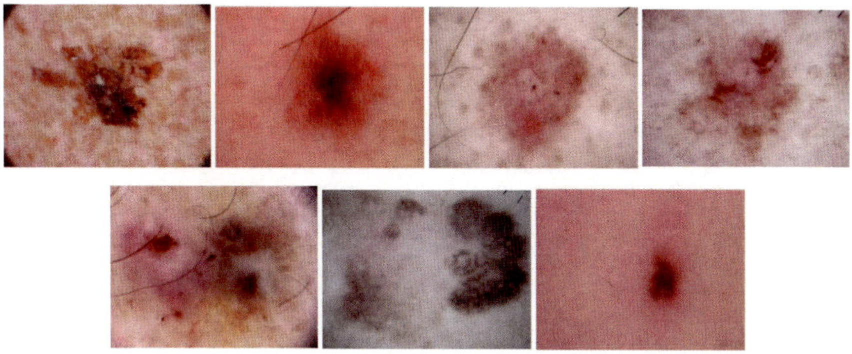

Fig. 5 Sample images from the ISIC dataset (arranged in ascending order of lesion type: melanoma, melanocytic nevus, basal cell carcinoma, actinic keratosis, benign keratosis, dermatofibroma and vascular lesion)

The first dataset which was 'ISIC 2018' contains 10,015 images of 7 types of skin lesion diseases namely: Benign Keratosis, Dermatofibroma, Vascular Lesion, Melanoma, Melanocytic Nevus, Basal Cell Carcinoma and Actinic Keratosis. These images were collected with approval of Medical University of Vienna and University of Queensland. All the images are in the standard size of 600 × 450 pixels in JPEG format [7, 8].

Table 1 summarizes the dataset 1 used for this project.

The second dataset which was 'ISIC 2019' contains 25,333 images of 8 types of skin lesion diseases namely: Dermatofibroma, Vascular Lesion, Squamous Cell Carcinoma, Melanoma, Melanocytic Nevus, Basal Cell Carcinoma, Actinic Keratosis, Benign Keratosis. All images are in the size of 1022 × 767 pixels in JPEG format [9].

Table 2 summarizes the dataset 2 used for this project.

A new dataset was created by combining both the datasets of ISIC 2018 and ISIC 2019. The seven common classes of skin lesions were retained and extra noise was

Table 1 Dataset information for ISIC 2018

Dataset	ISIC challenge 2018
Type	Dermoscopic
Image size	600 pixels × 450 pixels
Number of images	10,015
Image type	JPEG (RGB)
Class labels	0: Melanoma 1: Melanocytic Nevus 2: Basal Cell Carcinoma 3: Actinic Keratosis 4: Benign Keratosis 5: Dermatofibroma 6: Vascular Lesion

Table 2 Dataset information for ISIC 2019

Dataset	ISIC Challenge 2019
Type	Dermoscopic
Image size	1022 pixels × 767 pixels
Number of images	25,333
Image type	JPEG (RGB)
Class labels	0: Melanoma 1: Melanocytic Nevus 2: Basal Cell Carcinoma 3: Actinic Keratosis 4: Benign Keratosis 5: Dermatofibroma 6: Vascular Lesion 7: Squamous Cell Carcinoma

removed from the dataset. This enabled the models to learn more efficiently due to the abundance of samples now available per class.

Table 3 summarizes the final dataset used for this project.

Table 4 shows the number of images fed per class into the network.

Table 3 Dataset information for FINAL dataset

Dataset	FINAL
Type	Dermoscopic
Image size	600 pixels × 450 pixels (10,015) 1022 pixels × 767 pixels (25,333)
Number of images	35,348
Image type	JPEG (RGB)
Class labels	0: Melanoma 1: Melanocytic Nevus 2: Basal Cell Carcinoma 3: Actinic Keratosis 4: Benign Keratosis 5: Dermatofibroma 6: Vascular Lesion

Table 4 Number of images per lesion type

Lesion type	Number of images
Melanoma	5635
Melanocytic Nevus	19,580
Basal Cell Carcinoma	3837
Actinic Keratosis	1194
Benign Keratosis	3723
Dermatofibroma	354
Vascular Lesion	395

3.2 Method Overview

See Fig. 6.

3.2.1 Data Augmentation

Contemporary advances in deep learning models have been largely associated with large quantity and diversity of data. Having large data helps crucially in improving the performance of machine learning models. But obtaining such vast quantities of data is cost-intensive and tedious. Hence, we use the technique of Data Augmentation. It is a technique that enables us to considerably increment the diversity and quantity of data available, without actually aggregating new data. In order to generate new data through augmentation of images, various techniques such as cropping, padding, adding noise, brightness changing and horizontal flipping are commonly used to train large neural networks [10].

In this project, the training images are augmented in order to make the model robust to new data which in turn helps in increasing the testing accuracy.

Table 5 shows the augmentation techniques used on the dataset.

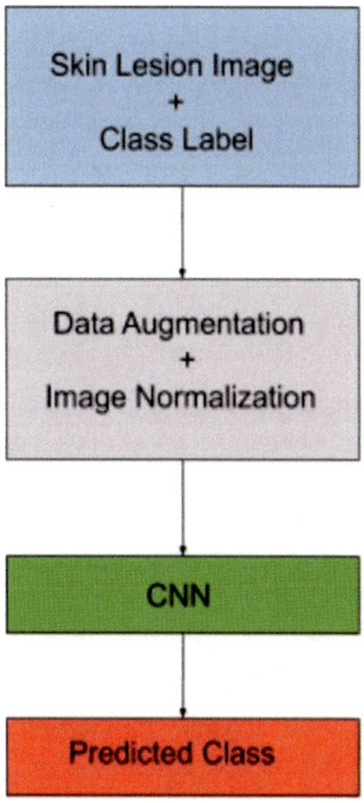

Fig. 6 Classification model overview

Table 5 Information about data augmentation parameters

Augmentation technique	Range
Zoom range	0.1
Rotation range	10
horizontal flip	False
Rescale	1./255
Width shift range	0.1
Height shift range	0.1

3.2.2 Image Normalization

Image Normalization is a technique used to normalize the pixel values of the image in a similar distribution [11]. It is beneficial to normalize images before feeding into the neural network as this helps in approaching the global minima at error surface at a faster rate while performing gradient descent. In a way, it helps the network to converge faster. Also, the computations become significantly less intensive for the machine to perform as all the pixel values are scaled.

3.2.3 Transfer Learning

Transfer Learning is a learning method in which a model trained for a particular task is reiterated as the origin for another model on a similar task. This approach is very mainstream in deep learning due to the vast computation resources and time consumed to train neural network models. In the case of problems in the computer vision domain, low-level features, such as shapes, corners, edges and intensity, can be shared across tasks, and thus enable knowledge transfer.

In this project, the CNN is trained by transfer learning from the pretrained weight of ImageNet classification. ImageNet Large Scale Visual Recognition Challenge (ILSVRC) is the classification challenge involving 14 million training images of approximately 20,000 classes [12]. Fine-tuning from the pretrained weight significantly increases the classifier training speed, overcomes the limit of small number of training data and makes the convergence more realistic to occur. The front layers of the CNN serve to detect the edges, points, corners and simple structures in the image and the pretrained weight enables the trained model has higher adaption speed.

The model is trained and tested on the state-of-art CNNs, namely Inception V3, ResNet50, VGG16, MobileNet and InceptionResnet to perform the seven-class classification of the skin lesion images.

3.2.4 Dropout

Deep Neural Networks (DNNs) may overfit a dataset with meager training samples. Collection or Ensemble of neural networks with different architectures are known to diminish overfitting, but still require us to train and maintain various models which becomes computationally intensive.

This is where Dropout comes into the picture. A single model is used to resemble having a vast number of distinct network architectures by arbitrarily dropping out nodes during training. This is how dropout works, and proposes a computationally viable and exceptionally effective method to cut down overfitting and enhance generalization error in DNNs of all types.

3.3 *Network Architecture*

3.3.1 Inception V3

Inception V3 is a widely acclaimed image recognition model that has been built by various researchers over time. Originally based on the paper "Rethinking the Inception Architecture for Computer Vision" by Szegedy, the model has attained an accuracy greater than 78.1% on the ImageNet dataset [13].

Figure 7 shows the model architecture of Inception V3 network.

Inception V3 proposed by Google is the third iteration in the series of Deep Learning Convolutional Architectures. The model was trained on 1000 classes from the ImageNet dataset which was itself trained on about 1 million images. It consists of the inception modules which apply the filters of multiple size on the same level of input. The high computation cost involved in the inception module is solved by applying 1×1 convolution to truncate the input into a smaller intermediate block, called the bottleneck layer. The multiple filters are applied on the bottleneck layer to significantly reduce the computation cost. The auxiliary classifiers in the Inception network contribute to the weighted loss function for regularization purpose. To utilize the ImageNet pretrained weight of the Inception V3 network, the input image has to be in the 299×299 size.

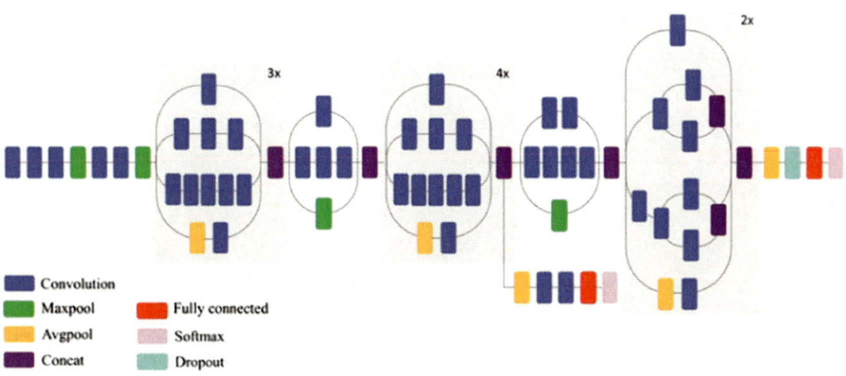

Fig. 7 Inception V3 architecture

Fig. 8 Residual block of the ResNet 50 model

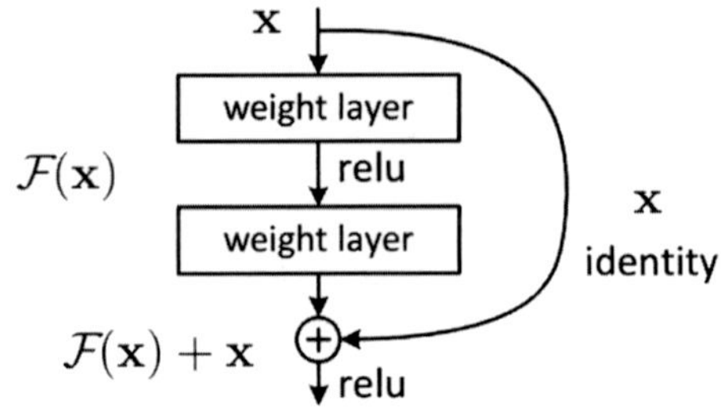

The key addition to the Inception V3 was Factorization. Factorization was introduced in the convolution layer to further reduce the dimensionality, so as to reduce the overfitting problem.

3.3.2 ResNet50

Residual Network (ResNet) is a typical model of neural network used as an integral part for various computer vision tasks. This ResNet model won the 2015 ImageNet challenge. The network is 50 layers deep and takes the input image size of 224 × 224 pixels [14].

Generally, in a deep CNN, many layers are stacked and trained. The network learns many low level, middle and high level features. In residual learning, rather than learning some features, we learn some residual. Residual can be simply interpreted the as reduction of features learned from input layer. It has been shown that training residual networks is easier compared to training simple deep CNNs. It also helps to tackle the problem of degrading accuracy.

A residual block of ResNet50 model is shown in Fig. 8.

3.3.3 VGG 16

VGG16 is a CNN model proposed by A. Zisserman and K. Simonyan from the University of Oxford in the paper "Very Deep Convolutional Networks for Large-Scale Image Recognition" [15].

Figure 9 shows the model architecture of VGG 16 network.

The input of this model has to be a 224 × 224 RGB image. The image is passed through a pile of convolutional layers, with kernel size or filter size of 3 × 3. The stride is set to 1 pixel; the padding of convolution layer input is set in a way such that the spatial resolution is preserved after convolution, i.e. the padding is 1-pixel for 3 × 3 conv. layers. Max-pooling is done over a 2 × 2 pixel window, with a stride of 2 pixels.

After the stack of convolutional layers, 3 Fully-Connected (FC) layers follow. The initial two layers have 4096 channels each, the third layer contains 1000 channels. The

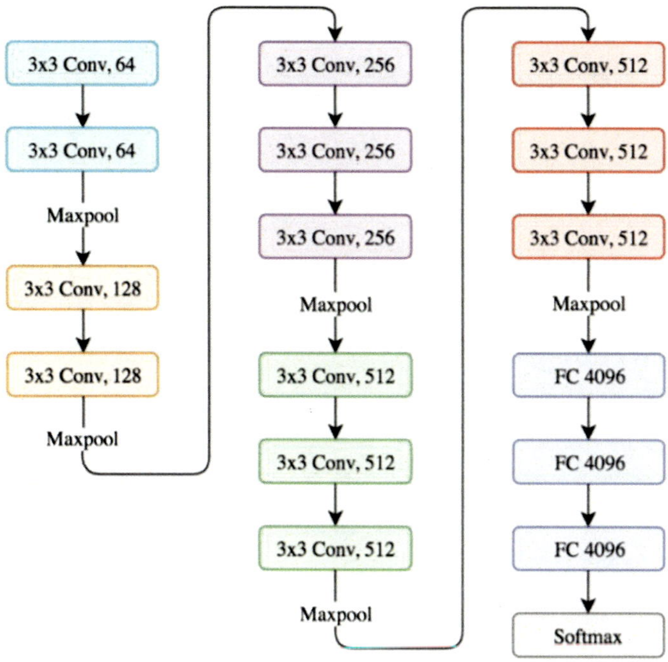

Fig. 9 VGG 16 architecture

final layer has the softmax activation function. All other hidden have the rectification (ReLU) non-linearity activation.

3.3.4 Inception-Resnet

Inception-ResNet-v2 is a CNN model that has been trained on the ImageNet database. It contains 164 layers and has the capability to classify images into 1000 categories, such as, mouse, many animals, keyboard, pencil etc. The network is robust and has learned excellent feature representations for a various kind of images. The input to the network has to be an image of size 299 × 299 [16].

Overall, Inception Resnet V2 has a similar structure and computation cost as Inception V4. It additionally introduces the residual connection to the submodules Inception Block A, B and C at the left side to enable to network to go deeper.

Figure 10 shows the model architecture of Inception-Resnet network.

3.3.5 MobileNet

MobileNets are based on the principle of streamlined architectures which use depth-wise separable convolutions followed by point-wise convolutions that considerably

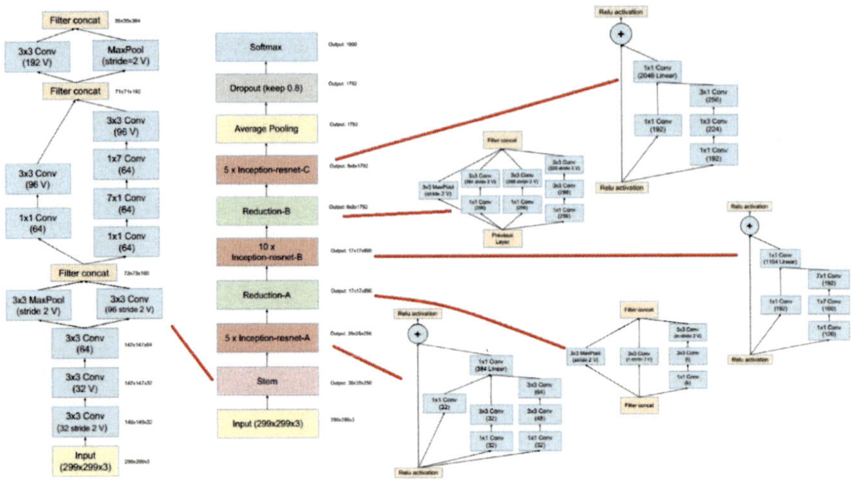

Fig. 10 InceptionResnet architecture

reduce the number of learnable parameters and assist in building light-weight deep neural networks [17]. Effectively, it decreases the total number of floating point calculations required which is supportive for embedded and mobile computer vision applications where there is shortage of computational power. The architecture was proposed by Google.

Figure 11 shows the depth-wise and point-wise convolution operations.

Figure 12 depicts the architecture of the MobileNet CNN model.

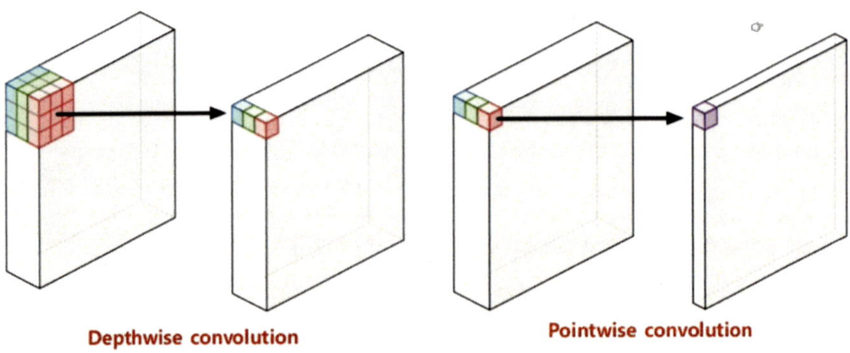

Fig. 11 Depthwise and pointwise convolution

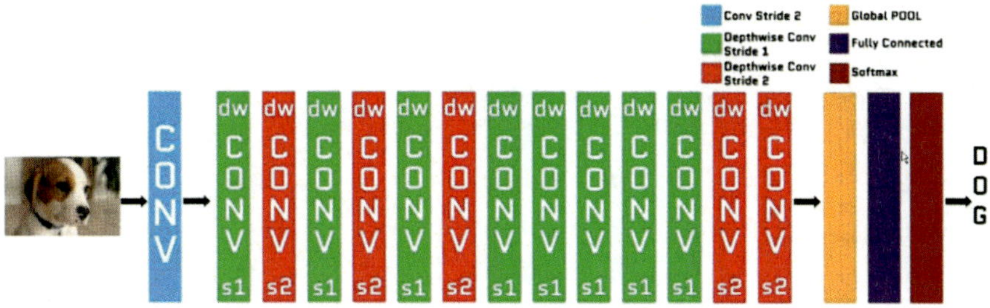

Fig. 12 MobileNet architecture

4 Results and Discussion

4.1 Performance Metrics

In a classification problem, only one metric such as Accuracy cannot help us evaluate the complete model efficiency effectively. Hence we measure the **Accuracy**, **Precision**, **Recall**, **F1 Score** and **Support** for every class of the skin lesion disease. We also plot the **Confusion Matrix** in order to check how well our model performs on every class.

We will now understand how the above mentioned metrics are calculated and what they mean [18].

Figure 13 is a confusion matrix where, **TP** = True Positive; **FN** = False Negative; **FP** = False Positive; **TN** = True Negative.

For example, if the skin lesion image is labelled with the melanoma and the model also predicts it as melanoma, this is considered as the **true positive** case. If the image is labelled with melanoma but it is classified as any of the other six classes, this is

Outcome of the diagnostic test	Condition (e.g. Disease) As determined by the Standard of Truth		
	Positive	Negative	Row Total
Positive	TP	FP	TP+FP (Total number of subjects with positive test)
Negative	FN	TN	FN + TN (Total number of subjects with negative test)
Column total	TP+FN (Total number of subjects with given condition)	FP+TN (Total number of subjects without given condition)	N = TP+TN+FP+FN (Total number of subjects in study)

Fig. 13 Terms for definition of classification metrics

the case of **false negative**. **False positive** case happens when the skin lesion image is indicated by the classification model to have melanoma but it actually belongs to any of the other six diseases. If a non-melanoma skin lesion image is suggested as non-melanoma by the classifier, it is the case of **true negative**.

4.1.1 Accuracy

Accuracy is the fraction of predictions our model has correctly guessed.

It is defined as:

$$Accuracy = \frac{TP + TN}{TP + TN + FP + FN}$$

4.1.2 Precision

Precision metric answers the following question: What proportion of positive identifications was actually correct?

It is defined as:

$$Precision = \frac{TP}{TP + FP}$$

4.1.3 Sensitivity/Recall

Sensitivity is also called as the True Positive Rate (TPR) or Recall. It answers the following question: What proportion of actual positives was identified correctly?

It is defined as:

$$Sensitivity = \frac{TP}{TP + FN}$$

4.1.4 F1 Score

F1 Score is also known as the F-score or F-measure. The F1 score is calculated as a weighted average of the precision and recall. Its best value is 1 and worst is 0. The contribution of recall and precision in the calculation of the F1 score are equal.

It is defined as:

Table 6 Hyperparameters used for training the CNN models

Optimizer	Adam optimizer
Learning rate	0.0001
Epochs	30
Loss function	Categorical cross-entropy
Batch size	64
Dropout	0.4

$$F_1 = 2 * \frac{precision * recall}{precision + recall}$$

4.1.5 Support

Support is defined as the number of actual occurrences of the class in the specified dataset.

4.2 Hyperparameters

Table 6 shows the hyperparameters used to train the CNN models.

4.3 Model Performance

4.3.1 ResNet50

This section represents the classification performance of the ResNet50 model. The input images were resized to 224 × 224 as this model requires the input to be in that format.

Table 7 shows the confusion matrix of this model.

Table 8 presents the classification report of this model.

The **average** metrics achieved by the model is presented in Table 9.

4.3.2 MobileNet

This section represents the classification performance of the MobileNet model. The input images were resized to 224 × 224 as this model requires the input to be in that format.

Table 10 shows the confusion matrix of this model.

Table 7 ResNet50 confusion matrix

		True class						
Predicted class		AKIEC	BCC	BKL	DF	MEL	NV	VASC
	AKIEC	**115**	16	79	7	11	10	0
	BCC	33	**574**	72	4	31	52	2
	BKL	11	17	**596**	1	29	36	0
	DF	1	5	6	**54**	6	5	0
	MEL	4	13	106	1	**828**	170	0
	NV	3	15	117	8	135	**3690**	2
	VASC	0	1	0	1	2	3	**72**

The bold numbers in the given table represent the correctly classified number of images belonging to a given class

Table 8 ResNet50 classification report

	Precision	Recall	F1 score	Support
Actinic Keratosis	0.69	0.48	0.57	238
Basal Cell Carcinoma	0.90	0.75	0.81	768
Benign Keratosis	0.61	0.86	0.72	690
Dermatofibroma	0.71	0.70	0.71	77
Melanoma	0.79	0.74	0.77	1122
Melanocytic Nevus	0.93	0.93	0.93	3970
Vascular Lesion	0.95	0.91	0.93	79

Table 9 ResNet50 average metrics

Accuracy	0.85
Precision	0.86
Recall	0.85
F1 score	0.85
Support	6944

Table 11 presents the classification report of this model.
The **average** metrics achieved by the model is presented in Table 12.

4.3.3 VGG16

This section represents the classification performance of the VGG16 model. The input images were resized to 224 × 224 as this model requires the input to be in that format.

Table 10 MobileNet confusion matrix

		True class						
Predicted class		AKIEC	BCC	BKL	DF	MEL	NV	VASC
	AKIEC	**119**	20	65	1	20	13	0
	BCC	22	**591**	57	7	33	57	1
	BKL	11	20	**582**	0	29	48	0
	DF	1	5	10	**42**	5	14	0
	MEL	7	18	85	0	**782**	230	0
	NV	4	26	126	2	53	**3753**	6
	VASC	0	1	2	1	2	7	**66**

The bold numbers in the given table represent the correctly classified number of images belonging to a given class

Table 11 MobileNet classification report

	Precision	Recall	F1 score	Support
Actinic Keratosis	0.73	0.50	0.59	238
Basal Cell Carcinoma	0.87	0.77	0.82	768
Benign Keratosis	0.63	0.84	0.72	690
Dermatofibroma	0.79	0.55	0.65	77
Melanoma	0.85	0.70	0.76	1122
Melanocytic Nevus	0.91	0.95	0.93	3970
Vascular Lesion	0.90	0.84	0.87	79

Table 12 MobileNet average metrics

Accuracy	0.85
Precision	0.86
Recall	0.85
F1 score	0.85
Support	6944

Table 13 shows the confusion matrix of this model.
Table 14 presents the classification report of this model.
The **average** metrics achieved by the model is presented in Table 15.

4.3.4 Inception V3

This section represents the classification performance of the Inception V3 model.

The input images were resized to 299 × 299 as this model requires the input to be in that format.

Table 13 VGG16 confusion matrix

		True class						
Predicted class		AKIEC	BCC	BKL	DF	MEL	NV	VASC
	AKIEC	**105**	44	48	4	19	18	0
	BCC	7	**656**	30	1	24	47	3
	BKL	9	15	**545**	2	30	89	0
	DF	1	7	5	**45**	2	14	3
	MEL	4	22	49	0	**824**	223	0
	NV	3	24	52	3	93	**3794**	1
	VASC	0	3	0	1	2	2	**71**

The bold numbers in the given table represent the correctly classified number of images belonging to a given class

Table 14 VGG16 classification report

	Precision	Recall	F1 score	Support
Actinic Keratosis	0.81	0.44	0.57	238
Basal Cell Carcinoma	0.85	0.85	0.85	768
Benign Keratosis	0.75	0.79	0.77	690
Dermatofibroma	0.80	0.58	0.68	77
Melanoma	0.83	0.73	0.78	1122
Melanocytic Nevus	0.91	0.96	0.93	3970
Vascular Lesion	0.91	0.90	0.90	79

Table 15 VGG16 average metrics

Accuracy	0.87
Precision	0.87
Recall	0.87
F1 score	0.87
Support	6944

Table 16 shows the confusion matrix of this model.
Table 17 presents the classification report of this model.
The **average** metrics achieved by the model is presented in Table 18.

4.3.5 InceptionResnet V2

This section represents the classification performance of the InceptionResnet model. The input images were resized to 299 × 299 as this model requires the input to be in that format.

Table 16 Inception V3 confusion matrix

		True class						
Predicted class		AKIEC	BCC	BKL	DF	MEL	NV	VASC
	AKIEC	**151**	27	35	0	18	7	0
	BCC	15	**667**	25	5	23	32	1
	BKL	12	18	**593**	0	34	33	0
	DF	1	5	6	**54**	2	8	1
	MEL	7	11	39	1	**954**	110	0
	NV	2	18	57	3	125	**3764**	1
	VASC	0	0	0	1	0	4	**74**

The bold numbers in the given table represent the correctly classified number of images belonging to a given class

Table 17 Inception V3 classification report

	Precision	Recall	F1 score	Support
Actinic Keratosis	0.80	0.63	0.71	238
Basal Cell Carcinoma	0.89	0.87	0.88	768
Benign Keratosis	0.79	0.86	0.82	690
Dermatofibroma	0.84	0.70	0.77	77
Melanoma	0.83	0.85	0.84	1122
Melanocytic Nevus	0.95	0.95	0.95	3970
Vascular Lesion	0.96	0.94	0.95	79

Table 18 Inception V3 average metrics

Accuracy	0.90
Precision	0.90
Recall	0.90
F1 score	0.90
Support	6944

Table 19 shows the confusion matrix of this model.
Table 20 presents the classification report of this model.
The **average** metrics achieved by the model is presented in Table 21.

4.4 Comparison and Discussion

Table 22 presents the average accuracy, precision, recall, F1 score and Support metrics among the seven disease classification achieved by all the final CNNs for comparison.

Table 19 InceptionResnet confusion matrix

		True class						
Predicted class		AKIEC	BCC	BKL	DF	MEL	NV	VASC
	AKIEC	**142**	42	35	1	11	7	0
	BCC	1	**717**	16	2	14	18	0
	BKL	3	23	**612**	1	18	32	1
	DF	1	4	5	**62**	1	4	0
	MEL	5	19	36	0	**943**	119	0
	NV	1	36	52	4	98	**3778**	1
	VASC	0	1	0	1	1	2	**74**

The bold numbers in the given table represent the correctly classified number of images belonging to a given class

Table 20 InceptionResnet classification report

	Precision	Recall	F1 score	Support
Actinic Keratosis	0.93	0.60	0.73	238
Basal Cell Carcinoma	0.85	0.93	0.89	768
Benign Keratosis	0.81	0.89	0.85	690
Dermatofibroma	0.87	0.81	0.84	77
Melanoma	0.87	0.84	0.85	1122
Melanocytic Nevus	0.95	0.95	0.95	3970
Vascular Lesion	0.97	0.94	0.95	79

Table 21 InceptionResnet average metrics

Accuracy	0.91
Precision	0.91
Recall	0.91
F1 score	0.91
Support	6944

Table 22 Average metrics achieved by all final CNNs

	Accuracy	Precision	Recall	F1 score	Support
ResNet50	0.85	0.86	0.85	0.85	6944
MobileNet	0.85	0.86	0.85	0.85	6944
VGG16	0.87	0.87	0.87	0.87	6944
Inception V3	0.90	0.90	0.90	0.90	6944
InceptionResnet	**0.91**	**0.91**	**0.91**	**0.91**	**6944**

The bold numbers in the given table represent the correctly classified number of images belonging to a given class

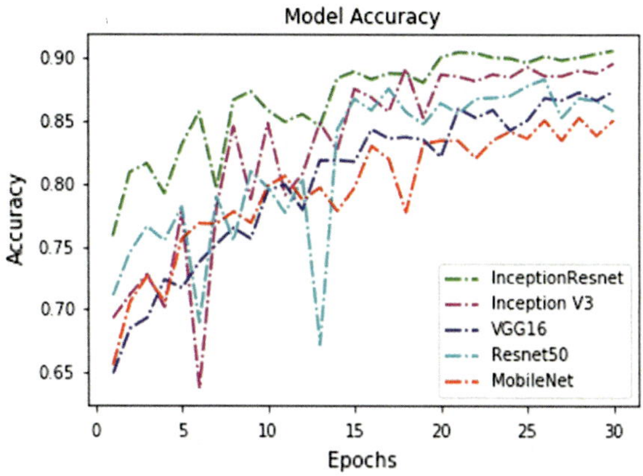

Fig. 14 CNN accuracy curves

Figure 14 shows the learning curve of all the CNN models.

From the Table 22 and Fig. 14, it is clearly visible that the Inception V3 and InceptionResnet CNN models have given the highest classification performance with Accuracies of **90** and **91%**. This model is robust enough to classify lesion images in the any of the seven types.

4.5 Conclusion

To restate, this project was conducted with the aim of developing convolutional neural network model to diagnose and detect skin cancer from lesion images. It also explored the data augmentation technique as a preprocessing step to strengthen the classification robustness of the CNN model.

The best model, namely InceptionResnet achieved an average accuracy of 91%.

References

1. G.P. Guy, C.C. Thomas, T. Thompson, M. Watson, G.M. Massetti, L.C. Richardson, Vital signs: melanoma incidence and mortality trends and projections—United States, 1982–2030. Morb. Mortal. Wkly. Rep. (2015)
2. "World Cancer Research Fund - Skin Cancer Statistics," 2018. [Online]. Available: https://www.wcrf.org/dietandcancer/cancer-trends/skin-cancer-statistics. Accessed: 28-Oct-2019
3. American Cancer Society, "Cancer Facts and Figures 2019." [Online]. Available: https://www.cancer.org/content/dam/cancer-org/research/cancer-facts-and-statistics/annual-cancer-facts-and-figures/2019/cancer-facts-and-figures-2019.pdf. Accessed: 29-Oct-2019
4. K. Kourou, T.P. Exarchos, K.P. Exarchos, M.V. Karamouzis, D.I. Fotiadis, Machine learning applications in cancer prognosis and predictio. Comput. Struct. Biotechnol. J. (2015)

5. E. Nasr-Esfahani et al., Melanoma detection by analysis of clinical images using convolutional neural network, in *Proceedings of the Annual International Conference of the IEEE Engineering in Medicine and Biology Society, EMBS* (2016)
6. A. Esteva et al., Dermatologist-level classification of skin cancer with deep neural networks. Nature (2017)
7. International Skin Imaging Collaboration, "ISIC 2018: Skin Lesion Analysis Towards Melanoma Detection," 2018. [Online]. Available: https://challenge2018.isic-archive.com/. Accessed: 29-Oct-2019
8. P. Tschandl, C. Rosendahl, H. Kittler, Data descriptor: the HAM10000 dataset, a large collection of multi-source dermatoscopic images of common pigmented skin lesions. Sci. Data (2018)
9. International Skin Imaging Collaboration, "ISIC 2019," 2019. [Online]. Available: https://challenge2019.isic-archive.com/. Accessed: 29-Oct-2019
10. D.A.V. Dyk, X.L. Meng, The art of data augmentation. J. Comput. Graph. Stat. (2001)
11. S.G.K. Patro, K.K. Sahu, Normalization: a preprocessing stage. IARJSET (2015)
12. O. Russakovsky et al., ImageNet large scale visual recognition challenge. Int. J. Comput. Vis. (2015)
13. C. Szegedy, V. Vanhoucke, S. Ioffe, J. Shlens, Z. Wojna, Rethinking the inception architecture for computer vision, in *Proceedings of the IEEE Computer Society Conference on Computer Vision and Pattern Recognition* (2016)
14. K. He, X. Zhang, S. Ren, J. Sun, Deep residual learning for image recognition, in *Proceedings of the IEEE Computer Society Conference on Computer Vision and Pattern Recognition* (2016)
15. K. Simonyan, A. Zisserman, VGG-16. arXiv Prepr. (2014)
16. C. Szegedy, S. Ioffe, V. Vanhoucke, A.A. Alemi, Inception-v4, inception-ResNet and the impact of residual connections on learning, in *31st AAAI Conference on Artificial Intelligence, AAAI 2017* (2017)
17. A.G. Howard et al., MobileNets. arXiv Prepr. arXiv1704.04861 (2017)
18. A. Baratloo, M. Hosseini, A. Negida, G. El Ashal, Part 1: Simple Definition and Calculation of Accuracy, Sensitivity and Specificity. (Emergency, Tehran, Iran, 2015)

Hardik Nahata is a Machine Learning enthusiast with a fusion of industrial and academic experiences. He is pursuing a B.Tech. in Computer Science and Engineering at the Institute of Aeronautical Engineering, Hyderabad, India. He has worked as a Research Assistant in the School of Computer Science and Engineering at Nanyang Technological University, Singapore for his Final Year Thesis. His research interests are Machine Learning for Healthcare, Deep Learning, Medical Image Analysis, Neural Networks and Computer Vision.

Satya P. Singh received Ph.D. degree in Electrical Engineering from Gautam Buddha University, India in 2017. Currently, he is a Post-Doc Research Fellow in the School of Computer Science and Engineering at Nanyang Technological University, Singapore. His research interests include Deep Learning for 3D, Artificial Intelligence in Healthcare, Analysis of MRI, fMRI, and DTI brain scans using AI.

Security of Healthcare Systems with Smart Health Records Using Cloud Technology

Priyanka Dadhich and Kavita

Abstract The information technology in healthcare become trustful and more reliable with the advancement in technology. The cloud computing is one of the advance information technologies which is comprises of huge and bulky computers and linked with online web and internet. This is also termed as distributed computing which transfer data from one source to another and can be access and controlled by authorized users in healthcare organization. The cloud computing and its related challenges builds a protected Electronic Health Record (EHR) in an exceedingly cloud computing setting has attracted heaps of attention healthcare business and educational community. The thought of Cloud computing is turning into a preferred data technology (IT) infrastructure for facilitating EHR or good health records sharing and integration. As advancement in technology comes with some limitation and data security, reliability and privacy are some of the hot issues which are existing in cloud computing. The proposed chapter includes various factors affecting cloud computing security and comparative analysis of existing algorithms for cloud computing security in healthcare.

Keywords Electronic health record · Information technology · Cloud computing

1 Introduction

1.1 Cloud Computing in Healthcare

Presently, healthcare domain rapidly expanding as a new commercial industry and contributing important part in global economy. In today scenario healthcare structure is heavily relay on IT infrastructure. Telemedicine, video-conferencing and robotic

---

P. Dadhich (✉) · Kavita
Jayoti Vidyapeeth Women's University, Mahlan, Jaipur, India
e-mail: dadhichpriyanka18@gmail.com

Kavita
e-mail: kavita_kulhari@yahoo.co.in; kavita.yogen@gmail.com

V. Jain and J. M. Chatterjee (eds.), *Machine Learning with Health Care Perspective*, Learning and Analytics in Intelligent Systems 13,
https://doi.org/10.1007/978-3-030-40850-3_9

operations in healthcare are the medical benefits of advance and modern technologies. In the modern era healthcare data portal and online apps assist doctors and healthcare professionals by providing information about patients and there health history in a secured way. New Emerging technologies like information analytics, cloud computing, and information science are utilized in social insurance to essentially improve a clinic's IT vitality productivity with the data in the most ideal manners to improve vitality proficiency for medicinal services in a financially savvy way.

"The rapid innovations in virtualization and distributed computing, as well as improved access to high-speed Internet, have accelerated interest in cloud computing. In the new information technology (IT), Cloud computing provides technique to perform the healthcare service in a tactful manner and decrease carbon dioxide path which also reduce the cost of information technology (IT), and on the basis of demand it deliver computing services" [1].

"Cloud Computing define both the term that is form of application and platform. In cloud computing platforms configures and reconfigures the servers which can be virtual or physical machines. In the Cloud Computing powerful servers and large data centers are used to define applications that can be accessible through the internet" [2].

"As per the National Institute of Standards and Technology (NIST)", "Cloud computing is an online application for sanctioning present, suitable, on-request arrange access to a joint pool of configurable resources of computing (like, services, applications and servers, webs, and storage) which will be promptly provisioned and available with nominal administrative work or service provider interface" [3].

Buyya has defined cloud computing as, "Cloud is a parallel and distributed computing system consisting of a collection of interconnected and virtualized computers that are dynamically provisioned and presented as one or more unified computing resources based on service-level agreements (SLA) established through negotiation between the service provider and consumers" [4].

The cloud computing enables communication at anytime and everywhere which is continuously available to share information from one platform to another. It creates a prospect for information portability within various medical devices exist in health organizations with another medical device deployed at other health organization. Cloud computing infrastructure enables to extract the various database of treatment, patients and diseases, that can be easily accessible to doctors to perform investigation and to understand statistical outcomes. The another system provides access which is connected with database server. This required lot of information, data and computing controlling the healthcare field. This technology provide patient historical data at the time of treatment. When patients moves from one department to another for several investigations and tests than it lead to high rate conversation of information between various medical departments. This technology enables clinicians to provide whole health information about the patients for performing comprehensive and precise treatment.

In present, Hospital Information Systems (HIS) is an essential part of healthcare sector, whereas Electronic Medical Records (EMR) are performed for computer aided operations and remotely located patient care. With the development of IT in healthcare digital data challenges are faced by the industry but also with the help of cloud computing they are able to offer various healthcare services that can be utilize for the better outcomes [5, 6].

2 Cloud Service Models

To operationalize the applications of cloud computing a medical clinics and hospitals must spare all medical data on clouds. Cloud comprises of numerous variants of service type models (like PaaS, SaaS and IaaS) and distribution type models (like community, private, hybrid and public clouds).

- **Infrastructure as a Service (IaaS)**: These model consist of networks, storage, processing and other computing infrastructure properties. This basic infrastructure can't manage and controlled by users, but they can control the functionings performed by the applications [7].
- **Platform as a Service (PaaS)**: The model employ specified coding languages and hardwares into the cloud computing infrastructure which empower the users to imply and develop various applications. These services are also can't controlled by users, but they control the functionings perform by applications.
- **Software as a Service (SaaS)**: The present cloud computing allows operators to access functions that going through Cloud setup from different end-user applications (usually via online browser). The operator doesn't operate or regulate the first Cloud structure or individual application capabilities apart from restricted user-specific function settings ("NIST Cloud Computing Reference Architecture Ver1.0") [8].

3 Deployment Models in Cloud Computing

The various methods of deploying cloud computing models:

- ***Private clouds*** are functionalized by its authorized users or by single organization. The basic infrastructure of private cloud computing may be available at both off-site or on-site, and it can be managed by third party the organization have full control over them. Whereas large organization or general public can access ***Public clouds*** which are controlled and accomplished by a Cloud service provider.
- ***Hybrid clouds*** are used to perform application and data portability which adjoin two or more (public or private) clouds that persist as a single entities and technology bound them altogether.
- ***Community clouds*** are formed when many organizations share their cloud computing infrastructure to a particular community. These clouds may be available at off-site or on-site and it can be managed by third party or organization have control over them (Fig. 1) [9].

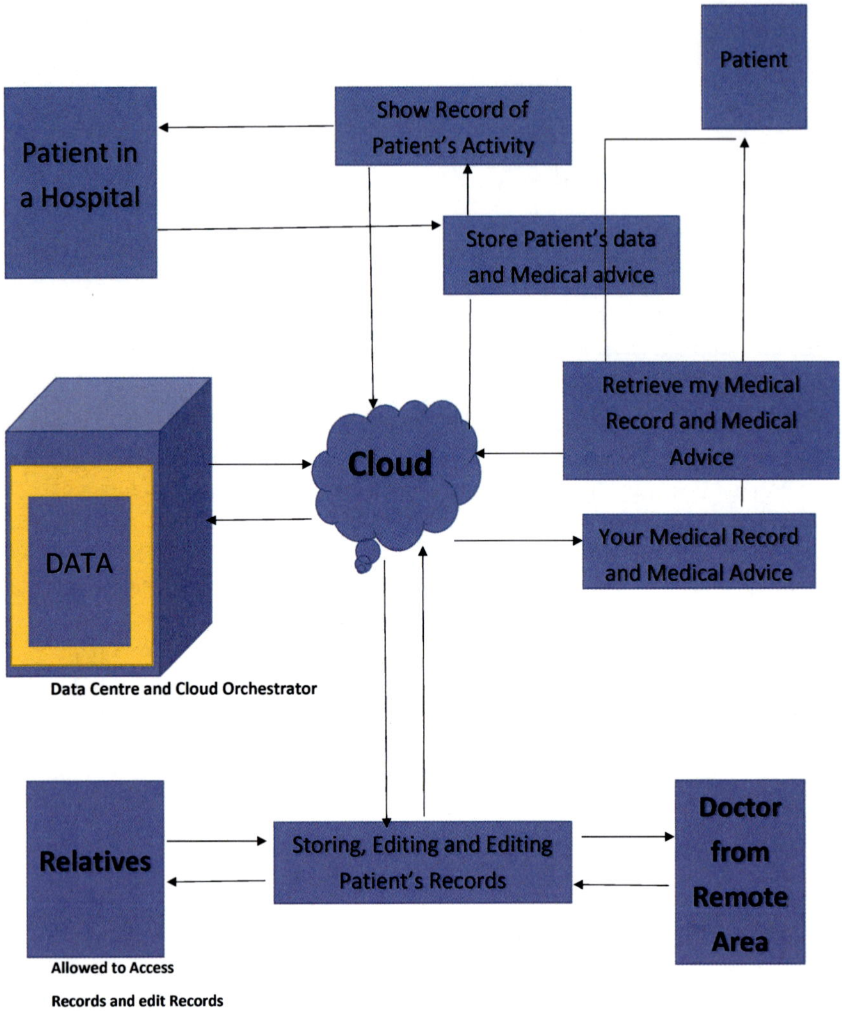

Fig. 1 Cloud computing flow diagram in healthcare systems

4 Cloud Computing Security

With the rapid growth of cloud services users store their sensitive data on clouds, so security of clouds is getting more important than that of earlier. The cloud data are immensely accessible by scalable data centers and can be accessed by anyone from anywhere. As the growth of clouds is increasing the growth of malicious or harmful activities like data theft is also increasing. Millions of users avail the services of the clouds everyday therefore they require persistent services and high safety. The main objective of cloud service is that the work done on the client side can be moved on the unseen collection of resources like hardware, software, and infrastructure services, over the internet. The Cloud Service Provider manages the database and applications on remote server for the users to provide them independence of using them from anywhere and any workplace. The cloud services allow the data centers to allow the organization to get their applications to easy to manage, run fast, easy to access, and less maintenance to meet the business demands [10].

Like we can access and manage the applications by using smart phones, tablets, and laptops without the requirements of storing the data at our end, by storing the data at only one location i.e. Clouds. To ensure data security, confidentiality, reliability of the data the service provider must provide better encryption mechanisms.

Cloud Security Issues: Cloud services can easily be accessed by the hackers or cyber attackers for many reasons. Some well-known cloud service providers such as Google, Microsoft, and Amazon have enough infrastructure to deflect and be tolerant against the cyber-attacks, but every cloud specially clouds for small organizations does not have such tolerant capability. If a cyber attacker has any idea about the capabilities of the clouds, it becomes very easy for them to crack the sensitive data from the clouds. Cloud computing security includes several major security issues like [11] :

(A) **Multi Occupancy**: The cloud service model is a kind of model which provides sharing of resources among multiple Independent users of that cloud. It becomes very easy for the competitor to co-exist on the same cloud. Such kind of situations enhanced the chances of data theft.
(B) **Data Loss and Leakage of Data**: Removal or modification of data on such medium can lead to data loss. Sometimes we store the data on unreliable medium, can make it easily available for the unauthorized users and they can make multiple attacks to affect the integrity of the data sets.
(C) **Easy Accessibility of the Clouds**: Clouds are easily accessible for any users from anywhere. The cloud service providers must use a registration system to make a validation control on the services.
(D) **Service Legal Agreement**: It is a legal document signed by cloud service provider and cloud user about terms and conditions of the services provided and availed. All the required security measures, services provided and legal measures to be taken in case of unsatisfactory performance must be mentioned in the agreement earlier at the time of start of the services [12–14].

4.1 Healthcare Data Security in the Cloud

There is not any requirement of infrastructure reference on cloud service on which these clouds are based. However, many pros and cons are linked by cloud applications. For the cloud-based resources, data protection and unauthorize access are one of the major risks linked with Cloud access. Both the Cloud Users and Cloud Providers faced various challenges associated with cloud computing which can be on both side

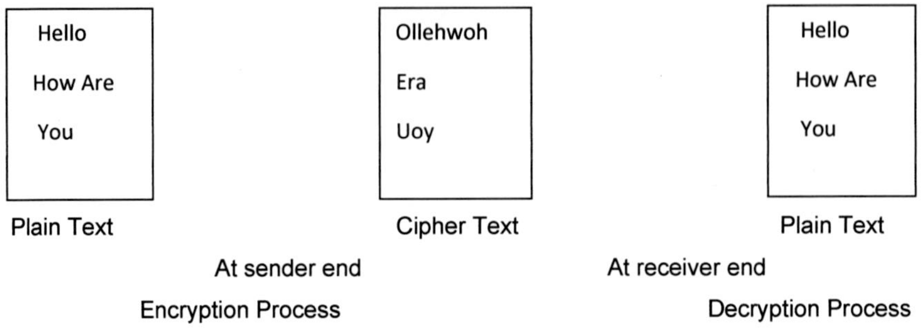

Fig. 2 Encryption decryption process

including web side or information side. The privacy protection for healthcare data is exist for a long time, but with large number of advantages and application benefits cloud computing are utilize more efficiently by healthcare providers [15, 16].

Cryptography plays an important role in data security and confidentiality on cloud computing. In the process of data storage and transmission cryptography convert data into unreadable data which looks waste to intruder. Cipher text is the unreadable coded form of data. The data will appear in its original form when it is received by its receiver, when data is decoded it is known as plain text. Encryption is the process of converting plain text into cipher text whereas reverse process of converting cipher text into plain text is recognized as decryption. At the sender's end encryption of text takes place whereas at receiver's end the decryption process is performed (Fig. 2) [17].

The cryptography algorithms are further classified into three algorithms; i.e. (a) Symmetric algorithms; (b) Asymmetric algorithms; (c) Hashing.

The application of algorithms is used to create fixed length signature and the encryption of data can be enabled by using hash function which is known as hashing. The different hash value is associated with the data. Once the data is encrypted in hashing that cannot be decrypted which is the major disadvantage of hashing. The symmetric and asymmetric algorithms perform important role to overcome from the limitation of hashing. The symmetric algorithm is also termed as "Secret Key Encryption Algorithm" that only apply one key for both decryption and encryption which termed as private key, where as in asymmetric algorithm for encryption and decryption both public and private keys are used, this algorithm is also termed as "Public Key Encryption Algorithm" [18].

The medical domain is everchanging towards a data-centric model, aided in part by open standards that support cooperation, collaborative workflows and information sharing. The wide range of healthcare services are provided by the cloud computing. All the stake holders in the healthcare industry whether they are healthcare institutions, insurance sectors, medical practices and research organizations can be connected on one platform with the help of cloud computing which enhanced computing resources at less capital expenditures [19].

In today's world many senior citizens have chronic and life-long medical conditions that require a high level of healthcare. Many aged persons are also living alone due to some social working conditions. They are not aware of their medical conditions and they also like to stay in home rather than health care centers. These patients need constant observation and consultation of healthcare providers like doctors and medical practitioners. In such cases patients' medical records with the prescription of the doctor must be keep away from unauthorized access and should be easily available when required the cloud computing with healthcare system is the best solution. Transferring patient data onto cloud-based systems is again a critical challenge faces by the IT healthcare mangers in respect to data security and confidentiality. To resolve the patient data privacy problem in cloud computing data encryption process is only the one solution that can be performed on cloud computing [20, 21].

5 Key Issues Pertaining to Cloud Computing Security in Healthcare Systems

Many IT professionals reflect that for best results and growth in business cloud computing can provide best solutions in each business sector. Alsanea and Barth [22] conduct a research study which revealed that 85.80% of responders are agree to adopt cloud computing technology, whereas majority of responders found cloud computing as useful tool, the major chunk of responders perceived that cloud computing provides quality of service and data security as a major application. "***Cloud Security Alliance (CSA)***", acknowledged the "Speedy increasing rate of cloud technology adoption as an important solution to the organizations for effective management of data due to its enormous number of benefits such as flexibility, scalability, and affordability" [23], stated that "Cloud computing technology also delivers more opportunities and gives advantages to both small and medium enterprise".

The fast-moving inventions containing cloud services has directed to abundant consequences in healthcare delivery. Various challenges are faced by existing e-healthcare systems that is related with assistance provided to clients, online connectivity, cost factor and disaster recovery.

Some key factors which affects the cloud computing security concerns in healthcare systems perspectives are as follows:

1. Hardware modularity
2. Software modularity
3. System complexity
4. System compatibility
5. Cost effectiveness
6. Network connection
7. Data security and Privacy of data
8. Training of the staff.

1. **Hardware modularity**: To augment the scalability of the healthcare applications within an organization enough hardware modularity is required. Health-related issues requires embedding essential devices to develop and debug. Therefore, healthcare workers must provide adequate hardware equipment which significantly lead to enhance the effectiveness of health care data throughout all the departments of hospitals and health care centers. It is guaranteed that the present hardware tools can be used in cloud environment and these hardware's can be used in hospitals also.
2. **Software Modularity**: The software modularity plays a significant role in the usage and adoption of any new technology. The user must be sure enough that existing platform can easily be handled by the cloud applications and can possibly be boost up the health care staff's authorization to operate the application available on cloud. If we have an up to date software, then we can assist and utilize the appropriate technology which can assist the organization to adopt new cloud services which can help to use the IT infrastructure.
3. **System Complexity**: As we move on to the higher level of the technology the overall relative complexity of the system increases, and the system also became complex to use as well as to understand. Majority of admin observed while using cloud-based applications in healthcare domains outcomes in some challenges for the health care workers and users who are lacking in technological proficiency and Information Technology specialty. HealthCare Workers must be provided easy to use systems which can easily be handled with simple knowledge and less expertise of technology. System complexity is frequently related with a fact that how the people perceive technology to develop relevance to the self-expertise among themselves.
4. **System Compatibility**: We can define the System compatibility as the capacity of two systems to work together without having to be transformed to do the same. Currently available cloud services must be compatible to process the Health-Care data records and their precision to deliver services to these mechanisms. If Cloud based health related data will be well-suited with healthcare workers requirement than they avail these services on their daily routine work basis and this will also enhance the use and reliability of Health clouds in medical care. Compatibility is an essential aspect which must be considered while we are going to adopt Health Clouds. Compatibility of Health clouds significantly affect the insight of users and Health Care professionals to make use of it.
5. **Cost Effectiveness**: One of the major preceding in using a new technology is the cost factor, and this can affect the pronouncement of the user about using it or not. The utilization and implementation of new technology may lead to major investments hence the cost increases accordingly. Cloud technology and use of health clouds is the better solution to store and access the health records without investing by individuals. It is comparatively inexpensive, and a cloud-based system will positively help the health care users either patients or doctors and hospitals perception about using it.
6. **Network Connection**: Lack of internet network or poor network connection is the major affecting key factor to use the new technology. We must face some

difficulties regarding internet services due to bad tele-communication capabilities as well as irregular supply of the power. In the cloud technology Network is necessary to connect with the Health centers as well as different Healthcare users within a Nation or even the state. Some complications are faced by Health Care workers and staff to access and share the health data records to various departments with the availability of limited accessibility and inadequate network connection. So, the quality of Network connectivity is an important factor which should be considered by health workers while adopting the new technology.

7. **Data Security and Privacy of Data**: A security threat in an organization might lead to loss of any person's information, privacy of data, or loss of other sensitive information. So, security and privacy of the patient's data is the key factor which is to be considered while adopting the Cloud Services. Health Care systems ensures and guarantee about privacy and security of patient's health-related records, as these records deals with sensitive information which should be accessible only if proper authentication is approved by the Health centers [24].
8. **Training of the staff**: Training is the resource that is provide by the organizations to their operating staff to achieve the proficiency required to operate and to efficiently utilize the technology. The training makes the users more confident and comfortable while using the cloud technology. Adequate Training programs must be provided by the management of the Health cloud service providers to the users in a timely manner. So the users got the knowledge about the services of the clouds and may feel comfortable while adopting the new technology. The role of proper training programs is to effectively use the cloud services in health care sector grounded on the needs demanded by Health-Care users as well as patients [25, 26].

6 Brief Description of Some Cloud Computing Algorithms and Their Comparison

Various symmetric and asymmetric algorithms are used to save the data on clouds, which provide high security to the data while transmission of the data over the network. Some of those are discussed below:

1. **DES Algorithm (Data Encryption Algorithm)**:

It is most popular security algorithm used for encryption the data. This is a symmetric algorithm which means that the same key is used to encrypt and decrypt the data from sender to receiver. This is used to encrypt the sensitive data. This algorithm uses a 64-bytes block of to encode the data at a time. In 16 rounds the encoding is done. One bit is used as a parity bit. In the working for encoding the 64 bits are divided into two parts left and right halves. Right half is mixed with function f where it is mixed with key. Then it is added to left half and at last both the halves are swapped with each other, except the last round [27, 28].

Here are the equations for the round i:
Equations for round i:

$$Li = Ri - 1 \text{ (Runs the steps except the last one)}$$
$$Li = Li - 1 \text{ XORed with } f(Ri - 1) \text{ (Mixing the right half with function f)}$$

In other words:

$$Ri - 1 = Li \text{ (Both the Right and Left Halves are swapped with each other)}$$
$$Li - 1 = Ri \text{ XORed with } f(Li)$$

Decryption is same as encryption.

Modes of operations for DES: There are three types of modes of operations for DES algorithm [29].

(i) ECB: Electronic Code Book mode: This type of mode encrypts and decrypts each 64 bits independently. The attacker could build the code book [30].
(ii) CBC: Cipher Block Chaining mode: This mode of mode encrypts the data in a way that each 64 bits is dependent on the previous one.

$$\text{Encryption:} Ci = EK(Pi \oplus Ci - 1)$$
$$\text{Decryption:} Pi = Ci - 1 \oplus DK(Ci)$$

(iii) CFB-OFB: This type of mode allows the encryption by byte-wise. This can be termed as Cipher Feed Back, Output Feed Back [31].

Advantages:

1. Secured algorithm, it's hard to attack.
2. It's easy to implement with both hardware and software.
3. Easy to analyze.

Disadvantages:

1. Key size is too small. So, it takes time to encrypt large voluminous data.

2. **Triple DES (Triple Data Encryption Algorithm)**:

This is known as symmetric algorithm. We run the DES algorithm three times in this algorithm. By applying DES algorithm with 3 different keys in sequence the size of encryption can be extended in Triple DES algorithm. Overall forty eight passes are runs across the DES algorithm, and 168 bytes present in the resultant key; and this may remain difficult to execute, it has also a 2 key option presented in Triple DES which operates via Encryption-Decryption-Encryption (EDE) method [32].

1. Encryption: This is the method of coding the data at sender site and key one is used to employed with content.

2. Decryption: This is the method of decoding of the data at receiver site, key two is applied for transforming encrypted to decrypted text.
3. Encryption: finally, followed by step two decrypted text is further encrypted by applying key two.

This type of 3-key process (that is bulkier, but highly secure), here 3 times data is encrypted successively. With key one data is encrypted, followed with key two data is again encrypted, at last messages and text encrypted with the help of key three [33].

Steps of 3-DES:

1. We run the DES algorithm three times: in ECB mode.
2. If the Key 2 (K2) is equal to Key 3 (K3), this is DES Backwards compatibility
3. Double-DES: Ci = EB(EA(Pi)) [Second round of DES]
4. Given P1, C1: Note that DB(C1) = EA(P1)
5. We will make a list of every Encryption Key [EK(P1)].
6. Try each L: if DL(C1) = EK(P1), then maybe K = A, L = B. (2^{48} L's might work.)
7. Test with P2, C2: if it checks, it was probably right.
8. Time roughly 2^{56}. Memory very large.

Advantages:

1. Significantly more secured as compared to DES.

Disadvantages:

1. It is slower process as compared to DES.
2. It becomes very lengthy to run DES three times so computation part is very difficult to implement.

3. **AES Algorithm (Advanced Encryption Algorithm)**:

This is considered as a very powerful algorithm as compared to its forerunners algorithms DES and Triple DES. This is critical and for data encryption it implies longer key. Data decryption is prompt, so this algorithm becomes more secured and trusted to share sensitive information on network via using firewall, routers and security protocols. This algorithm works on 128 bits and can be expanded up to 256 bits. Original name for this algorithm is Rijndael, which works with different block size and key size [32].

Enciphering rounds are involved in the AES algorithm. The sequence of several process steps completes one round, that involve pairing of transposition, substitution and coordinating associations between the output and inputted plaintext.

"The total 10 rounds are applicable and applied with 128-bit keys, 12 rounds for 192-bit keys and 14 rounds for 256-bit keys which vary the number of rounds with number of keys". The Advanced Encryption Standard text uses menu-based replacement of text by columns and rows.

AES algorithm has been explored for its vulnerabilities and since its recognized as the new standard. Some novel methods are applied for identifying innovative approaches to strike the algorithm Brute-force attacks is one of them. Some attacks are effective against the less complicated versions of Rijndael (AES) exploitation shorter keys and fewer rounds. AES is widely recognized algorithmic program to be a reliable and sensible approach [33].

The encryption process of AES algorithm can be further divided into four sub processes which are as follows:

1. **Byte substitution**: This is the initial step of Encryption process; the 16 info bytes are supplanted by watching them and put in a fixed table. The consequence of this procedure at that point displayed in a lattice containing four lines and four segments.
2. **Shift Rows**: Every rows of the matrix are shifted to left. Then new entries are reinserted on the right side of the row that are termed as 'fall off'. Shifting of the rows can be done as mentioned below:

    - Initial row remains unmovable it remains fixed.
    - Subsequent row is move from position of 1 byte towards left.
    - Next 3rd row is moved from position 2 towards left.
    - 4th or last row is moved from the position 3 to the left.
    - As a end output, we found a fresh matrix which is comprises of sixteen bytes and moved respectively [34].

3. **Mix Columns**: Now Encoding is done at column level; with the application of mathematical function every column with four bytes can be transformed. The further task replaces the original column by incorporating 4 bytes of one column as input and for output 4 new bytes entirely. As an outcome of this process one more different matrix is found which comprising sixteen new bytes. In the last round this process does not take place.
4. **Add Round Key**: Here in this step, the received new sixteen bytes of the matrix contemplated as 128 bytes and binary XOR operation performed in the process. The final output in last series comprises of ciphertext. This step can be further repeated with sixteen bytes resulted from 128 bytes [35].

After encryption the decryption takes place at the receiver's site. The decryption process is just reverse of the encryption process. The steps of the decryption are:

1. Round key is Added
2. Mixing of the Columns
3. Shifting of the Rows and
4. Substitution of the Bytes.

Advantages:

1. It can be considered as most secured algorithm because it is implemented for both hardware and software.
2. High length key sizes such as 128, 192, and 256 bits is used for encryption. So, it becomes robust against hacking.

3. Most widely used security protocol so can be used for numerous applications such as wireless communication, e-business, financial transactions etc.
4. For 128 bits, about 2^{128} attempts needed to break so, it makes difficult to hack the sensitive private data.

Disadvantages:

1. Algebraic structure is used so it becomes simpler.
2. Every block is always encrypted in same way.
3. Hard to implement with software.

4. **Proposed Security Algorithm**:

Several security algorithms are available for encryption/decryption of the data for providing security in Health care systems. For Health Care security a proposed algorithm can be used, which provides high security and it also makes difficult encryption codes which is difficult to hack. The proposed algorithm uses ten rounds to encryption, and it can also be applied for large block sized data up to 256 bits [36].

The steps of the proposed algorithm are as follows:

1. Create cipher key using place value swapping pattern.
2. Convert the cipher key into 64-bit value.
3. From the cipher key, set of round keys are derived.
4. With block data (Plain text), State array is initialized
5. Add the initial Round key to the starting State Array.
6. Followed by BODMAS Encryption.
7. The block to be encrypted is just a sequence of 128-bits. First convert the 128-bits into 16-bytes. At the start of the encryption, the 16-bytes of the data.
8. Convert the data in 128-bit pattern.
9. To alter the State Array encryption process is comprises of several steps. The following 2 operations are involved in these steps:

    a. XOR Round Key
    b. Shift Rows.

10. After this we get 128-bit data.
11. Then, the encrypted text is converted into Sixty-four-bits (eight Bytes) from one hundred twenty-eight-bits (sixteen-Bytes).
12. Then, sixty-four-bit encrypted text block is handed over to an Initial Permutation (IP) Function.
13. On encrypted text Initial Permutation is performed.
14. Next, the Initial Permutation (IP) produces permutation block.
15. Now implement short multiplication formulae on permutation block.
16. Then convert the derived block into 128-bit by using the encrypted data which is received in step 10.

By this way this algorithm can be considered as better solution for the high security issues for healthcare systems [37].

Advantages:

1. Uses high length key size as 128–256 bytes.
2. Have a different approach because it uses Algebraic expressions to encrypt the code.
3. Extreme tough in decryption because algebraic expressions work on assumed random data samples.
4. Only authorized person will be able to decrypt the code not by using formulae, but the person will use API only.
5. Faster approach because a little block cipher is there.
6. It has no lengthy codes and processes.
7. It has better use of Arithmetic and Logical Unit (ALU).
8. It is Easy to process and tough to hijack.

7 Conclusion

Cloud computing have become indispensable ingredients of the future information infrastructures. Evaluation of Cloud service performance is crucial and beneficial to both service providers and service consumers. Cloud computing not only helps healthcare stakeholders to solve many of their existing problems but also to deliver quality healthcare services in a timely and cost-effective manner. The adoption of cloud computing is also new challenges which faces problem like data security and privacy, lack of trust, organizational acceptance and unavailability of system development standards etc.

The suitable analysis, planning and measures should be taken into consideration before shifting to the cloud environment and hence the present proposed security algorithm is based on the performance analysis of various existing cloud computing security algorithms to find out the most suitable algorithm for healthcare industry for providing secured data with high efficiency and minimum cycles of encryption.

References

1. M. Masrom, A. Rahimli, W.N.B.W. Zakaria, S.M. Aljunid, Understanding the problems and benefits of using cloud computing in Malaysia healthcare sector. Int. J. Comput. Appl. **4**(1) (2016). Retrieved from: https://pdfs.semanticscholar.org/7426/0a0aae6671db660f0834d3b1c8795f35d723.pdf
2. G. Boss, P. Malladi, D. Quan, L. Legregni, H. Hall, Cloud computing (2007). www.ibm.com/developerworks/websphere/zones/hipods/. Accessed 20 May 2010
3. P. Mell, T. Grance, The NIST definition of cloud computing. Natl. Inst. Stand. Technol. **53**(6), 1–7 (2009)
4. R. Buyya, C.S. Yeo, S. Venugopal, J. Broberg, Y. Brandic, Cloud computing and emerging IT platforms: vision, hype, and reality for delivering computing as the 5th utility. Future Gener. Comput. Syst. **25**, 599–616 (2009)

5. N. John, S. Shenoy, Health cloud—healthcare as service (HaaS), in *Proceedings of International Conference on Advances in Computing, Communications and Informatics (ICACCI)*, 2014, pp. 1963–1966
6. K. Al-Begain, M. Zak, W. Alosaimi, C. Turyagyenda, Security of the cloud, in *Emerging Research in Cloud Distributed Computing Systems*, ed. by S. Bagchi (IGI Global, Hershey, PA, 2015), pp. 363-404. https://doi.org/10.4018/978-1-4666-8213-9.ch012
7. R. Asija, R. Nallusamy, Healthcare SaaS based on a data model with built-in security and privacy, in *Cloud Security: Concepts, Methodologies, Tools, and Applications*, ed. by I. Management Association (IGI Global, Hershey, PA, 2019), pp. 744–759. https://doi.org/10.4018/978-1-5225-8176-5.ch037
8. High Performance Architecture and Grid Computing International Conference, HPAGC 2011 (Chandigarh, India, July 19–20, 2011). Proceedings
9. F. Stoicuta, I. Ivanciu, E. Minzat, A.B. Rus, V. Dobrota, An OpenNetInf-based cloud computing solution for cross-layer QoS: monitoring part using iOS terminals (10th International symposium on electronics and telecommunications, Timisoara, 2012), pp. 167–170
10. S. Carlin, K. Curran, Cloud computing security. Int. J. Ambient Comput. Intell. (IJACI) **3**(1), 14–19 (2011). https://doi.org/10.4018/jaci.2011010102
11. P.T. Vaikunth, P.S. Aithal, Cloud computing security issues—challenges and opportunities. Int. J. Manage. Technol. Soc. Sci. (IJMTS) **1**(1), 33–42 (2017). https://doi.org/10.5281/zenodo.569920
12. N. Sultan, Making use of cloud computing for healthcare provision: opportunities and challenges. Int. J. Inf. Manag. **34**(2), 177–184 (2014). ISSN: 0268-4012
13. V.R. Pancholi, B. Patel, *Enhancement of cloud computing security with secure data storage using AES* (2016)
14. F. Ogigau-Neamtiu, Cloud computing security issues. J. Defense Resour. Manage. **3**(2), 141–148 (2012)
15. Md. Fakhrul Alam Onik, S.S. Salman-Al-Musawi, K. Anam, N. Rashid, A secured cloud based health care data management system. Int. J. Comput. Appl. **49**(12) (2012). Retrieved from: https://pdfs.semanticscholar.org/7426/0a0aae6671db660f0834d3b1c8795f35d723.pdf
16. S. Kannan, S. Ramakrishnan, Performance analysis of cloud computing in healthcare system using tandem queues. Int. J. Intell. Eng. Syst. **10**(4) (2017). Retrieved from: http://oaji.net/articles/2017/3603-1498897498.pdf
17. M. Panhwar, S. Ali Khuhro, G. Panhwar, Ghazala, K. Ali, SACA, *A Study of Symmetric and Asymmetric Cryptographic Algorithms* (2019)
18. A. Er. Pansotra, S. Er. Preet Singh, Cloud security algorithms. Int. J. Secur. Appl. **9**(10), 353–360 (2015)
19. Y. Al-Issa, M. Ashraf Ottom, A. Tamrawi, eHealth cloud security challenges: a survey. J. Healthc. Eng. **2019**, 1–15 (2019)
20. E. Jaul, J. Barron, Age-related diseases and clinical and public health implications for the 85 years old and over population. Front. Public Health **5**, 335 (2017). https://doi.org/10.3389/fpubh.2017.00335
21. R. Anitha, S. Mukherjee, Data security in cloud for health care applications, in *Advances in Computer Science and its Applications*, vol 279, ed. by H. Jeong, M.S. Obaidat, N. Yen, J. Park (Springer, Berlin, Heidelberg, 2014), Lecture Notes in Electrical Engineering
22. M. Alsanea, J. Barth, Factors affecting the adoption of cloud computing in the government sector: a case study of Saudi Arabia. Int. J. Cloud Comput. Serv. Sci. **x**(x), 1–16 (2014). https://doi.org/10.11591/closer.v3i6.6811
23. R. Amponsah, J. Panford, J. Hayfron-Acquah, Factors affecting cloud computing adoption in a developing country–ghana: using extended unified theory of acceptance and use of technology (UTAUT2) model. Int. Res. J. Eng. Technol. (2016)
24. M. Ahmadi, N. Aslani, Capabilities and advantages of cloud computing in the implementation of electronic health record. Acta Informatica Med. **26**(1), 24–28 (2018). https://doi.org/10.5455/aim.2018.26.24-28

25. A.M. Kadhum, M.K. Hasan, Assessing the determinants of cloud computing services for utilizing health information systems: a case study. Int. J. Adv. Sci. Eng. Inform. Technol. **7**, 503–510 (2017)
26. A. Meri et al., Success factors affecting the healthcare professionals to utilize cloud computing services. Asia-Pacific J. Inf. Technol. Multimed. **6**(2), 31–42 (2017). e-ISSN: 2289-2192
27. M.H.M. Zaharuddin, R.A. Rahman, M. Kassim, Technical comparison analysis of encryption algorithm on site-to-site IPSec VPN, *International Conference on Computer Applications and Industrial Electronics, Kuala Lumpur*, 2010, pp. 641–645. https://doi.org/10.1109/ICCAIE.2010.5735013
28. J.O. Grabbe, *The DES algorithm illustrated*, 2010.
29. Federal information processing standards publication (FIPS 197), *Advanced Encryption Standard (AES)*, 2001.
30. S.K. Rahimi, F.S. Haug, *Distributed database management systems: a practical approach*
31. D. Mukhopadhyay, B.A. Forouzan, *Cryptography and network security*, 2nd edn. (Mcgraw Hill Education, 2011)
32. G. Singh, S. Kinger, Integrating AES, DES, and 3-DES encryption algorithms for enhanced data security. Int. J. Sci. Eng. Res. **4**(7) (2013)
33. S. Bansal, G. Jagdev, Analyzing working of DES and AES algorithms in cloud security. Int. J. Res. Stud. Comput. Sci. Eng. (IJRSCSE) **4**(3), 1–9 (2017). ISSN 2349-4840 (Print) & ISSN 2349-4859 (Online). http://dx.doi.org/10.20431/2349-4859.0403001
34. K. Arul Jothy, K. Sivakumar, M.J. Delsey, Enhancing the security of the cloud computing with triple AES, PGP over SSL algorithms. IJESRT Int. J. Eng. Sci. Res. Technol. **7** (2018) ISSN: 2277-9655
35. N.N. Pathak, M. Nagori, Enhanced security for multi cloud storage using AES algorithm. Int. J. Comput. Sci. Inf. Technol. **6**(6) (2015). ISSN: 0975-9646
36. Data encryption standard (DES). Retrieved from: https://www.geeksforgeeks.org/data-encryption-standard-des-set-1/
37. P. Gowthami, M. Nishanthi, B. Indhumathi, M. Navaneethakrishnan, An aggregated aes algorithm for preventing keyleakage problem in cloud computing. Int. J. Recent Res. Sci. Eng. Technol. **5**(4) (2017, April)

Priyanka Dadhich is a research scholar at Jayoti Vidyapeeth Women's University, Mahlan, Jaipur. She is also working as assistant professor at S.S. Jain Subodh Girls P.G. College, Sanganer, Jaipur. She is very hard working and dedicated to the work assigned to her. She has several publications and about 10 citations also. She has more than 15 years of experience in academics and research. She is specialized with C language, Data Structure, DBMS, SAD, Computer Graphics, MIS, Simulation and Modeling.

Dr. Kavita Ph.D. (Computer Science) and she received her M.C.A. degree in computer science from Modi Institute of Technology and Science, Lakshmangarh, Sikar. Presently working as an Associate Professor at Jagan Nath University, Jaipur. She has nine years of teaching experience in the field of Computer Science and supervising research scholars in the field of E-commerce, mobile Commerce, Data Mining etc. Dr. Kavita has published several research papers in the reputed 'National and International Journals' and National and International Conferences. She is also author of a book cloud computing from International publisher SCHAND publication.

Intelligent Heart Disease Prediction on Physical and Mental Parameters: A ML Based IoT and Big Data Application and Analysis

Rohit Rastogi, D. K. Chaturvedi, Santosh Satya and Navneet Arora

Abstract Nearly 17.5 million deaths from cardiovascular disease occur worldwide. Currently, India has more than 30 million heart patients. People's unconscious attitudes towards health are likely to lead to a variety of illnesses and can be life threatening. In the healthcare industry, large amounts of data are frequently generated. However, it is often not used effectively. The data indicates that the generated image, sound, text, or file has some hidden patterns and their relationships. Tools used to extract knowledge from these databases for clinical diagnosis of disease or other purposes are less common. Of course, if you can create a mechanism or system that can communicate your mind to people and alert you based on your medical history, it will help. Current experimental studies use machine learning (ML) algorithms to predict risk factors for a person's heart disease, depending on several characteristics of the medical history. Use input features such as gender, cholesterol, blood pressure, TTH, and stress to predict the patient's risk of heart disease. Data mining (DM) techniques such as Naive Bayes, decision trees, support vector machines, and logistic regression are analyzed in the heart disease database. The accuracy of various algorithms is measured and the algorithms were compared. The result of this experimental analysis is a 0 or 1 result that poses no danger or danger to the individual. Django is used to run a website.

R. Rastogi (✉)
ABES Engineering College, Ghaziabad, UP 201009, India
e-mail: rohit.rastogi@abes.ac.in

D. K. Chaturvedi
Dayalbagh Educational Institute, Agra, UP 282005, India
e-mail: dkc.foe@gmail.com

S. Satya
Indian Institute of Technology, Delhi 110016, India
e-mail: ssatya@rdat.iitd.ernet.in

N. Arora
Indian Institute of Technology, Roorkee 247667, India
e-mail: navneetroorkee@gmail.com

V. Jain and J. M. Chatterjee (eds.), *Machine Learning with Health Care Perspective*, Learning and Analytics in Intelligent Systems 13,
https://doi.org/10.1007/978-3-030-40850-3_10

Keywords Stress · Internet of Things (IoT) · Mental health · Meditation · Tension type headache (TTH) · Naïve Bayes (NB) · Support vector machine (SVM) · Machine learning (ML) · Logistic regression · Mental and physical scores · Decision tree (DT) · Connected devices/smart devices · Big data (BD) tools · Big data analysis (BDA)

1 Introduction

We are told to live in the "information age". A lot of data is generated every day. However, it is not used more effectively. Effective tools to extract knowledge from these databases are not very popular for the diagnosis of clinical diseases and other purposes.

The main challenge that medical institutions (health centers, hospitals) are addressing is that they are affordable, high quality facilities and more people could afford. Qualitative services refers checking patients properly and providing the most effective treatment. Decisions can lead to risk consequences that may not be accepted by the general public.

Hospitals also need to reduce the cost of various clinical trials. Hospitals can achieve this by designing appropriate decision support systems or computer-based information [1, 2].

Today, almost all hospitals use some sort of hospital computer-based information system to manage patient information. These systems typically store large amounts of data that can be used to train the machine, thus providing sufficient results at a low cost. Unfortunately, this information is not used wherever possible. This can be thought of as a treasure trove of hidden information that is often overlooked. This raises an important issue:

"How can intelligent clinical decisions translate into information that helps doctors make better clinical decisions?" [3, 4].

1.1 Machine Vision

In today's world, machine vision can be seen as an important tool to cure the disease along with the medical science and in future machine vision could be seen as very important technology which can combine the hard work of doctors and surgeon and the scientific technology which could improve the efficiency of treating any disease and reduce the effective cost of any treatment. From the feedback of various patients from past 2 years it has been observed that medical technology's importance and need is increasing day by day. And from the world wide it can be seen as the fifth industry which has the most number of visitors [5].

This technology includes works as use of cameras during endoscopy during surgery to have a crisp and clear image during operating, use of scanner which could improve the quality of dentures and scanners are also used to detect skin cancer. It is fast as well as accurate as it takes around 100 images in just 1 s which reduces a maximum amount of time during operating which is consumed in scanning. In future we would be able to see more camera based medical technologies due to which the patients need not to go to the doctor every time, he/she could get prescribed even at home, there would be a physical distance between the patient and doctors. And if we are talking about machine vision then we shouldn't forget about artificial intelligence and machine learning [6, 7].

These two are very essential in today's medical science. AI has become so smart that you don't need to code to search any specific image, because of its knowledge and the ability to identify patterns the AI find the images when any similar image is presented.

1.2 Medical Images

It is a way through which the images of human body part for any clinical purpose is required for the study of anatomy and physiology of humans or operating upon it. Its main purpose is to get the information of the part which is under the skull and bones for diagnosing and treating disease. The medical imaging uses technologies like X-ray radiography, magnetic resonance imaging, medical ultrasonography or ultrasound, endoscopy, elastography, tactile imaging, thermography, medical photography. It helps the doctor to understand the problems and the damage of internal organs so that they could operate well and could provide a better and efficient treatment. The process of medical images is hustle free and it doesn't require any special preparations and in cases like cancers medical imaging can serve as a boon for the patient [8, 9].

One of the technology is ultrasound which helps the doctor to look internal body structure clearly for example tendons, muscles and other internal organs. The medical images are used in various surgeries. The high resolution images allow the doctor to have a look at real time progress of surgery and work according to it. It may look like it is very costly and the surgery which involve medical images are little bit costly, but the situation is just opposite as the doctor just take the image of the affected region and plans the treatment for it.

This process reduces the efforts of doctor due to which the overall cost of the surgery gets reduced. And it is considered to be the safer technology because there are chances that human could interpret the wrong disease but because of the medical imaging all the work become safer and smooth. In previous time it is difficult to share the medical files over a long distance and because of this there were chances of the patient of not getting the proper treatment or could had died, but after medical imaging it become very easy to share the reports from one place to another and get the best treatment [10–12].

1.3 Analysis of Medical Images

In the previous segment we have discussed about medical imaging, machine vision and how much they are useful for treating the disease effectively, efficiently and accusatively. In this section we will talk about the analysis of medical image and particular for headache and for which medical imaging is known as Neuroimaging [13].

So for Neuroimaging the technologies which are used are Computed tomography also known as CT and Magnetic resonance imaging which is generally known as MRI.

We have already discussed about TTH and migraine that they have frequent and severe attacks and now we have two technologies through which we could analysis their medical images [14, 15].

1.3.1 Computed Tomography (CT)

This process provides the help to diagnostic or the therapist to look at the internal changes of the brain or the positioning of muscles when the attack occurs. There are various 3D high resolution images which makes it easy for the doctor to look at the issue clearly, the doctor has the clear view of the person's anatomy. But there are some risks in CT scan such as radiation exposure which could lead to cancer. As the radiation which is emitted from the CT scan emits the effective amount of 3–5 mSv to take the CT of a normal head [16, 17].

1.3.2 Magnetic Resonance Imaging (MRI)

This medical image process helps the diagnosis but it is slightly differ from CT as there is no emission of radiations in it. It uses a strong magnetic field for the imaging of human anatomy. If we compare MRI from CT we come to know that it provide better result and there will be increase in the contrast in between the soft tissues which are present in the body. And this is because of the high magnetic field. The environment which is magnetic in nature because of this special care is provided to the patient and is it recommended not to wear anything which could be attracted by the magnetic field as if it does then it harm the machine as well as the patient.

1.4 Application of Machine Vision

We have already seen it previous analysis that in what ways machine vision is helpful to us for treating complex disease and surgeries, but in this section we are going to see that how machine vision could be helpful in treating stress. Yes it could look weird that how a machine could treat something which is not physical. In today's time this is possible with the help of Artificial Intelligence [18, 19].

1.4.1 Use of AI in Treating Stress

Stress and anxiety has become very common now a days. Every other person is suffering with it. But everyone takes it very lightly because of the assumption that it will be fix accordingly with time but they didn't not have clue that it if it is not treated at the right time then it could lead to various serious disease. And because of the busy schedule no one have proper time to visit psychiatrist on regular because so to the problem of this solution scientists and various doctors took help of Artificial Intelligence [20, 21].

The patient just need counselling from someone who could understand him. The new tools that are equipped with AI have capabilities to communicate with the patient directly either with the help of chatbots or with the help of echo tools or we could say virtual doctors.

1.4.2 Use of AI with Biofeedback

As we have already discusses that how biofeedback is used in treating tension type headache. It is a great method through the doctors could able to make patients understand about the complex problems and structure of their brain. And after a biofeedback therapy, patients should communicate with AI technologies that are helpful in treating stress. This makes patients to communicate more and more and helps in reducing the stress level and if this works efficiently, then there will be no need to gulp many medicines and having expensive surgeries [22, 23].

1.5 BD and IoT Applications

As we are growing with technologies we experience that the different technologies are making out tasks easier and cheaper. And some of those technologies are reducing the overall cost of treatments and surgeries. These technologies include the devices which can have continuously monitoring system for patients, system that provides automatic therapies and for that patient need not go to the doctor, he just have to use these technologies and will be prescribed according to his problem. These technologies are faster as they have internet access and could have the real time status of the health of the person. For this IoT and BD are using fast [24, 25].

IoT (Internet of Things) is the network of physically connected devices and various other devices which are embedded with different software, GPS connectivity, which enables to record the real time data. Its impact on medical science will be huge and will be very important. The speed by which this technology is growing we could make a estimate that large number of tasks of medical science will be done by IoT which will create a billion market and reduce the effective cost of various treatments, and saving more lives.

Let's take an example and understand that how this will change the procedure in medical science. Suppose a person is a diabetic patient, he have his id card, and when it is scanned, it will link to the cloud server which is storing his previous details such as lab records, medical and prescription history [26–28].

It might look really simple and easy but this task is little bit difficult and a game changer in the field of medicine. After implementing of this in a very less time all the records which are hundreds and thousand years old will get digitized, and the information will be easily sharable. The challenges that come across implementing these technologies is communication, as there are so many devices which are enabled with sensors which will record data, and sometimes they will talk to server in some language. While the each manufacturer has their own protocol due to which sensors with different manufacturer can't speak to each other much [29].

The environment thus created which is coupled with private data, there might be possibility that the data might get steal and IoT could get failed. The data which will be inbuilt into the sensors will be done by the drivers. And that's how BD comes into the picture in medical science. The BD works mainly on 3V's

- **Volume**—Data in large volume
- **Variety**—The variation in data that had been recorded
- **Velocity**—The speed by which the data will get analyzed.

As defined the medical science is becoming the emerging user of BD. Some most difficult high dimensional documents data sets include X-rays, MRIs, some wave analysis for example EEG and ECG. The thing that data analysis should be given to its users is the constant updates based on the knowledge that had been gained, while keeping all the data at one place [30, 31].

Therefore these two technologies IoT and BD will change the medical science in coming years. And it will reduce the treatment duration and effective cost and the medical science will be the combination of business methods and real time decision.

1.6 Manage Lifestyle: Solution to the Issue Raised

1.6.1 Control the Stress Level

The only and best possible solution to reduce tension type headache is by not thinking about the problem much and for this you must have to plan things accordingly from previous itself, start organizing yourself in proper manner, take proper time to take mental as well as physical rest and if you ever find yourself stuck in any difficult or stressful situation always to keep back from it as soon as possible [32, 33].

1.6.2 Go Hot or Cold

Another effective way to reduce tension type headache is by relaxing the muscles and for this try to apply hot or cold pack (any one which you would prefer), this process sore the muscles and reduce extent of tension type headache.

If you want to apply hot pack then you can use a heating pas which is set on a hot table and start compressing on the affected area by slowly padding. And if you want to apply cold one then for that wrap the ice in a cloth and apply it on the effective area and try to avoid rashes [34, 35].

1.6.3 Perfect Your Posture

If you feel lazy to perform the methods to reduce the level of tension type headache then this the simplest and easiest way to reduce it. And you could reduce it just by correcting your posture a little bit. As correct posture doesn't allow muscles to get tense. What you should keep in mind when standing is to keep your shoulders at the height of your head. Pull the abdomen and the butter. When sitting, make sure your thighs are parallel to the ground and your head is not bent forward [36].

1.7 Tension Type Headache

TTH is the most basic form of headache occurring in about three quarters of the world. Tension type headache can usually last from 31 min to 8 days. The pain is generally mild and moderate in most of the times and sometimes it is worst. There are no diagnosis tests to conform Tension type headache (TTH). Galvanic skin resistance (GSR) therapies are used to treat these headaches and are found very effective [37].

At times, it can be very severe that it becomes difficult to distinguish between tension type headache and migraine attack. The most commonly used preventive measures include medications such as allopathic treatments and non-medication treatments such as biofeedback therapies. The term "tension type headache" (TTH) has been declared by the International Classification Headache Diagnosis I (ICHD I). The terms "tension" and "type" represents its different meanings and reflect that some sort of mental or muscular tension could cause a impact.

However, studies at large extent at this topic shows that there somehow a doubt about its neurobiological nature. It is one of the common form for headache and in normal cases people took it lightly and they don't consult to the doctor and they treat themselves with various drugs in most cases people got cured but sometime the drugs have chronic impact on the health which could be severe [38, 39].

1.7.1 Causes

The main reasons due to which TTH happens, could be the contraction between the neck and scalp. The victims of this in majorities are females. In many researches, it is also concluded that if you are sitting in only one position for a longer period of time and if there is strain in neck and head then TTH could occur. The people who suffer from this problem are mainly from IT industries because of working on the computer for longer period of time.

Other circumstances may include consumption of heavy caffeine, alcohol, taking any amount of physical or mental pressure, chronic fever, and intake of cigarettes [40, 41].

1.7.2 Symptoms

We have discussed about the causes that how TTH could may occur, but in this section we will discuss that what are the possible ways or signs which confirms that one is suffering with TTH.

- The first sign could be if you are feeling constantly dull and in pain.
- If you wrap a tight band around your head.
- If you are feeling pain all over the head not only on particular points.

In these conditions the problem one may feel could be as difficulty in sleeping. There is no problem such as nausea and vomiting by TTH [42, 43].

1.7.3 Tests and Treatments

If the pain is mild or moderate and after applying the home remedies the pain gets settle down then there is no need to see any doctor. The treatment of this disease could be easily done by oneself by keeping track of the amount and duration of pain on a daily basis and if you see serious issue in those record then reach to the doctor immediately. Its treatment also includes of some drugs such as aspirin, acetaminophen to prevent pain. In this disease narcotic pain relievers are not prescribed [44].

Just make sure to have a proper treatment because if you don't do it properly then there are chances that you will be again affected by it very soon [43].

1.7.4 Alternative Medicine

There are some non-conventional therapies are listed below:

Acupuncture—This is one the easiest way to get rid of tension type headache. It provides relief in the chronic headache. The practitioners of thus treatment treat the patient by using very thin needles which produces some amount of pain and discomfort but relief the headache for very long period.

Massage—It helps to reduce the stress level and relieve tension. It is mainly used to give relaxation to tightened and sensitive portion in the back part of head, some part of neck and shoulder [45, 46].

1.7.5 Deep Breathing, Biofeedback and Behavior Therapies

Various types of relaxation therapies are helpful in dealing with tension headaches, which include deep breath and biofeedback.

Coping and Support

It makes you very anxious and depressed and affects relationships, productivity and quality of life. Living with extreme pain is very difficult [3].

Here are some suggestions are listed below to avoid all this.

Talk to a Counselor or Therapist

Talk therapy helps you in a significant way to cope up with the effect of extreme pain [47, 48].

Join a Support Group

Support groups are a good option for tackling this issue. Group members usually know the latest treatments. And you will feel better after you get a good company [49–51].

2 Literature Survey

2.1 TTH and Stress

There are various researches are carried out on Heart Diseases, stress and illness. Budzynski and his colleagues were the initials to publish demonstration for biofeedback for tension type headache. They design EMG model for biofeedback and decides some protocol that would be for TTH. The main mode to treat people who are affected from TTH is Pharmacotherapy, it is very effective in decreasing the extent and duration of TTH. But it also have drawback as it is all over used as antidepressant medicine and those have risk of having adverse effect.

Other than that it has potential risk for analgesic medication because of the overuse. In 2008 there was an article which state the efficiency of biofeedback in treating TTH. Galvanic skin resistance (GSR) therapies are used to treat these headaches and are found very effective. At times, it can be very severe that it becomes difficult to differentiate among tension type headache and migraine attack.

The most commonly used preventive measures include medications such as allopathic treatments and non-medication treatments such as biofeedback therapies. The term "tension type headache" (TTH) has been declared through the International Classification Headache Diagnosis I (ICHD I). The terms "tension" and "type" represents its different meanings and reflect that there are sort of mental or physical tension could cause an impact. However, studies at large extent at this topic shows that there somehow a doubt about its neurobiological nature.

2.2 *Heart Disease Prediction*

In the research manuscript title as Intelligent Heart Disease Risk Predictor System Using Data Mining Techniques, the authors Sellappan Palaniappan and his colleague [52]. They proved the fact that stakeholder data generated by the healthcare industry is not used for better decision making. Finding masked patterns, ideas, and correlations is often not found. In this situation, new data mining techniques are useful. Research using various data mining techniques such as Naive Bayes technique, neural network, decision tree, etc. created an example of a heart disease risk prediction system.

Another important experimental analysis conducted by the author is described in Marjia Sultana and her team members [53] and a research paper on the analysis of data mining techniques for predicting the risk of heart disease. I found that the main cause of death was the heart. Diseases around the world are difficult to predict because they require high expertise and accuracy. In this article, we will discuss some of the features of an individual's medical history entry to predict heart disease based on several data mining techniques.

They have used Weka software to predict heart disease risk using W48, J48, SMO, Bayesnet, multilayer perceptron, KStar software. Use the collected and standard data to compare and measure the performance of each algorithm by adding prediction accuracy results, AUC values, and ROC curves. Based on Bayes Net and SMO performance techniques, J48 and KStar algorithms outperform multi-layer perceptrons.

The Authors: Soodeh Nikan team member in a study of a ML program to predict the risk of CAD, they presented the fact that it is the leading cause of death in the world of coronary artery disease (CAD). In this paper, they proposed an algorithm that uses ML algorithms to predict the risk of coronary atherosclerosis. REMI techniques are being investigated to find and estimate missing values in the atherosclerosis database.

The proposed algorithm was calculated using UCI and STULONG databases. Two performance classification models assess predictors of heart disease risk and compare them with previous studies. The results showed that the improvement in the rate of risk prediction accuracy of the proposed method in relation to other tasks in the experiment was investigated. The effect of residual values on the prediction algorithm is also evaluated and compared to other traditional methods, and the proposed REMI approach is more effective.

3 Product Perspective

Here we shall define the user and admin interfaces. Also we shall explain the software, hardware, memory and communication interfaces of our product (web app).

3.1 System Interfaces

3.1.1 User Interface

- He will enter his details upon which the risk for heart disease will be predicted.
- User can sign up to make an account where he could know how many times he has predicted the risk and what were the results.

3.1.2 Admin Interface

- He can add new entries into existing dataset.
- He can access the details of the users and can access the database.

3.1.3 Interfaces

- The default page is the login page where users can sign in, sign in, sign in, and sign in.
- After logging in, each user is redirected to a prediction page where they can enter details and predict the outcome. The user can now be redirected to the profile page or logged in again.
- The profile page displays the user profile and profile picture, as well as past predictions (if any) already made by the user.

3.1.4 Hardware Interfaces

The web application has no extra hardware requirements.

3.1.5 Software Interfaces

OS: Ubuntu 17.04, Windows 10.
Web browser: Mozilla Firefox, Google Chrome.
MySQL database (v5.7.19): Connect to the background database.
Run the Anaconda-ML code.
For Django and Python-Web applications.

3.1.6 Communications Interfaces

A web browser with good internet connection.

The minimum internet speed for uploading, processing and details is 1 Mbps.

3.1.7 Memory Constraints

At least 4 GB of RAM to run the project.

3.2 Product Functions

- This product performs user login, login, and administrator login.
- Administrators have the task of importing new data into and accessing records.
- Users can enter values for various parameters based on their risk factors is calculated.

3.2.1 User Characteristics

- Users who want to predict risk need to be aware of the features they need Incorrect attribute value prediction pages can mislead users.
- The value of each attribute must be in the correct format and scope.

3.2.2 Constraints

- The dataset used is small.
- The database used is a regular database that can store a limited amount of data.
- This product makes it clear that there is no "name" input field, so only user data is used for forecasting.
- Therefore, privacy is protected. However, since the program uses open source libraries, skeptical developers can check the source code directly.
- The system must be reliable.
- If the request is not processed, an error message is displayed.
- The web page will load in seconds.
- The algorithm should be chosen with the highest accuracy of the web application.

3.3 UML Diagrams

We represent here the Use Case, Sequence, Architecture, Dataflow (DFD Level 0, 1 and 2), Activity and ER-diagrams to represent product functionality for proposed web app.

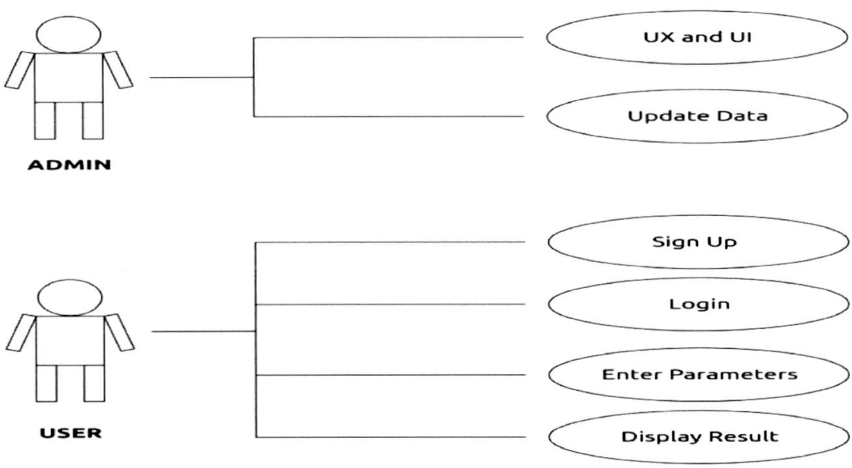

Fig. 1 Use case diagram of heart disease predictor

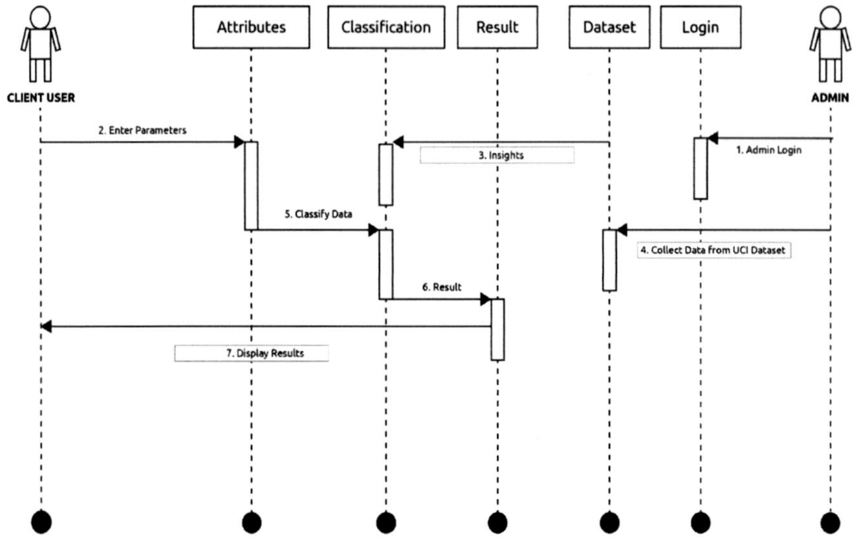

Fig. 2 Sequence diagram of heart disease predictor

3.3.1 Use Cases

Use Case Diagram
Figure 1 is a diagram of the project. There are two types of users. One is an administrator accessing the database, and the other is a user or doctor who can log in or register by entering details and predicting risk.

3.3.2 Sequence Diagram

Figure 2 the sequence diagram shows how to perform the activity and how to respond to the user. This shows how the data is sent to the model and how the data is sorted.

3.3.3 Architecture Diagram

Figure 3 the system architecture shows that there is a model trained using experimental data, and that the trained model is used to classify the data into a set of values provided by the user.

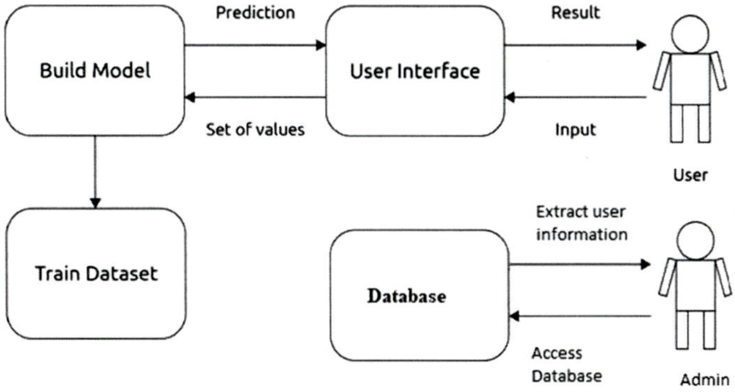

Fig. 3 Architecture diagram of heart disease predictor

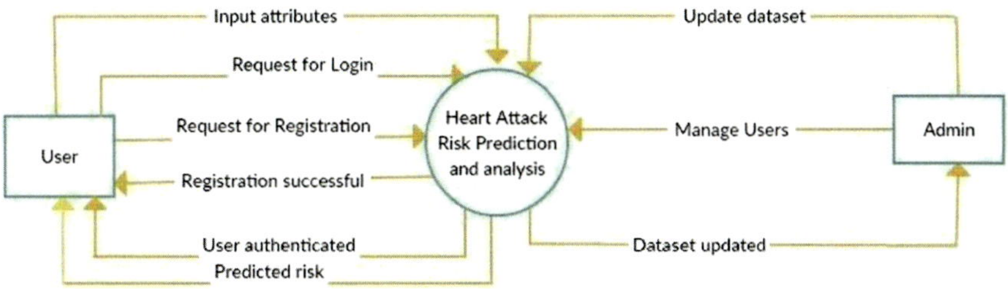

Fig. 4 Data flow diagram (level 0)

3.3.4 Data Flow Diagram

DFD Level 0

Figure 4 represents how the data flows in our system. When a user sign in or logs in the user is returned by the appropriate response. Any admin can manage a user data, update the database or delete a wrong entry.

DFD Level 1

Figure 5 a level 1 data flow diagram is shown. Shows details of the data flow and various entity and system functions.

DFD Level 2

Figure 6 the level 2 data flow diagram shows how requests and responses are sent from the system to the user and vice versa.

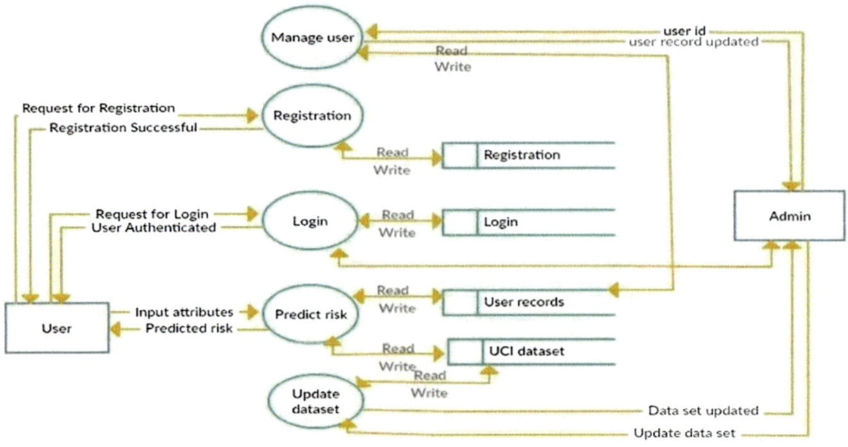

Fig. 5 Data flow diagram (level 1)

3.3.5 Activity Diagram

Figure 7 displays an activity diagram. Describes the dynamic aspects of the system. Represents a flow from one activity to another.

3.3.6 ER Diagram

Figure 8 represents how our database is designed from different entities, the attributes of the entities and how different entities are related to each other.

4 Experimental Setup and Product Configuration

Firstly we focus on Software and Hardware Requirements.

4.1 Hardware Requirements

- Processor: core i5
- RAM: >4 GB.

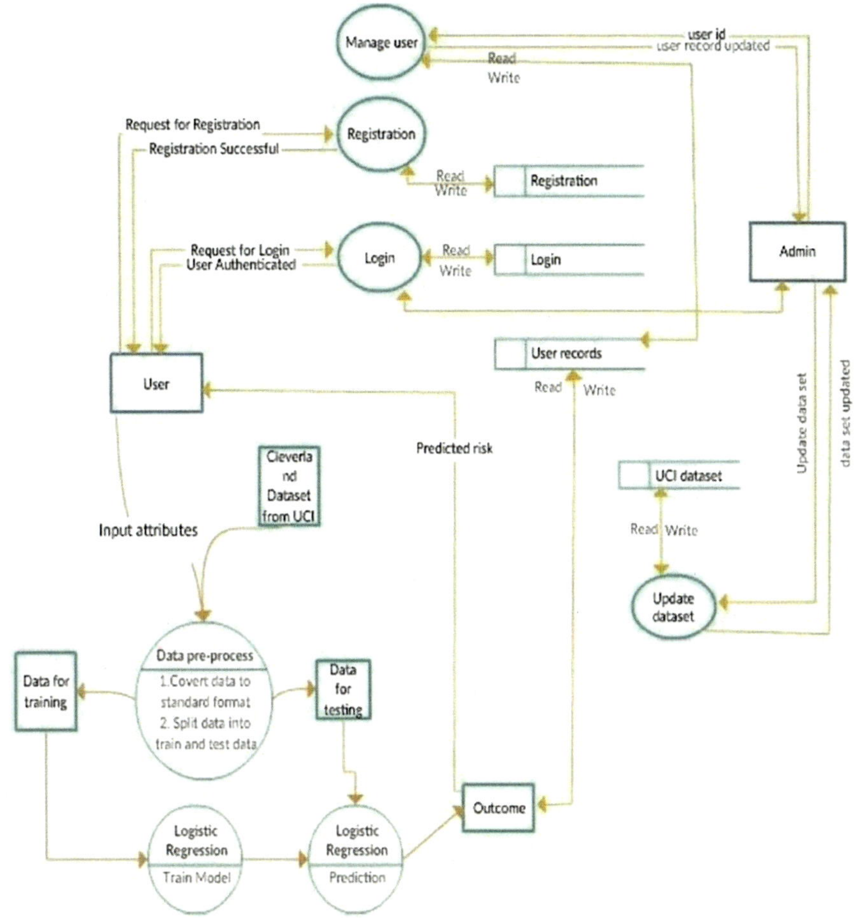

Fig. 6 Data flow diagram (level 2)

4.2 Software Requirements

- Operating System: Windows 7 or above.
- Web Browser: Google Chrome, Mozilla Firefox.
- MySQL Database (v5.7.19): For connectivity with background databases.
- Anaconda: Software to run machine learning code.

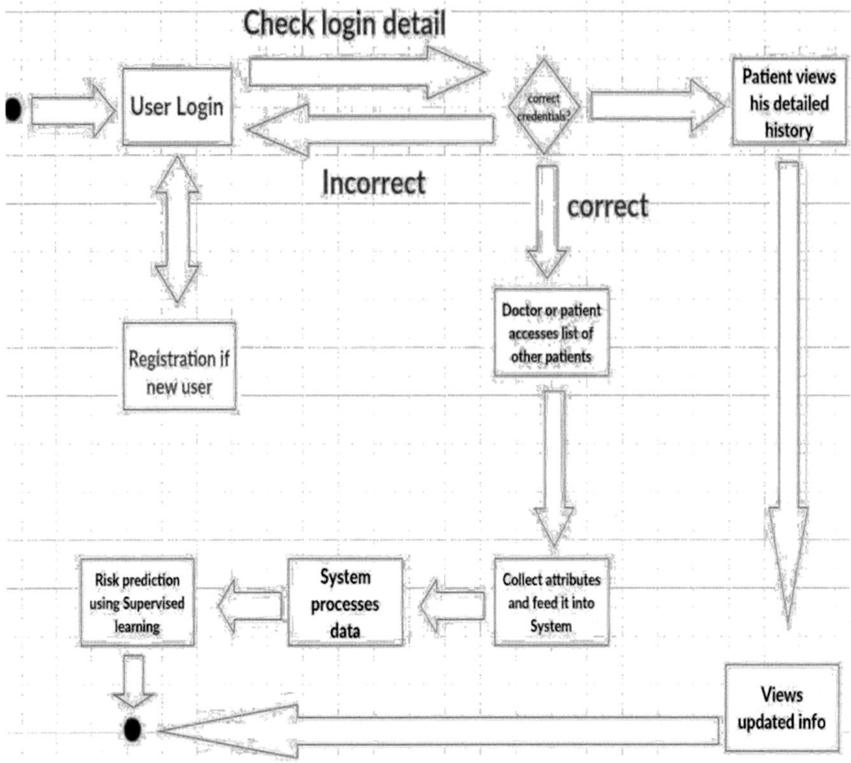

Fig. 7 Activity diagram of heart disease predictor

- Django: Web browser.
- User Interface: Adobe Illustrator.

4.3 Implementation Activities

This whole experimental research work and product development and implementation can be broadly divided into two parts: Implementation of Prediction engine and implementation of web application.

4.3.1 Implementation of Prediction Engine

The prediction engine is implemented using various Python libraries such as scikit-learn, panda, numpy. Use various characteristics, such as cholesterol, age, and blood sugar, to predict whether you are at risk for heart disease.

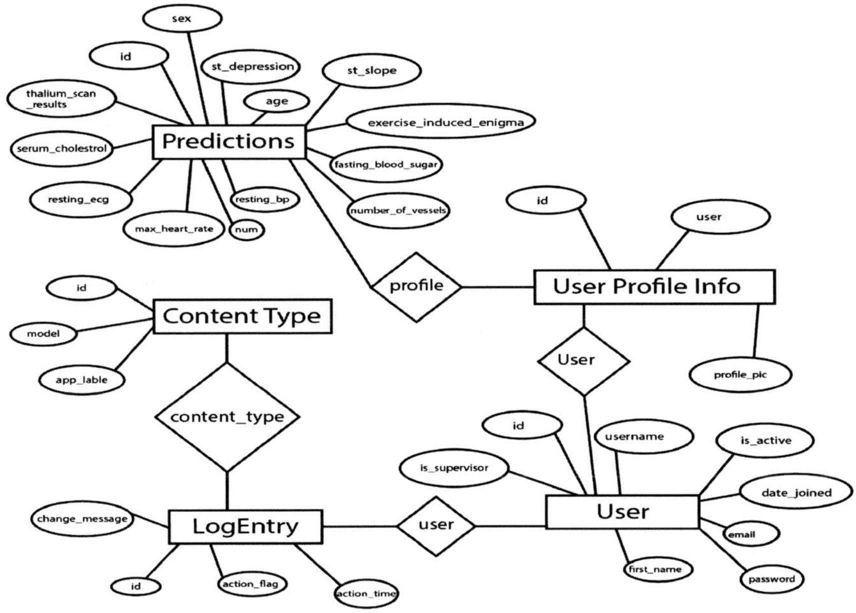

Fig. 8 ER diagram of heart disease predictor

This section describes the dataset used in forecast engine development, followed by the implementation of the forecast engine iteration.

The "Cleveland Heart Disease" Dataset

The UCI ML repository is used to create the Cleveland Heart Disease Database. The UCI database contains 351 data sets managed by the University of California, Irvine. This allows users to browse the dataset, download the dataset, and assign data to the data (as per Fig. 9).

The Cleveland data set contains 303 records. The dataset contains 76 attributes, but this project considers only a subset of 14 attributes. Table 1 shows the selected characteristics.

4.4 Development

The prediction engine is developed in four stages. The libraries used to run the prediction engine are:

- numpy: To manipulate an array
- Panda: To manipulate CSV files and data frames

Fig. 9 UCI data repository—home page

- matplotlib: To create a graph using pyplot, define the parameters using rcParams and color them with cm.rainbow
- Warning: Ignores warnings that appear in the notebook due to past/future depreciation
- train_test_split: To split a data set into training data and test data
- StandardScaler: Scales all functions to improve machine learning model and dataset compatibility.

4.4.1 First Increment

The first iteration was done using the "Jupyter notebook". First goal:

- To visualize a dataset
- To find relationships between features
- To find prediction accuracy with different algorithms
- To create an overall prediction workflow.

Figure 10 shows how the different attributes in the UCI dataset are distributed.

4.4.2 Second Increment

This plugin has been removed by the Jupyter notebook. The results of this increase are as follows:

Table 1 Cleveland dataset database description

S. No.	Field	Description	Range and values
1	Age	Age of the patient	0–100 in years
2	Sex	Gender of the patient	0–1 (1: male, 0: female)
3	Chest pain	Type of chest pain	1–4 (1: typical angina; 2: atypical angina; 3: non-anginal; 4: asymptotic)
4	Resting blood pressure	Blood pressure during rest	mm Hg
5	Cholesterol	Serum cholesterol	mg/dl
6	Fasting blood sugar	Blood sugar content before food intake if >120 mg/dl	0–1 (0: false; 1: true)
7	ECG	Resting electrocardiographic results	0–1 (0: normal; 1: having ST-T wave)
8	Max heart rate	Maximum heart beat rate	Beats/min
9	Exercise induced angina	Has pain been induced by exercise	0–1 (0: no; 1: yes)
10	Old peak	ST depression induced by exercise relative to rest	0–4
11	Slope of peak exercise	Slope of the peak exercise ST segment	1–3 (1: up sloping; 2: flat; 3: down sloping)
12	Ca	Number of vessels colored by fluoroscopy	0–3
13	Thal	Defect type	3: normal 6: fixed defect 7: reversible defect
14	Num	Diagnostics of heart disease	(0: <50% narrowing; 1: >50% narrowing)

- The code was extracted from the notebook, the visual was deleted, the prediction code was extracted using a different algorithm, and a function was created.
- A function is created to save and download the trained model.
- Performance was developed to represent different types of scores for trained models.

4.4.3 Third Increment

The third increase introduces a web API that predicts heart disease. In this section, you will use Django for the web framework and the form for form validation.

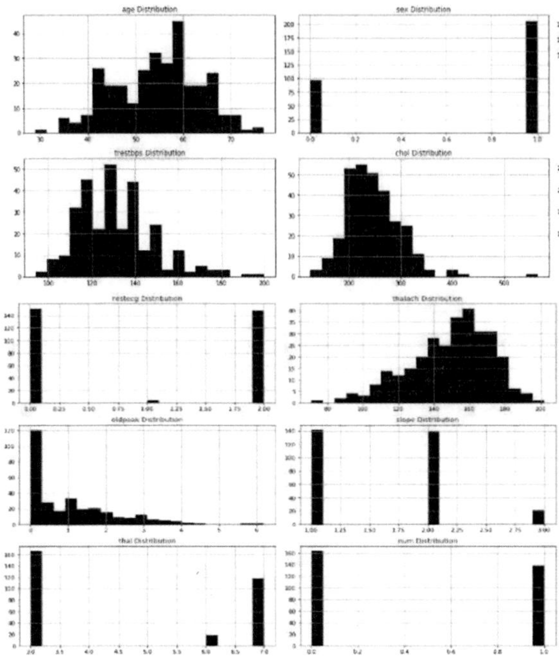

Fig. 10 Histograms of heart disease dataset

5 Results, Interpretation and Discussion

5.1 Implementation Snapshots of Interfaces

The various snapshots of our system are as follows (as per Figs. 11, 12, 13, 14 and 15):

- Login Page
- Signup Page
- Predict Heart Disease Page
- Filled Prediction page
- Prediction result Page
- About Us Page
- Admin Login
- Admin Dashboard.

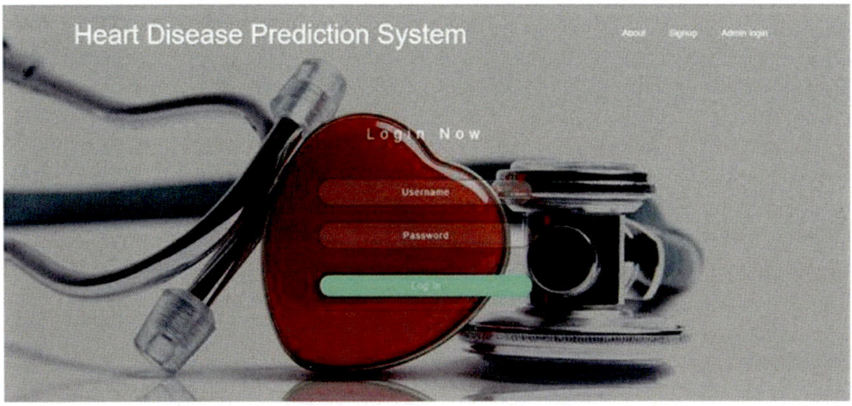

Fig. 11 Login page of web app for heart disease predictor

Fig. 12 Sign up page of web app for heart disease predictor

5.2 Graphical Analysis and Results Discussion

Figures 16, 17, 18 and 19 show the ROC curves for the different algorithms we used.

The area under the ROC curve is a measure of how to distinguish parameters between the two diagnostic groups. The figure shows that SVM performance is better than other algorithms because the SVM has the highest AUC, followed by a logistic regression algorithm.

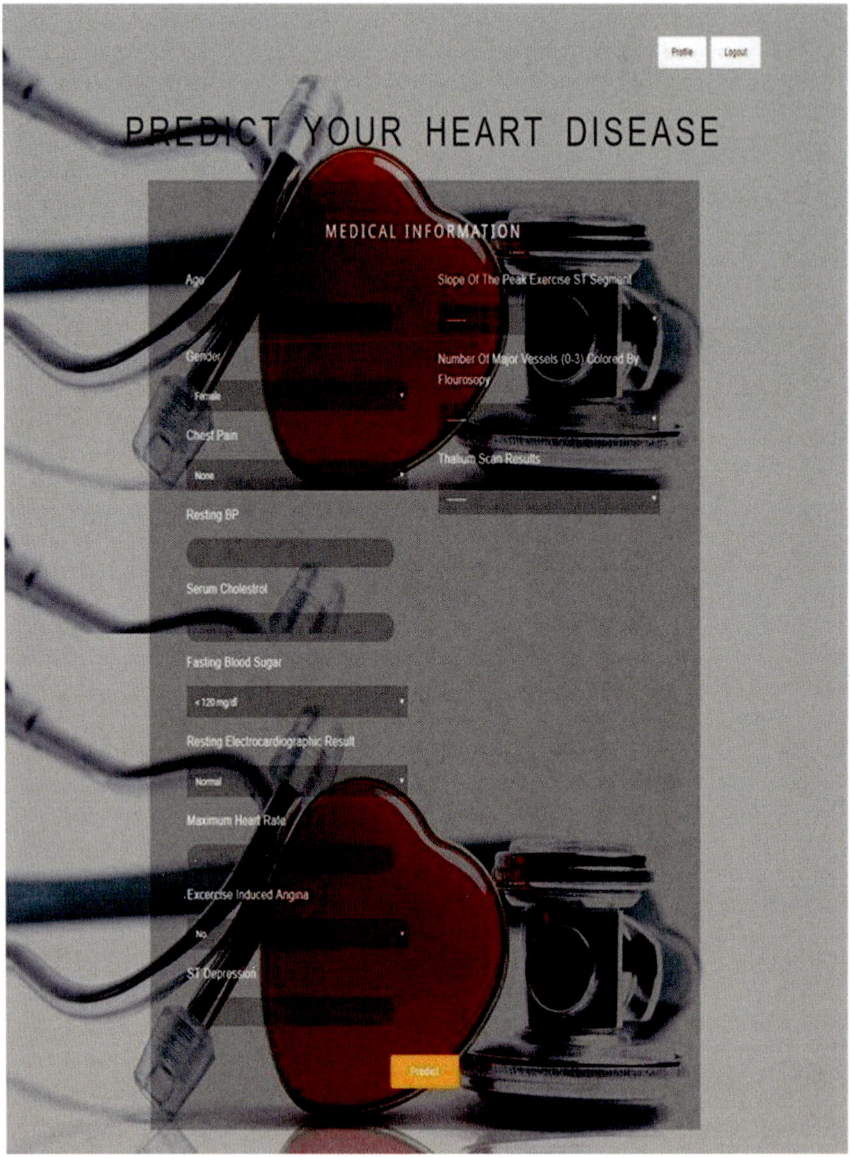

Fig. 13 Predict heart disease page

5.3 Performance Evaluation

Table 2 shows the evaluation of ML algorithms and a comparison of various algorithms based on their accuracy.

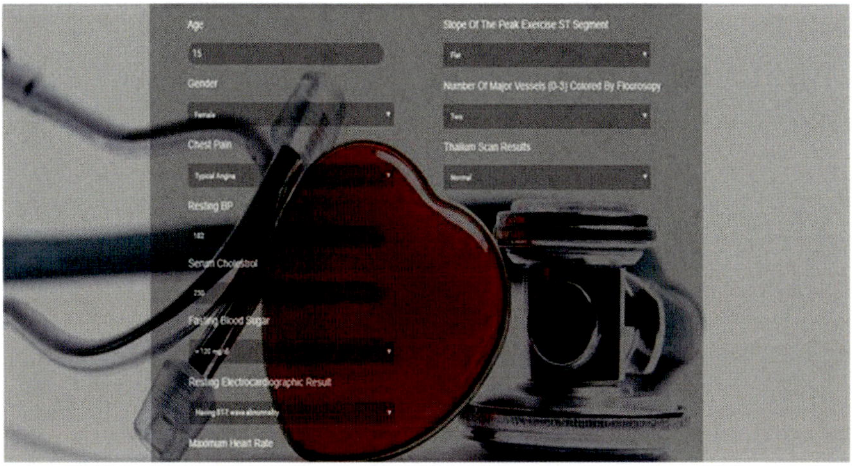

Fig. 14 Filled prediction page of web app for heart disease predictor

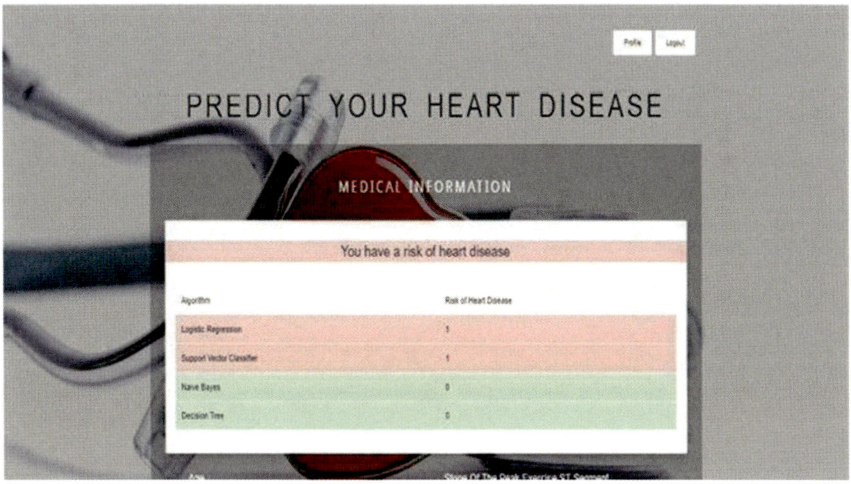

Fig. 15 Prediction results page of web app for heart disease predictor

6 Limitations of System

6.1 Limited Dataset

Using ML requires a large amount of data. There is little data on heart disease. Also, the number of samples without illness is higher than the number of samples actually ill.

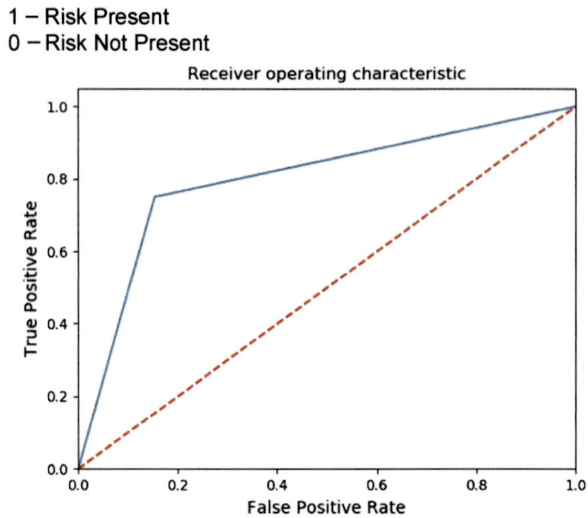

Fig. 16 Decision tree ROC curve

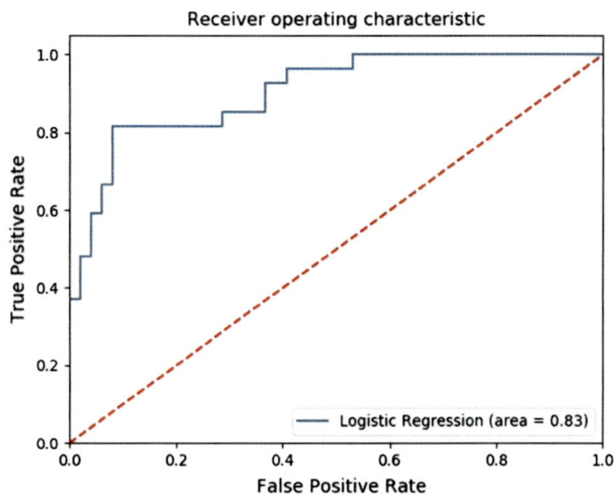

Fig. 17 Logistic regression ROC curve

6.2 Lack of Suggestions Based on Risk

This Research work and experimental simulation helps the user to detect whether there is a risk of heart disease or not, but it does not give any suggestions to the user as to what he/she should do in case of a positive result of heart risk [54, 55].

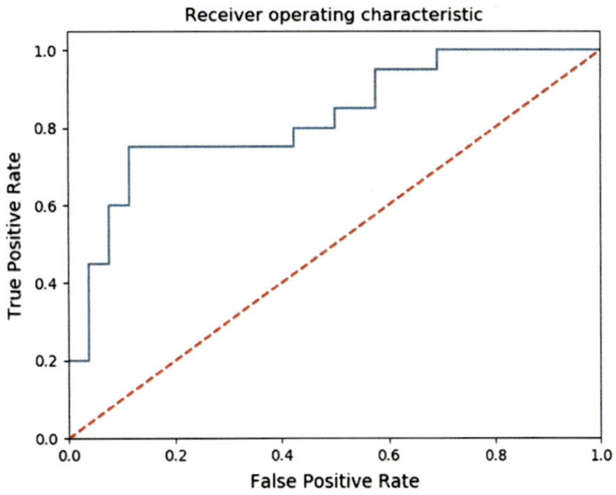

Fig. 18 Naïve Bayes ROC curve

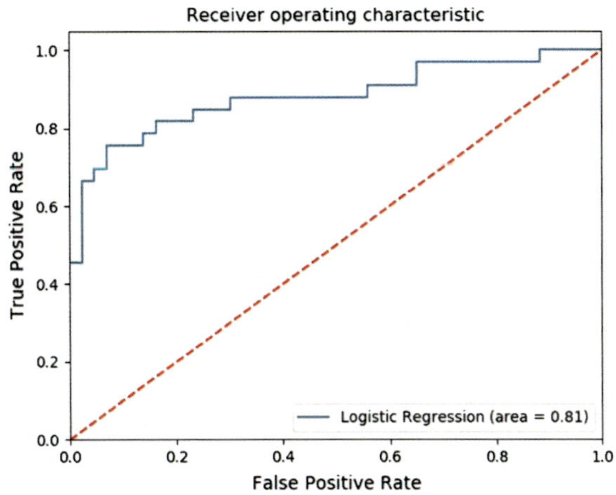

Fig. 19 SVM ROC curve

Table 2 Performance evaluation of ML algorithms

Algorithms	Accuracy (%)
Logistic regression	83.38
Naïve Bayes	80.73
Decision tree	80.54
Support vector classification	87.89

6.3 *Lack of Trust on Machines*

There's skepticism among people in trusting machines to make life-death decisions. A lack of trust in machines, prioritizing individuals over statistics are some of the social issues that might arise from this simulation.

6.4 Assistive Tool

Changing this decision-making process from a doctor to an algorithm can lead to claims of failure due to improper treatment due to the wrong algorithm. In many countries, doctors are responsible for and may be liable for negligence in patient care. Therefore, this project should only be considered as an auxiliary tool and should be fully trusted only after future validation by integrating more practical data [56].

7 Recommendations

The purpose of our research is to predict the risk of heart disease using data mining techniques and ML algorithms. We designed a drug analysis system that can improve drug discourse and reduce costs as well. This can help reduce the risk that doctors will "predict patients" based on certain health parameters.

The main purpose of this study is to help non-specialists make the right decisions about the risk of heart disease in patients. It aims to combine medical decisions with computer-based data to reduce medical errors, improve patient safety and improve overall results [57].

This not only reduces medical costs, but also strengthens the medical perspective. This means having enough knowledge and better data. Large companies can invest heavily in research to address potential activities and risks. This type of work can collect all available data as the basis for reasonable assumptions about the future.

8 Novelty and Advantages of Experimental Work and Product

8.1 System Faster Diagnosis

Doctors rely on general knowledge for treatment. If there is a lack of public knowledge, we will summarize the study after studying several cases. However, this process is time consuming, but ML can make a diagnosis faster because it can diagnose heart disease faster.

8.2 Reduced Medical Errors and Misdiagnoses

Any medical professionals which are working in the field of heart disease know that they can predict the chance of heart attack up to 67% accuracy. Nowadays for more accurate prediction of heart disease doctors need a support system. By using ML for medical diagnosis, this simulation has the potential to greatly reduce the number of medical errors and misdiagnoses [58, 59].

8.3 Easy to Use

This project is simple and easy to use; as here the user must provide the medical details and based on the features extracted the algorithm will then detect the risk of heart disease. As here algorithm does the task hence a well-trained model is less bound to make errors in predicting the heart disease, in short accuracy is improved and thereby it also saves time [60].

8.4 No Human Intervention Required

To detect the heart disease, the user must provide medical details such as age, cholesterol, etc. and here the algorithm will provide the results based on the features extracted and hence here chances of error been made are very minimum since there is no human intervention and it also saves lot of time for the patients or doctors and they can further proceed for treatments or other procedures must faster [61, 62].

9 Future Scope and Possible Applications

Once a particular form of heart disease is diagnosed, the patient has several treatment options. Data extraction is very useful. This is a good data set. Therefore, different models need to be combined with more complex models to improve the accuracy of predicting the risk of heart disease. This system is very intelligent as more data is added to the database [63, 64].

Many improvements can be made to improve scalability and accuracy. Due to time constraints, the following tasks cannot currently be performed, but should be performed in the future. Using different interpretation methods, different types of decision trees increase information and profitability, meaning multiple classification voting methods. I want to discover various rules such as related laws, logistic regression, and clustering algorithms.

10 Conclusions

- Admin Individual who is responsible for carrying the overall process of maintaining and smooth functioning of the system.
- Patient Individual who will be tested for cardiac arrest and feeds updated information into the system.
- Doctor Individual who looks over the progress of individual patients, and suggest appropriate action over the case.
- Absolute risk is the risk of illness over a period of time. We all have an absolute risk of developing various diseases such as heart disease, cancer, and stroke.
- Therefore, feedback, including enhancement and correction feedback, is an important element of clinical education. This study focuses on the GP, the population that needs to be studied, especially because it provides a lot of feedback on medical education.
- Pulse rate is measured as arterial pulse rate per unit time. Usually 60–80 min per minute for adults.
- Heart rate, number of ventricular contractions per unit time (usually per minute).

Appendix

Implementation—Web Application
Included here is the implementation of web application through feature driven development. Each activity of feature driven development is discussed with artifacts produced during that activity.

The web application has been developed using Django framework of python. It utilizes the MVT (Model View Template) architecture.

Development of Overall Model
During this activity of feature driven development, software requirement specification document was prepared for capturing the requirements. ER Diagram and requirement specification document was designed. After that, for the completion of this activity, a domain object model was prepared along with the overall application architecture.

Functional Specifications
Included in this section are the functional/non-functional requirements of the systems along with the use-cases and wireframes.

Functional Requirements

- The system allows users to predict heart disease.

- The system allows users to create an account and login.
- The system allows the users to update their profile and password.
- The system provides login for admin.
- The system should allow administrator to monitor and remove inappropriate datasets and code.

Non-functional Requirements

- The website should be responsive and have consistent across different screen sizes and resolutions.
- The website should provide user information about different values used during the prediction.

Architecture

The major components of Django project architecture are models, views and templates along with urls.py, settings.py, admin.py and forms.py.

Our Simulation architecture has been described as follows:

Figure 20 shows the architecture of our project and how different files are distributed in different directories.

Models

A model is a definitive data source that includes the underlying context and behavior of the data. A model is usually a table in a database. Each part of the table in the database is a model property. Django provides a set of automatic programming interfaces (APIs) for database applications for the convenience of the user.

Fig. 20 Simulation of heart disease predictor architecture

View

The file view is short. This file contains a Python function that receives a web request and returns a web response. The answer is an XML document or HTML content or a "404 error". The function logic of the function is not optional until it returns the desired response. To link a view function to a specific URL, you must use a structure called URLconf that maps the URL to a display function.

Template

Django templates are simple text files that can generate text-based templates such as HTML and XML. The template contains tags and variables. When evaluating the pattern, the variable is replaced with the result. Pattern logic is controlled by tags. You can also change variables using filters. For example, a small filter can convert a variable from uppercase to lowercase.

Figures 21 and 22 show how website and Analysis works. It shows how a request is sent and response is sent back.

The working of a template is as follows. A request for a resource is made by any user from the template. Than Django works as a controller and check the availability of the resource in the URL. If the URL matches with any view than Django returns it to the template as a response.

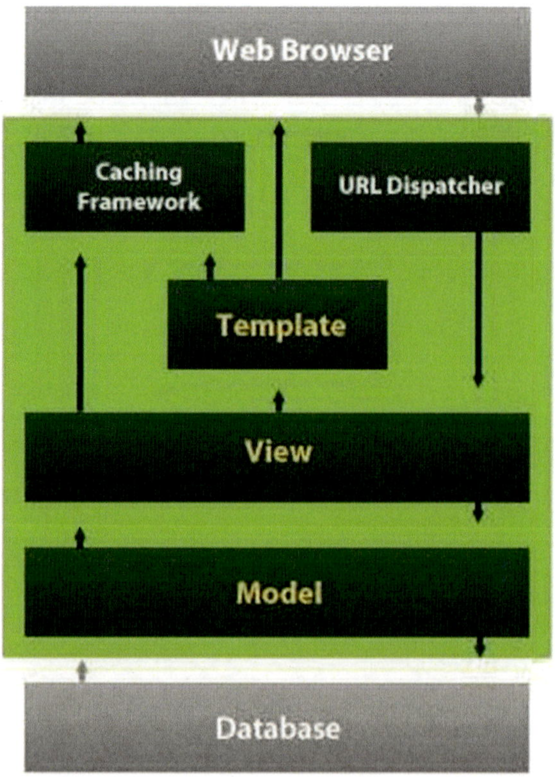

Fig. 21 Experimental working of heart disease predictor

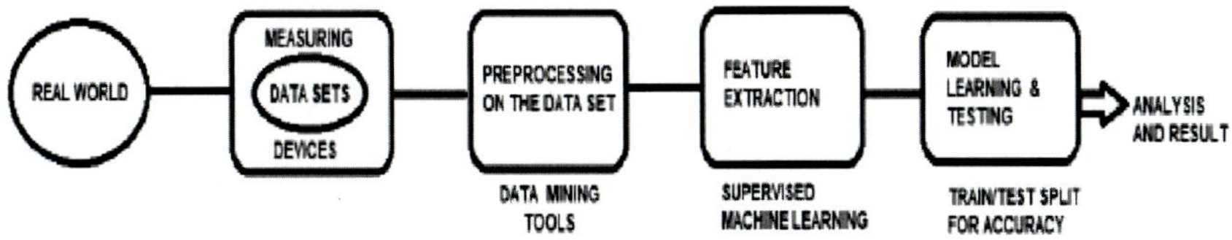

Fig. 22 The flow chart of prediction system

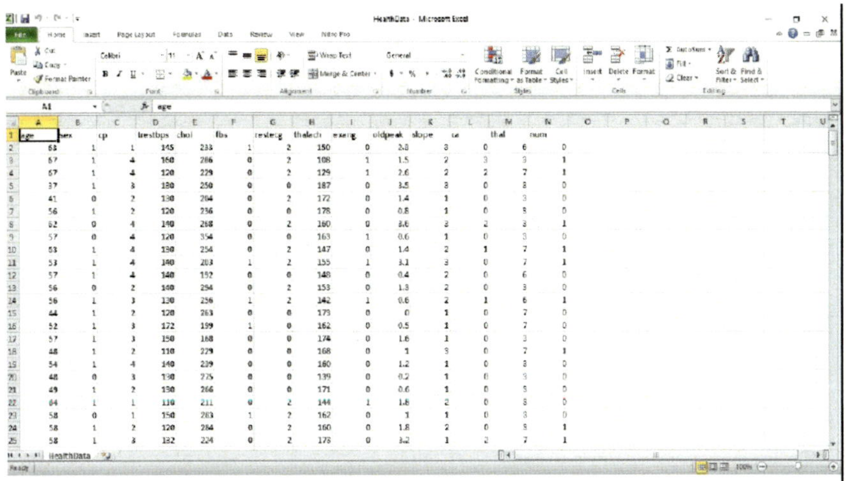

age	sex	cp	trestbps	chol	fbs	restecg	thalach	exang	oldpeak	slope	ca	thal	num
63	1	1	145	233	1	2	150	0	2.3	3	0	6	0
67	1	4	160	286	0	2	108	1	1.5	2	3	3	1
67	1	4	120	229	0	2	129	1	2.6	2	2	7	1
37	1	3	130	250	0	0	187	0	3.5	3	0	3	0
41	0	2	130	204	0	2	172	0	1.4	1	0	3	0
56	1	2	120	236	0	0	178	0	0.8	1	0	3	0
62	0	4	140	268	0	2	160	0	3.6	3	2	3	1
57	0	4	120	354	0	0	163	1	0.6	1	0	3	0
63	1	4	130	254	0	2	147	0	1.4	2	1	7	1
53	1	4	140	203	1	2	155	1	3.1	3	0	7	1
57	1	4	140	192	0	0	148	0	0.4	2	0	6	0
56	0	2	140	294	0	2	153	0	1.3	2	0	3	0
56	1	3	130	256	1	2	142	1	0.6	2	1	6	1
44	1	2	120	263	0	0	173	0	0	1	0	7	0
52	1	3	172	199	1	0	162	0	0.5	1	0	7	0
57	1	3	150	168	0	0	174	0	1.6	1	0	3	0
48	1	2	110	229	0	0	168	0	1	3	0	7	1
54	1	4	140	239	0	0	160	0	1.2	1	0	3	0
48	0	3	130	275	0	0	139	0	0.2	1	0	3	0
49	1	2	130	266	0	0	171	0	0.6	1	0	3	0
64	1	1	110	211	0	2	144	1	1.8	2	0	3	0
58	0	1	150	283	1	2	162	0	1	1	0	3	0
58	1	2	120	284	0	2	160	0	1.8	2	0	3	1
58	1	3	132	224	0	2	173	0	3.2	1	2	7	1

Fig. 23 Dataset used and analysis by ML for disease prediction

Figure 23 shows the Cleveland data set used and machine learning algorithm applied by python software.

Documentation (User Manual):

For Users

Steps that how the system is executed:

Firstly the user has to connect his phone to internet by the medium of mobile data pack or by the medium of Wi-Fi.

Then the user has to download or clone the project. Steps to run

- create virtual env ex. mkvirtualenv mytest_env (optional)
- pip install -r requirements.txt
- python manage.py migrate
- python manage.py runserver.

Steps to execute Registration and Login

- Then the user must register into the application using his basic personal details like username and email.
- If the user is already registered, then the user can log into the application using his login credentials including username and password.

Prediction of Heart Risk

- Now to predict the risk of heart disease, user can enter the values of various parameters on the basis of which his risk factor will be calculated.
- After entering all the values, click on Predict button.
- The page will be reloaded and the result will be shown according to 4 ML.
- If result is 1, user has risk of heart disease. If result is 0, user does not have a risk of heart disease.
- If two or more models give result as 1, the user has a risk of heart disease.
- The user can also view his profile and previous predicted results by clicking on Profile tab.

References

1. M. Ashina, L.H. Lassen, L. Bendtsen, R. Jensen, J. Olesen, Effect of inhibition of nitric oxide synthase on chronic tension-type headache: a randomized crossover trial. Lancet **353**(9149), 287–289 (1999)
2. S. Chauhan, R. Rastogi, D.K. Chaturvedi, N. Arora, P. Trivedi, Framework for use of machine intelligence on clinical psychology to study the effects of spiritual tools on human behavior and psychic challenges, in *Proceedings of NSC-2017 (National System Conference)*, DEI, Agra, 1–3 December 2017
3. R. Rastogi, D.K. Chaturvedi, S. Satya, N. Arora, V. Yadav, S. Chauhan, P. Sharma, SF-36 scores analysis for EMG and GSR therapy on audio, visual and audio visual modes for chronic TTH, in *Proceedings of the ICCIDA-2018 on 27 and 28th October 2018. CCIS Series* (Springer, Gandhi Institute for Technology, Khordha, Bhubaneswar, Odisha, India, 2018)
4. J.-B. Waldner, Nanoinformatique et intelligence ambiante, *Inventer l'Ordinateur du XXIeme Siècle* (Hermes Science, London, 2007), p. 254. ISBN 978-2-7462-1516-0
5. M. Gulati, R. Rastogi, D.K. Chaturvedi, P. Sharma, V. Yadav, S. Chauhan, M. Gupta, P. Singhal, Statistical resultant analysis of psychosomatic survey on various human personality indicators: statistical survey to map stress and mental health, *Handbook of Research on Learning in the Age of Transhumanism* (IGI Global, Hershey, PA, 2019), pp. 363–383 (chapter 22). https://doi.org/10.4018/978-1-5225-8431-5.ch022. ISSN: 2326-8905|EISSN: 2326-8913
6. G. Bronfort et al., Non-invasive physical treatments for chronic/recurrent headache. Cochrane Database Syst. Rev. (3), CD001878 (2004). https://doi.org/10.1002/14651858.cd001878.pub2. pmid15266458
7. A. Sharma, R. Rastogi, D.K. Chaturvedi, S. Satya, N. Arora, P. Trivedi, A. Singh, A. Singh, Intelligent analysis for personality detection on various indicators by clinical reliable psychological TTH and stress surveys, in *Proceedings of CIPR 2019 at Indian Institute of Engineering Science and Technology*, Shibpur on 19–20th January 2019. AISC Series (Springer, 2019a)
8. Commission of the European Communities, Internet of Things—An Action Plan for Europe, *COM*-278 (18 June 2009)
9. D.K. Chaturvedi, Human rights and consciousness, in *International Seminar on Prominence of Human Rights in the Criminal Justice System (ISPUR 2012)*, Organized Ambedkar Chair, Dept. of Contemporary Social Studies & Law, Dr. B.R. Ambedkar University, Agra, 30–31 March 2012, p. 33
10. D.M. Biondi, Physical treatments for headache: a structured review. Headache **45**(6), 738–746 (2005). https://doi.org/10.1111/j.1526-4610.2005.05141.x.pmid15953306

11. D.K. Chaturvedi, M. Arya, Correlation between human performance and consciousness, in *IEEE-International Conference on Human Computer Interaction*, Saveetha School of Engineering, Saveetha University, Thandalam, Chennai, IN, India, 23–24 August 2013
12. E. Van der Zee, H. Scholten, Spatial dimensions of big data: application of geographical concepts and spatial technology to the Internet of Things. Stud. Comput. Intell. **23**, 156–178 (2014)
13. D.K. Chaturvedi, R. Satsangi, The correlation between student performance and consciousness level, in *International Conference on Advanced Computing and Communication Technologies (ICACCT™-2013)*, Asia Pacific Institute of Information Technology SD India, Panipat (Hariyana), Souvenir, 16 November 2013, pp. 66, 200–203
14. P. Sharma, R. Rastogi, D.K. Chaturvedi, S. Satya, N. Arora, V. Yadav, S. Chauhan, Analytical comparison of efficacy for electromyography and galvanic skin resistance biofeedback on audio-visual mode for chronic TTH on various attributes, in *Proceedings of the ICCIDA-2018 on 27 and 28th October 2018. CCIS Series* (Springer, Gandhi Institute for Technology, Khordha, Bhubaneswar, Odisha, India, 2018)
15. M.E. Lenaerts, Burden of tension-type headache. Curr. Pain Headache Rep. **45**(2), 57–69 (2006)
16. R. Rastogi, D.K. Chaturvedi, N. Arora, P. Trivedi, V. Mishra, Swarm intelligent optimized method of development of noble life in the perspective of Indian scientific philosophy and psychology, in *Proceedings of NSC-2017 (National System Conference)*, DEI, Agra, 1–3 December 2017
17. N. Bessis, C. Dobre (eds.), *Big Data and Internet of Things: A Roadmap for Smart Environments* (Springer, Cham, 2014), pp. 137–168. ISBN 9783319050294
18. A.P. Verhagen, L. Damen, M.Y. Berger, J. Passchier, B.W. Koes, Lack of benefit for prophylactic drugs of tension-type headache in adults: a systematic review. Fam. Pract. **4**(3), 156–178 (2010)
19. D.K. Chaturvedi, Science, religion and spiritual quest, *Edited book on Linkages between Social Service, Agriculture and Theology for the Future of Mankind* (DEI Press, 2004), pp. 15–17
20. P. Vyas, R. Rastogi, D.K. Chaturvedi, S. Satya, N. Arora, P. Singh, Statistical analysis for effect of positive thinking on stress management and creative problem solving for adolescents, in *Proceedings of the 12th INDIACom*, 2018, pp. 245–251. ISSN 0973–7529 and ISBN 978-93-80544-14-4
21. Y. Nestoriuc, W. Rief, A. Martin, Meta-analysis of biofeedback for tension-type headache: efficacy, specificity, and treatment moderators. J. Consult. Clin. Psychol. **76**(3), 379–396 (2008). https://doi.org/10.1037/0022-006x.76.3.379.pmid18540732
22. J.-L. Gassée, Internet of Things: The "Basket of Remotes" Problem (12 January 2014)
23. D.K. Chaturvedi, S. Rajeev, The correlation between student performance and consciousness level. Int. J. Comput. Sci. Commun. Technol. **6**(2), 936–939 (2014). ISSN: 0974-3375
24. R. Want, B.N. Schilit, S. Jenson, *Enabling the Internet of Things* (IEEE Computer Society, IEEE, 2015), pp. 28–35
25. D.K. Chatruvedi, Lajwanti, Correlation between energy distribution profile and level of consciousness. Shiakshk Parisamvad Int. J. Educ. SPIJJE **4**(1), 1–9 (2014). ISSN: 2231-2323
26. M. Gulati, R. Rastogi, D.K. Chaturvedi, S. Satya, N. Arora, P. Singhal, Statistical resultant analysis of spiritual & psychosomatic stress survey on various human personality indicators, in *International Conference Proceedings of ICCI 2018*, 2018
27. Yaniv Chen, Advances in the pathophysiology of tension-type headache: From stress to central sensitization. Curr. Pain Headache Rep. **56**(1), 43–49 (2009)
28. A. Agrawal, R. Rastogi, D.K. Chaturvedi, S. Sharma, A. Bansal, Audio visual EMG & GSR biofeedback analysis for effect of spiritual techniques on human behavior and psychic challenges, in *Proceedings of the 12th INDIACom*, 2018, pp. 252–258. ISSN 0973–7529 and ISBN 978-93-80544-14-4
29. S. Greengard, *The Internet of Things* (MIT Press, Cambridge, MA, 2015), p. 90. ISBN 9780262527736
30. D.K. Chaturvedi, M. Arya, A study of correlation between consciousness level and performance of worker. Ind. Eng. J. **6**(8), 40–43 (2013). D.K. Chaturvedi, Lajwanti, Dayalbagh way of life for better worldliness. Quest J. J. Res. Hum. Soc. Sci. **3**(5), 16–23 (2015). ISSN(Online): 2321-9467

31. W.M. Kang, Y.S. Moon, J.H. Park, An enhanced security framework for home appliances in smart home. Human-centric Comput. Inf. Sci. **7**(6) (2017). https://doi.org/10.1186/s13673-017-0087-4
32. L. Bendtsen, R. Jensen, Treating tension-type headache—an expert opinion. Expert Opin. Pharmacother. **12**(7), 1099–1109 (2011)
33. D.K. Chaturvedi, J.K. Arora, R. Bhardwaj, Effect of meditation on chakra energy and hemodynamic parameters. Int. J. Comput. Appl. **126**(12), 52–59 (2015)
34. V. Yadav, R. Rastogi, D.K. Chaturvedi, S. Satya, N. Arora, M. Gupta, S. Chauhan, P. Sharma, Chronic TTH analysis by EMG & GSR biofeedback on various modes and various medical symptoms using IoT. Advances in ubiquitous sensing applications for healthcare. Big Data Analytics for Intelligent Healthcare Management (2019). ISBN: 9780128181461
35. Q.F. Hassan, *Internet of Things A to Z: Technologies and Applications* (Wiley, Hoboken, NJ, 2018), pp. 41–44. ISBN 9781119456759
36. D.M. Simpson, M. Hallett, E.J. Ashman, C.L. Comella, M.W. Green, G.S. Gronseth, M.J. Armstrong, D. Gloss, S. Potrebic, J. Jankovic, B.P. Karp, Headache and disease. Naumann **12**(2), 23–39 (2016)
37. D.K. Chaturvedi, Relationship between chakra energy and consciousness. Biomed. J. Sci. Tech. Res. **15**(3), 1–3 (2019), https://doi.org/10.26717/bjstr.2019.15.002705. ISSN: 2574-1241
38. P. Singh, R. Rastogi, D.K. Chaturvedi, N. Arora, P. Trivedi, P. Vyas, Study on efficacy of electromyography and electroencephalography biofeedback with mindful meditation on mental health of youths, in *Proceedings of the 12th INDIACom*, 2018, pp. 84–89. ISSN 0973–7529 and ISBN 978-93-80544-14-4
39. L.J. Kricka, History of disruptions in laboratory medicine: what have we learned from predictions? Clin. Chem. Lab. Med. (2018). https://doi.org/10.1515/cclm-2018-0518 (inactive 2018-11-27). PMID: 29927745
40. S. Derry, P.J. Wiffen, R.A. Moore, Aspirin for acute treatment of episodic tension-type headache in adults. Cochrane Database Syst. Rev. **12**(6), 45–57 (2017)
41. Richa, D.K. Chaturvedi, S. Prakash, The consciousness in Mosquito. J. Mosquito Res. **6**(34), 1–9 (2016). ISSN: 1927-646X
42. V. Singh, R. Rastogi, D.K. Chaturvedi, S. Satya, N. Arora, H. Sirohi, M. Singh, P. Verma, Which one is best: electromyography biofeedback efficacy analysis on audio, visual and audio-visual modes for chronic TTH on different characteristics, in *Proceedings of ICCIIoT-2018,* 14–15 December 2018 at NIT Agartala, Tripura (ELSEVIER-SSRN Digital Library, 2018). ISSN 1556-5068
43. R. Rastogi, D.K. Chaturvedi, S. Satya, N. Arora, S. Chauhan, An optimized biofeedback therapy for chronic TTH between electromyography and galvanic skin resistance biofeedback on audio, visual and audio visual modes on various medical symptoms, in *National Conference on 3rd MDNCPDR-2018* at DEI, Agra on 06–07 September 2018
44. H. Verma, R. Rastogi, D.K. Chaturvedi, S. Satya, N. Arora, H. Saini, K. Mehlyan, Y. Varshney, Statistical analysis of EMG and GSR therapy on visual mode and SF-36 scores for chronic TTH, in *Proceedings of UPCON-2018* on 2–4 November 2018 MMMUT, Gorakhpur, UP, 2018
45. R. Walls, R. Hockberger, M. Gausche-Hill, Rosen's emergency. Med Concepts Clin Pract **13**(3), 37–47 (2017)
46. H. Saini, R. Rastogi, D.K. Chaturvedi, S. Satya, N. Arora, H. Verma, K. Mehlyan, Comparative efficacy analysis of electromyography and galvanic skin resistance biofeedback on audio mode for chronic TTH on various indicators, in *Proceedings of ICCIIoT-2018*, 14–15 December 2018 at NIT Agartala, Tripura (ELSEVIER-SSRN Digital Library, 2018). ISSN 1556-5068
47. Richa, D.K. Chaturvedi, S. Prakash, Role of electric and magnetic energy emission in intra and interspecies interaction in microbes. Am. J. Res. Commun. **4**(12), 1–22 (2016). ISSN: 2325-4076
48. D.K. Chaturvedi, Lajwanti, T.H. Chu, H.P. Kohli, Energy distribution profile of human influences the level of consciousness, in *Towards a Science of Consciousness, Arizona Conference Proceeding*, Tucson, Arizona, 2012

49. J. Smith, *Ferri's Clinical Advisor* (Elsevier, Philadelphia, 2019), p. 1348. ISBN 978-0-323-53042-2
50. V. Yadav, R. Rastogi, D.K. Chaturvedi, S. Satya, N. Arora, I. Bansal, Intelligent analysis for detection of complex human personality by clinical reliable psychological surveys on various indicators, in *National Conference on 3rd MDNCPDR-2018* at DEI, Agra on 06–07 September 2018
51. P.M. Gadient, J. Smith, The neuralgias: diagnosis and management. Curr. Neurol. Neurosci. Rep. **14**, 459 (2014)
52. S. Palaniappan, R. Awang, Intelligent heart disease prediction system using data mining techniques, in *IEEE/ACS International Conference on Computer Systems and Applications*, Doha, 2008, pp. 108–115. https://doi.org/10.1109/aiccsa.2008.4493524
53. M. Sultana, A. Haider, M.S. Uddin, Analysis of data mining techniques for heart disease prediction, in *3rd International Conference on Electrical Engineering and Information Communication Technology (ICEEICT)*, Dhaka, 2016, pp. 1–5. https://doi.org/10.1109/ceeict.2016.7873142
54. V. Yadav, R. Rastogi, D.K. Chaturvedi, S. Satya, N. Arora, V. Yadav, P. Sharma, S. Chauhan, Statistical analysis of EMG & GSR biofeedback efficacy on different modes for chronic TTH on various indicators. Int. J. Adv. Intell. Paradig. **13**(1), 251–275 (2018). https://doi.org/10.1504/ijaip.2019.10021825
55. M. Weiser, The computer for the 21st century. Sci. Am. **265**(3), 94–104 (1991). https://doi.org/10.1038/scientificamerican0991-94
56. R.S. Raji, Smart networks for control. IEEE Spectrum **31**, 49–55 (1994)
57. M. Gupta, R. Rastogi, D.K. Chaturvedi, S. Satya, N. Arora, H. Verma, P. Singhal, A. Singh, Comparative study of trends observed during different medications by subjects under EMG & GSR biofeedback, in *ICSMSIC-2019*, ABESEC, Ghaziabad, 8–9 March 2019. IJITEE **8**(6S), 748–756 (2019). https://www.ijitee.org/download/volume-8-issue-6S/
58. P. Singhal, R. Rastogi, D.K. Chaturvedi, S. Satya, N. Arora, M. Gupta, P. Singhal, M. Gulati, Statistical analysis of exponential and polynomial models of EMG & GSR biofeedback for correlation between subjects medications movement & medication scores, in *ICSMSIC-2019*, ABESEC, Ghaziabad, 8–9 March 2019. IJITEE **8**(6S), 625–635 (2019). https://www.ijitee.org/download/volume-8-issue-6S/
59. P. Magrassi, T. Berg, A world of smart objects. Gartner research report, R-17-2243 (12 August 2002)
60. H.C. Tsai, H. Cohly, D.K. Chaturvedi, Towards the consciousness of the mind, in *Towards a Science of Consciousness, Dayalbagh Conference Proceeding*, Agra, India, 2013
61. H. Saini, R. Rastogi, D.K. Chaturvedi, S. Satya, N. Arora, M. Gupta, H. Verma, An optimized biofeedback EMG and GSR biofeedback therapy for chronic TTH on SF-36 scores of different MMBD modes on various medical symptoms, *Hybrid Machine Intelligence for Medical Image Analysis. Studies in Computational Intelligence*, vol. 841 (Springer, Singapore, 2019) (chapter 8). https://doi.org/10.1007/978-981-13-8930-6_8. ISBN: 978-981-13-8929-0
62. S. Nikan, F. Gwadry-Sridhar, M. Bauer, Machine learning application to predict the risk of coronary artery atherosclerosis, in *International Conference on Computational Science and Computational Intelligence (CSCI)*, Las Vegas, 2016, pp. 34–39. https://doi.org/10.1109/csci.2016.0014
63. P. Magrassi, Why a universal RFID infrastructure would be a good thing. Gartner research report, G00106518 (2 May 2002)
64. A. Singh, R. Rastogi, D.K. Chaturvedi, S. Satya, N. Arora, A. Sharma, A. Singh, Intelligent personality analysis on indicators in IoT-MMBD enabled environment, in *Multimedia Big Data Computing for IoT Applications: Concepts, Paradigms, and Solutions* (Springer, Singapore, 2019), pp. 185–215 (chapter 7). https://doi.org/10.1007/978-981-13-8759-3_7

Rohit Rastogi received his B.E. degree in CSE from C. C. S. Univ. Meerut in 2003, the M.E. degree in CS from NITTTR-Chandigarh, Punjab Univ. in 2010. Currently he is pursuing his Ph.D. from Dayalbagh Educational Institute, Agra, India. He is a Associate Professor in CSE Dept. in ABES Engineering. College, Ghaziabad, India. He has been awarded in different categories for improved teaching, significant contribution, human value promotions and long service etc. He is strong believer that transformation starts within. He keeps himself engaged in various competitive events, activities, webinars, seminars, workshops, projects and various other teaching Learning forums.

D. K. Chaturvedi is working in Dept. of Elect. Engg., Faculty of Engg., D.E.I., Dayalbagh, Agra since 1989. Presently he is Professor. He did his B.E. from Govt. Engineering College Ujjain, M.P. then he did his M.Tech. and Ph.D. from D.E.I. Dayalbagh. He is gold medalist and received Young Scientists Fellowship from DST, Government of India in 2001–2002 for post doctorial research at Univ. of Calgary, Canada. He is fellow of The Institution of Engineers (India), Aeronautical Society of India, IETE, SMIEEE, USA and Member of IET, U.K., ISTE, Delhi, ISCE, Roorkee, IIIE, Mumbai and SSI etc. and The IEE, U.K.

Santosh Satya has been Professor in Center for Rural Development and Technology in Indian institute of Technology, Delhi. She is Hardworking and reputed lady in the field of Rural development. She has 75 research works with 2238 citations and 33,379 reads. She has keen interest in human Health and psychology domain.

Navneet Arora is serving as Professor at Indian Institute of Technology, Roorkee. He received Best Paper Presentation award ACMME 2016, Kuala Lumpur, Malaysia, Adjudged Institute Star Performer-2004, at IIT Roorkee, best Application Paper Award for Res. Work from an Int. Advisory Board, U.S.A., 1997, University Gold Medal from University of Roorkee, for standing First in M.E., 1990. GATE Scholarship awarded by I.I.T. Kanpur, in 1988, National Merit Scholarship Awarded by MHRD, supervised many Ph.D. and M.E. Thesis and Research Publications, Sponsored Research Project and Industrial Consultancy.

Medical Text and Image Processing: Applications, Issues and Challenges

Shweta Agrawal and Sanjiv Kumar Jain

Abstract Text and image analysis are playing very important role in healthcare and medical domain. The whole clinical process is getting affected positively by text and image processing. Many datasets, algorithms, models and tools are available for extracting useful information and for applying natural language processing, machine learning and deep learning algorithms. But there exist many challenges in healthcare data for successful implementation of text and image based machine learning models, which include: (i) storage and retrieval of high resolution images, (ii) scarcity of data (iii) dataset generation and validation, (iv) appropriate algorithms and models for extracting hidden information from images and texts, (v) use of modern concepts like deep neural networks, recurrent neural network, (vi) data wrangling and (vii) processing capacity of processors. This chapter will: (i) establish a background for the research work in the area of machine learning and deep learning, (ii) provide the brief about various types of medical texts and images, (iii) discuss the various existing models and applications of natural language processing, machine learning, computer vision and deep learning in medical domain and (iv) discuss various issues and challenges on applying natural language processing, machine learning and deep learning on medical data.

Keywords Machine learning · Deep learning · Medical imaging · Information extraction · Medical text · Electronic health records

1 Introduction

Availability of massive multimodal data in the field of healthcare motivated researchers to develop and apply AI techniques in this domain. In past decades role of

S. Agrawal (✉)
Professor CSE, SIRT, SAGE University, Indore, Madhya Pradesh, India
e-mail: shweta.cse@sageuniversity.in

S. K. Jain
Medi-Caps University, Indore, Madhya Pradesh, India
e-mail: sanjivkj@gmail.com

V. Jain and J. M. Chatterjee (eds.), *Machine Learning with Health Care Perspective*, Learning and Analytics in Intelligent Systems 13,
https://doi.org/10.1007/978-3-030-40850-3_11

AI techniques in healthcare domain grown very rapidly [1–3]. A recent report of *The Lancet*, state that a dermatologist may take decades to review over 200,000 images of skin lesions, and a computer by using AI techniques can review the same images in few days [4]. It is beyond imagination that how fast and precise an AI assisted doctor could be as compare to human physicians. Medical text and image processing is becoming highly popular among researchers because of availability of wider range and vast volume of data and applications [5–8]. Machine Learning (ML) sub domain of AI is becoming the essential component of many real life applications and research projects. Medical diagnosis and treatment is one of the very important research areas of ML. The power of ML in disease diagnosis, prediction, sorting and in classifying medicinal data is empowering medical practitioners and improving the speed of decision making in the clinical history. This domain has a lot of challenges also, because the applications are going to affect human life, hence the accuracy and precision are very important factor, while preparing application models for this domain. The major components of healthcare data analysis are medical images and texts. Text and image analysis play vital role in the entire clinical process starting from diagnostics, treatment, requirement of surgical procedures, predictions, suggestions and follow up studies. It is desirable and appreciable to produce information and knowledge from huge Electronic Health Records (EHR). EHRs are allowing health experts to access multimodal medical data which include images, clinical text, history, drug prescriptions, visiting notes and many more; hence to mine EHR data precisely with the help of ML is becoming a very useful application [9]. The support system and recommendation system designed with the help of EHR are proving revolutionary changes in healthcare industry.

1.1 ML in Medical Text and Image Processing

ML is a field in computer science which allows computers to learn with the experiences and need not require explicit program for individual task. It incorporates the concepts of computational statistics and data mining. It is a discipline which emphasis on, to make computers learn from data, as human beings learn from experiences. To get the inference and relationship from the available data it utilizes methods of statistics.

Block diagram for the process flow in ML is shown in Fig. 1. The raw data is accumulated and stored in acquire data stage. Data sets are selected and prepared according to the specified process/observations in prepare data stage. In the next stage decision features are extracted. Training of partial data set is done with an appropriate ML training algorithm. Evaluation of the trained model is done in next step for the assessment of efficiency of ML model. After that the tested model is deployed for prediction/detection/assessment or analysis of real-world related process data sets.

The ML can be classified into three categories:

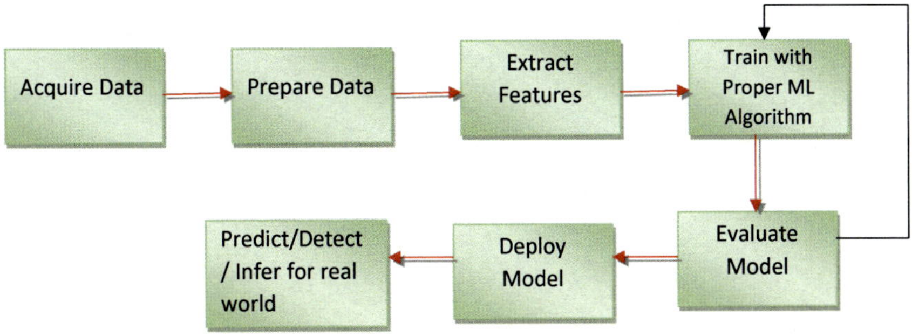

Fig. 1 Machine learning process flow

(i) **Supervised ML**: In supervised ML available data is labelled, means targets are known. Supervised ML can be further classified as Classification and Regression. In classification problems, output variable represents a category, such as given tumour image represents "cancerous" or "non-cancerous" tumour and in regression problems output variable represents a real value, such as "How much the body temperature, blood pressure". Examples of supervised ML in healthcare may include disease prediction and diagnosis like detection of a lung nodule from a chest X-ray automatically, training a model to relate a person's characteristics like height, weight, smoking status and family history to a certain outcome which may be like inception of diabetes within five years, or chances of heart blockage.

(ii) **Unsupervised ML**: In unsupervised ML [10] available data is unlabelled means targets are not known. In this learning method computers try to find patterns or groups which are occurring naturally or lie within the data. In this method different algorithms are used to make groups, clusters, categories based on the similarities in data. The applications of unsupervised ML in healthcare may include clustering of tissue samples based on similarity in gene expression, testing of novel drugs, activity monitoring and unusual activity detection for elderly homes and many more.

(iii) **Reinforcement Learning (RL)**: RL describes the process of taking suitable actions of a software agent in an environment, so as to get maximum cumulative reward. Agent decides an action at each time step based on its current state and feedback received from environment. The goal of the agent is to learn an optimal path to maximize the accumulated reward and to reach the goal. It means we don't give direct instructions to agent regarding the sequence of action they should take; instead they learn which actions are the best through trial-and-error interactions with the environment. Adaptive learning feature of RL is unique and make it distinct from traditional supervised learning methods, which require a list of correct labels, or from unsupervised learning approaches which try to find hidden structures within data. These features make RL a right solution for the problems of healthcare domain where the decision making is based on a sequential procedure [11]. Normally a medical treatment is based on a sequence of decisions to determine the right results such as patient history, type of treatment, drug dosage, or timing for re-examination at an instance based on current health status of an individual patient, by considering the goal of improving the patient's health conditions.

1.2 Deep Learning (DL) in Medical Text and Image Processing

An ML system requires domain expertise and a lot of techniques to convert the available unprocessed data into a form, which would be suitable for learning model like classifier, for extracting patterns and generating inference. Traditional ML techniques normally implement linear conversion of the input data and have limited ability to process natural data in their raw form [12]. DL is different from traditional ML in representation of input data to the training model. Feature extraction is not required from input data in DL. DL models are also called deep neural networks because of more than two hidden layers as compared to traditional artificial neural networks (ANNs) [13] which are usually limited to three layers (Input layer, one hidden layer and output layer). DL models also differ in number of interconnections as compare to ANN; hence have the capability to learn meaningful and useful concepts from the inputs. DL architectures have hierarchical learning structure and due to this have ability to incorporate heterogeneous data available in different- different data types.

Block diagram for the process flow in DL is shown in Fig. 2. All the process is same as ML approach, but there is no feature extraction block. The DL algorithm itself performs the task.

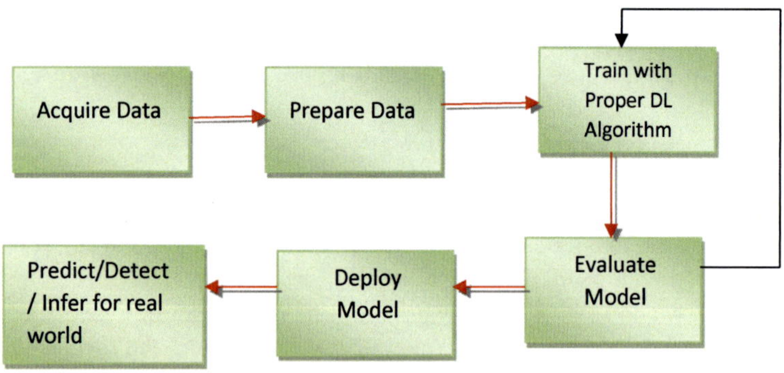

Fig. 2 Deep learning process flow

1.2.1 Convolution Neural Networks (CNN)

In deep neural network there exist one input layer, one output layer and many hidden layers. DL architectures are very complex and inefficient due to the interconnections of nodes of one layer to the next layer. By reducing number of interconnections using domain expertise, the performance can be improved. A CNN is a type of ANN with very few interconnections among the layers. CNNs are well suitable for image oriented task due to reduced features and nodes [14]. A CNN consists of multiple components which are:

(i) Convolution layers: This layer performs convolution operation with the help of small set of filters on the inputs received from its previous layers, for example a filter for detecting horizontal lines may be used to detect horizontal lines in an image.
(ii) Activation layers: The feature map created by convolution layer is fed into nonlinear activation functions. By this entire neural network can be approximated almost in a non linear function [15]. The most popularly used activation function is rectified linear unit (ReLU).
(iii) Pooling layers: Function of this layer is to reduce the amount of parameters, spatial size of the representation and number of computations in network. Generally used pooling methods are max pooling and average pooling. In max pooling most activated parameters for representing the feature are considered while average pooling considers average presence of a feature [16].
(iv) Dropout regularization: This method improves performance of CNN tremendously. In this [17] some number of nodes from a layer are randomly dropped out means temporarily removed from the network including their incoming and outgoing connections.
(v) Batch normalization: These layers produce normalized activation maps by applying the statistical formulas like mean and standard deviation. Due to this layer network periodically changes its activations to zero mean and unit standard deviation. This layer speeds up training, and reduces dependency on parameter initialization [18].

1.2.2 Recurrent Neural Network (RNN)

A RNN is a type of ANN where nodes are connected in temporal sequence and create a directed graph, hence exhibit temporal dynamic behaviour. RNNs have memory element to store their internal state which helps in sequential processing of inputs. Due to this feature RNNs are useful for the applications like handwriting recognition [19] or speech recognition [20, 21]. Unlike traditional neural network RNN converts the independent activations into dependent activations by assigning same weights and biases to all the layers. RNNs are applied on medical series data for detecting patterns have obtained very good accuracy.

1.3 Natural Language Processing (NLP) in Medical Text Processing

NLP is a technique in which machines are programmed to understand and interpret human language. The NLP applications in medical domain increased tremendously in last decades. In NLP text is being reformatted for subsequent analysis with the help of ML or artificial intelligence techniques. In medical domain text may be in form of clinician documentation, patient history, laboratory reports, billing documentation, transcripts of patient or even social media discussions.

NLP has been used for decision-making in medical domain in developing tools for risk stratification, identifying postoperative complications of a surgery with the help of physician notes, to provide training to patients for identifying syndromes, to identify biomedical concepts by radiology reports [22], nursing documentation [23], and discharge summaries [24] etc.

NLP is used to translate words or phrases into concepts or knowledge. Mapping from words or phrases to concepts involves steps:

(i) Tokenization: breaking a sentence into words or tokens,
(ii) Lemmatization: figure out the most basic form or *lemma* of each word in the sentence,
(iii) Identifying stop words: There are a lot of words like 'and' 'the' 'or' appears very frequently in a text and create noise. These words should be filtered out before applying any statistical analysis, and
(iv) Mapping: Mapping of each lemma into one or more concepts.

1.4 Computer Vision for Medical Image Processing

Computer Vision [25] algorithms aim to give computers a visual understanding of the world. It is one of the main components of machine understanding and learning. The three main components of computer vision are: (i) Image acquisition, (ii) Image processing, and (iii) Image analysis and understanding (creating knowledge). Image acquisition is the process of converting real world analog images into binary form. Image processing involves characterization of images by edges, distances, point features or segments and image analysis includes classification, recognition, prediction and matching. It involves extensive use of matrix representation and statistical algorithms. For image analysis ML or DL algorithms are applied. Examples of images analysis are 3D scene mapping, object detection, object counting, face recognition, face identification and many more. Orlando Health Winnie-Palmer Hospital for Women & Babies is an example of computer vision's applications in healthcare. In this, a computer vision tool with the help of artificial intelligence developed by Gauss is used for assessing blood loss during surgical process of childbirth.

2 Medical Images and Texts Datasets

Medical texts are the script written by medical practitioners which communicate the details regarding present and past status of a patient. These reports mention the reason behind the patient's visit to the doctor, diagnostic procedures done, treatments given, tests suggested, findings of tests and many other information. The medical text include history reports, physical examination reports, consultation reports, operative reports, admission reports, discharge summary, Colonoscopy reports, radiology reports, pathology reports, laboratory reports and conversation chats etc. Medical Imaging is a type of tests used to capture images of inner parts of the body. It helps to screen possible health conditions in prior manner, diagnose the likely cause of a particular disease and monitor health conditions that have been diagnosed, or the effects of treatment for them. Types of medical images include: Computer Tomography (CT), Positron Emission Tomography (PCT), Magnetic Resonance Imaging (MRI), Ultrasound and X-ray [26].

2.1 Types of Medical Texts

2.1.1 History Reports

A history report consists of the information about a person's health. It provides a background to the physician about how patient's body react to the particular medication, previous ailments, what could be the best treatment, how much time it would take to cure a particular disease, what may be the probability of reoccurrence of the disease and many more. A medical history report have many subsets which include surgical history, information about medical and medication allergies, results of physical exams, medicine prescribed and taken, diet habits, daily routine and family history. Family history shows health information of close family members like parents, grandparents, children, brothers and sisters.

2.1.2 Physical Examination Reports

This process utilizes observation, palpation, percussion, and auscultation methods to assess anatomic findings in a patient. The obtained information integrated with the patient's history and pathological reports. The carefully performed physical examination can yield 20% necessary data to diagnose the patient. In physical examination information can be collected by variety of sources like: verbal communication, body movements, behaviour, gait, and changes in the observable features like: pigmentary changes such as jaundice, cyanosis, and paleness may be noted. Patient's habits, interest and routine can be observed from his personal objects, exterior social and family environment [27].

2.1.3 Consultation Report

This report is used to converse the observations and suggestions by a physician to a patient. This report may consist patient's detailed medical history including earlier medications, social and family history, allergies, need for consultation, earlier and current symptoms described by the patient, observations from physical examination, tests prescribes, laboratory reports, conclusion drawn by consultant highlighting the patient's diagnoses and recommended treatment.

2.1.4 Operative Reports

An operative report is a document that describes the indication, procedure, specimen removed and any complications of the procedure. This report also consist preoperative and postoperative diagnoses and details of surgeon and assistants. Some structured tools for creating operative reports are also available [28] which generates systematic electronic operative reports.

2.1.5 Discharge Summary

This document indicates care plan of a patient after getting discharge from a hospital to assist the family members. A discharge summary with all the relevant information is necessary for proper care and safety of patient during the shifting between hospital and home or other hospital. The Joint Commission standards [29], shows that each medical institute's discharge summary should contain following components: (i) need for admission in hospital, (ii) disease diagnosed (iii) procedures and treatment provided, (iv) condition of patient at the time of discharge, (v) instructions for patient and family members, and (vi) signature of attending physician.

2.1.6 Admission Reports

Admission reports consist the information like: reason for admission or chief complaint, history of present illness, review of symptoms, allergies, current status of patient, physical examination reports and the initial instructions for that patient's care.

2.1.7 Clinical Reports

A clinical case report is a detailed academic report containing symptoms, diagnosis, treatment given, and follow-up of a patient. These reports usually describe a novel disease and helps in medical progress and new idea generation in medicine. Some clinical reports are prepared for extensive review of the relevant literature on the specified topic. The clinical reports are published to (i) show an unusual association between diseases and symptoms, (ii) unusual event during the observation period of a patient, (iii) findings related to genesis of a disease or unfavourable effect, (iv) some unique or rare seen feature of a disease, (v) illustration of new theory and many more [30, 31].

2.1.8 Laboratory Reports

These are the reports obtained in a clinical laboratory [32] where tests are performed on clinical samples to get information about the health of a patient in order to diagnose a particular disease, to check the response of a treatment, and to prevent disease. Medicinal laboratories are divided into two categories anatomic pathology and clinical pathology. In Anatomic pathology diagnosing of disease had done by the examination of organs and tissue samples while in clinical pathology diseases diagnosis is done through analyzing body fluids in a lab.

2.2 Types of Medical Images

Visual representation of the inner parts of the body and functions of some organs are termed as medical imagining. Medical images are capable to unveil the parts of body which are covered by skin, bones or other parts of body. These images are also capable to show the functioning of a body part, detecting any abnormalities, risk of occurrence of a disease and many more. Different category of Medical Imagining include (i) Digital radiography, (ii) Digital Mammography, (iii) Digital Fluoroscopy, (iv) Magnetic Resonance Imaging, (v) Magnetic Resonance Spectroscopy, (vi) Nuclear Medicine, (vii) Single Photon Emission Computed Tomography (SPECT), (viii) SPECT and X-ray CT, (ix) Positron Emission Tomography, (x) Positron Emission Tomography and X-ray CT, (xi) Computed Tomography, (xii) Ultrasound, (xiii) Photoacoustic Imaging, (xiv) Echocardiography, (xv) Functional Near-Infrared Spectroscopy, and (xvi) Magnetic Particle Imaging.

2.2.1 Digital Radiography

Uniform beams of x-rays incident on the patient's affected area. The beams are modulated by the tissues, hence identify and convert the x-ray detector information into two-dimensional greyscale image.

2.2.2 Digital Mammography

An x-ray projection imaging procedure specially used for identifying the abnormalities in the breast. In digital mammography dedicated x-ray systems and digital detectors are used in combine manner.

2.2.3 Digital Fluoroscopy

The acquirement sequence of x-ray projection image in real time is known as Fluoroscopy. It is utilized in dynamic assessment procedures of patients for interventional and investigative radiology. For observation and assessment of patient anatomy videos of X-ray pictures are captured stretched from 1 to 60 frames per second.

2.2.4 Magnetic Resonance Imaging

Magnetic Resonance Imaging (MRI) is an image technique which exploits a powerful magnetic field (order of 1.5–3 Tesla) procedure for proton magnetization in human body tissues. The subsequent emission of Radio Frequency (RF) signal from protons after energy assimilation due to the RF excitation, which are identified and processed to expose the tissue magnetic features in the form of an image on greyscale. Differences in tissue contrast are produced from specific pulse strings and for an explicit MRI study normally, a number of sequences are obtained.

2.2.5 Magnetic Resonance Spectroscopy

Magnetic Resonance Spectroscopy (MRS) determines the characteristics of native tissue metabolites in lone or multiple voxels (quantity elements). It is a MR image scheme in deciding tissue areas called non-invasive "biopsy". The solutions are considered vendor-specific regarding data storage into Picture Archiving and Communication System (PACS) for spectroscopy data. For MRS, an interorbital distance is specified in the digital image and communication in medicine criterion.

Safety training for hospital personnel functioning in the environs of MRI systems for knowledge of magnetic resonance is essential, covering informatics and PACS specialists. For the protection of patients as well as working personnel they must have knowledge of some features like fringe fields, heating, RF excitation, hazards of ferromagnetic materials, effects of intense magnetic fields and several others.

2.2.6 Nuclear Medicine (Molecular Imaging)

Tagging of radioactive materials with tracers is used for molecular imaging which gets disseminated into patient after injection. With the help of scintillation camera, images of emission are captured. This are generated through localization activity over explicit time- duration.

2.2.7 Single Photon Emission Computed Tomography

Multiple angles planar images are recorded from SPECT. From the imaged volume, it utilizes them to restructure slices of tomography.

2.2.8 SPECT and X-Ray CT (SPECT/CT)

SPECT/CT is a methodical structure where attenuation and relationship of metabolic movement correction is provided for fused data of CT and SPECT images through anatomy of high resolution.

2.2.9 Positron Emission Tomography (PET)

To generate annihilation photons positron emitters are used in PET imaging. The elements redistributed arc in human body after being attached to an agent. The determination of tomographic slice is achieved from an activity of line integral formed from photons of two oppositely directed gamma rays.

2.2.10 Positron Emission Tomography and X-Ray CT (PET/CT)

Measurements for attenuation correction are provided through CT system part for accuracy and high resolution. It generates Metabolic/Physiological information about structural and anatomical through image fusion. Using CT a huge amount of data in nuclear science is created now-a-days for further recognition and analysis.

2.2.11 Computed Tomography (CT)

The thin-slice protrusion data is acquired utilizing a detector array and rotary x-ray tube in CT scanners. After that algorithms regarding computer reconstruction are used for several tomographic images of the anatomical volume. The contrast resolution is essentially reliant on the amount of x-ray photons/voxels. Techniques for CT acquisition depend on the patient dimension and also on study type.

2.2.12 Ultrasound (US)

Body tissues greyscale anatomic images utilizing acoustic properties are generated in the Ultrasound technique by means of transporting small and high frequency pulses of sound into a particular volume. The significant capabilities of Ultrasound lies in the assessment of flow of blood by Doppler signal investigation and colour flow imaging. US are useful in different observation techniques viz. vascular laboratory (Vascular surgery), Gynaecology and obstetrics (imaging), Radiology (biopsy, breast), Ophthalmology, Cardiology.

2.2.13 Photoacoustic Imaging

Photoacoustic imaging is a hybrid biomedical imaging newly developed modality relying on the photoacoustic effect. It is helpful in optical deep imaging diffusive and/or quasi-diffusive systems thus merging the benefits of optical assimilation contrast with an ultrasonic spatial resolution. Latest studies have revealed that photoacoustic imaging can be utilized for monitoring tumours angiogenesis, detection of skin melanoma, mapping of blood oxygenation and imaging of brain functionality.

2.2.14 Echocardiography

Echocardiogram is the technique for imaging of heart using ultrasound. This method utilizes Doppler, 2D and 3D imaging to generate images of the heart and envisage the blood flowing in all of the four valves of heart. Echocardiography permits detailed configurations of the heart, collectively with size of chamber, functioning of heart, the heart valves, and the pericardium.

2.2.15 Functional Near-Infrared Spectroscopy

FNIR is a latest non-invasive imaging process used for the objective of functional neuro-imaging. It is the technique widely used for the imaging of brain related functionalities.

2.2.16 Magnetic Particle Imaging (MPI)

Magnetic Particle Imaging utilizes iron oxide Nano-particles in super-paramagnetic state, also a recent and progressive investigative imaging technique. It is exploited for super-paramagnetic state iron oxide Nano-particles tracking. The chief benefit is the high specificity and sensitivity. Also shows advantage of almost constant signal strength with depth of tissue. MPI shows its strength in research regarding imaging of Neuro-perfusion, cell tracking and cardiovascular performance.

2.3 Medical Image Data Sets

Many researchers and organizations have prepared the repositories for medical datasets. The table list some repositories for medical dataset.

S. No.	Dataset source	Description
1	National Institute of Health Clinical Center (NIH)National Institutes of Health—Clinical Center [33]	Approximately 100,000 Chest X-ray images, meta data and diagnosis
2	The Cancer Imaging Archive (TCIA) [34]	Medical images of cancer includes: • Breast MRI • Lung PET/CT • Lung Image Database Consortium (LIDC) • Neuro MRI • Virtual Colonoscopy • Osteoarthritis Initiative (MIA) • Reference Image Database to Evaluate Response (RIDER) • PET/CT phantom scan collection • CT Colongraphy
3	National Biomedical Imaging Archive (NBIA) [35]	Provides Image archives which can be used for development and validation of analytical software tools
4	Open Access series of Imaging Studies (OASIS) [36]	This database Provides Cross-sectional MRI Data of Young, Middle Aged, Nondemented and Demented Older Adults, Longitudinal MRI Data of Nondemented and Demented Older Adults, Longitudinal Neuroimaging, Clinical, and Cognitive Dataset for Normal Aging and Alzheimer's Disease
5	The Federal Interagency Traumatic Brain Injury Research (FITBIR) [37]	MRI, PET, Contrast, and other data on a range of Traumatic brain injury (TBI): a major medical problem for both military and civilian populations
6	STructured Analysis of the Retina(STARE) [38]	The full set of 400 retina images
7	Alzheimer's Disease Neuroimaging Initiative (ADNI) [39]	MRI and PET images, genetics, cognitive tests, CSF and blood biomarkers as predictors of the Alzheimer's Disease
8	Duke Center for In Vivo Microscopy [40, 41]	Provides CT, MRI, X-Ray, Confocal, optical, ultrasound and SPECT images which have very good resolution

(continued)

(continued)

S. No.	Dataset source	Description
9	MIDAS	It provides flexible, intelligent data storage system. The medical platforms available at MIDAS include [42] National Alliance for Medical Image Computing (NAMIC) [43] Provides neurosience and neurological disorders data Imaging Methods Assessment and Reporting (IMAR) [44] Provide MRI images for image segmentation, registration, and computer-aided diagnosis methods
10	Digital Retinal Images for Vessel Extraction (DRIVE) [45]	Helps in segmentation of blood vessels in retinal images
11	Digital Database for Screening Mammography (DDSM) [46]	Provides mammographic images to researchers
12	Public Lung Database To Address Drug Response [47, 48]	This database provide annotated CT image scans that highlights many of the key issues in measuring large lesions in the lung
13	The Osteoarthritis Initiative (OAI) [49]	Provides data related to prevention and treat knee osteoarthritis
14	SCR database: Segmentation in Chest Radiographs [50]	The SCR database provides data for comparative studies on segmentation of the lung fields, the heart and the clavicles in chest radiographs
15	Japanese Society of Radiological Technology (JSRT) Database [51]	This database contains digital images that shows with and without chest lung nodules
16	Standard Diabetic Retinopathy Database (DIARETDB1) [52–54]	This database provides digital images for benchmarking diabetic retinopathy detection
17	Cornell Visualization and Image Analysis (VIA) group [55]	Image databases containing lung CT images in together with documentation of abnormalities by radiologists
18	Omni Medical search [56]	Medical image and study databases and library
19	Spineweb [57]	A platform for getting spinal images for image analysis
20	Facebase [58]	Repository for datasets by organisms, experiment type, age stage, mutation, genotype and more

2.4 *Medical Data Processing Tools and Technologies*

2.4.1 Medical Text Processing Tools and Algorithms

Clinical Language Annotation, Modelling, and Processing Toolkit (CLAMP) [59] is a broad range NLP software for clinical text that is used for automatic encoding and recognizing useful clinical information from patient's history reports. MetaMap [60, 61] is a Natural language processing tool developed at National Library Medicine and freely available. This tool can be used to find metathesaurus from given text. Medical Language Extraction and Encoding System (MedLEE) [62] is a clinical NLP tool designed to assess and understand clinical text. An NLP tool is developed at Columbia University by Carol Friedman et al. It is from one of the most basic and wide-ranged clinical NLP tools to process radiology notes [63]. The tool was further extended to understand other clinical domains also like pathology reports [64] and on patient's discharge summaries [65]. The cTAKES [66] is an NLP based tool for processing clinical texts. It is utilizing the Unstructured Information Management Architecture (UIMA) framework and the OpenNLP natural language processing toolkit. The BioLark [67] for automatic recognition of terms using semantic web technologies. It comprises ontology focusing on representation and modelling of disorder, text mining and concept recognition. An ML based tool with manual rules has been developed by Lobo et al. [68]. Example of one more tool is OBO annotator used for Open Biological and Biomedical Ontologies [69]. Classification algorithm has been used to infer the sentiments of a patient based on their comments whether they are positive or negative [70], Kiroku's [71] is an application to pick up context in a conversation between doctor and the patient, it can write clinical notes also automatically. MDOps [72] reduces documentation time drastically; it can prepare clinical notes using iPhone or iPad. The various applications of clinical text and image processing include clinical decision support system [73, 74], analyzer for patients records [75], for public health bio surveillance [76], and in biomedical research [77, 78].

CNN Based Medical Text Processing Tools

CNNs are applied: (i) on a biomedical article classification model which identifies the hallmarks of cancer given in an article abstract [79], (ii) for attention mechanism to extract drug interaction [80], (iii) on biomedical articles to retrieve the relevant articles [81], (iv) for providing support to the patients based on message forwarded on portal [82], (v) for short text classification on radiology practices through protocol determination [83].

RNN Based Medical Text Processing Tools

RNNs are used (i) for preparing model to predict punctuation in medical reports [84], (ii) for biomedical articles to understand and suggest treatments with the help of relational and contextual similarities amongst the text [85], (iii) for extracting clinical concepts with the help of EHR reports [86], (iv) for designing the tool 'Doctor AI' which has used patient history to predict and diagnose medications and follow-ups [87], (v) for the Tool DeepCare [88] based on RNN with long short term memory (LSTM) which provides a dynamic network to analyze current state of disease and predicts future medical results. DeepCare was evaluated for diabetic and mental health patients for disease modelling, progression, recommendation and for future risk and (vi) for the tool RETAIN: A predictive model for clinical tasks [89].

2.4.2 Medical Image Processing Tools and Algorithms

The Visualization Toolkit (VTK) is open source software developed for processing 3D computer graphics, images and visualization [90]. Insight Segmentation and Registration Tool Kit (ITK) is an open source tool which provides huge collection of software for image analysis [91]. GIMIAS provides a workflow-oriented environment and mainly focuses on biomedical imagining. It is specially designed for solving individualized simulation and biomedical image computing problems. Problem specific plug-ins can be added in GIMIAS [92]. Classification algorithms have been used to predict that a tumour is benign or malignant [93], to classify skin lesions based on malignancy [94], to discover risk factors in retinal fundus photographs [95].

Liu s et al. predicted Alzheimer disease through brain MRI scans by applying CNN [96]. Prasoon, A. et al. used CNN for predicting osteoarthritis risk with the help of low-field knee MRI scans [97]. CNN is used for segmentation of multiple sclerosis lesions present in multi-channel 3D MRI [98] and for diagnosing benign and malignant breast nodules in ultrasound images [99]. CNN is used to analyse retinal fundus photographs for identifying diabetic retinopathy by Gulshan et al. [100]. Prediction of heart failure due to congestion and skin pulmonary disease through four-layer CNN is proposed by Liu et al. [101]. Sahba et al. used RL for the segmentation and classification of prostate ultrasound image [102, 103]. Liu and Jiang [104] used a Trust Region Policy Optimization (TRPO) for segmenting and classifying joint surgical gestures. Alansary et al. used deep RL for anatomical landmark localization of fetal 3D ultrasound image [105].

3 Challenges

The key challenges of text analysis include (i) identification of important clinical terms stated in the digital text, for example treatments, disorders, symptoms etc.

(ii) locating key entities, (iii) lack of clinical text due to patient and medical institutes privacy, (iv) unstructured format, (v) lot of repeated words and ambiguity, (vi) misspellings, (vii) Use of abbreviations, (viii) variety of communication styles and (ix) availability of appropriate models, tools and algorithms for text processing.

The key challenges of medical image analysis include requirement of high resolution images, for achieving better accuracy and best results. For analytics and machine learning the number of data samples should also be sufficient to train, test and validate a machine learning model. All these factors generate the challenges like (i) storage of huge amount of high resolution medical images, (ii) generation and validation of high resolution medical image dataset, (iii) use of virtual reality in medical visualizations, (iv) Appropriate algorithms for extracting useful information from medical images, (v) Appropriate tools and technologies for getting hidden information from images, and (vii) processing capacity of processors.

Few common challenges may be further elaborated as:

(i) Data quality: Medical domain data is very diverse, noisy, ambiguous, complex and imperfect, unlike other domains where the data could be well arranged and clean. Training a good ML and DL model with these huge and diverse data sets leads to many challenges and attracts research communities to address the issues, like sparsity of data, missing values and redundancy in data. Due to complexity and heterogeneity of medical data models can be prepared only by considering some aspects of healthcare system [106].

(ii) Temporality: The patient's response for a particular treatment may change over time and may be progressive also. Many existing ML and DL models are designed by considering static vector-based inputs, which are ignoring the changes over time. Development of time sensitive DL model is a complex task but RNNs and other architectures which are coupled with memory like LSTM will play significant role for such issues.

(iii) Domain complexity: Unlike other application domains like face recognition, speech recognition, the problems in medical imagining and text processing are more complicated. Because of highly varied data and for most of the diseases the information regarding their cause, progress, treatment, and response is not known properly and completely.

(iv) Interpretability: Designing of ML and DL models for the application areas; where the black box implementation of model is not a problem is achieving great success. For example in image annotation in which user itself can validate the tags assigned to the images. But in medical domain quantitative algorithmic performance and reason behind selection of a particular algorithm is equally important. It is very crucial to convince a medical professional for the actions suggested and recommended by a recommender system for example a medication provided by a recommender system may lead to side effects [107].

(v) Feature Collection: Feature capturing is an important aspect of ML and DL algorithms. The feature collection from patients and hospitals leads to many challenges. For collecting more and more information the medical data should be collected from any possible data source which may include EHRs, social media [108, 109], wearable devices, surroundings, online and offline surveys, online forums, medical organizations, genome profiles. The integration of this highly heterogeneous data, feature selection, storage and data wrangling are the important and very challenging research topic. A lot of efforts for combining heterogeneous medical data are required.

(vi) Data Volume: Data collection in medical domain is still a challenging task. World Health Organization's (WHO) December 2017s report states that approximately half of the world's population is not getting primary and essential health services. Consequently, getting as many patients as we want for data collection is not possible. To get the right inference for any disease and diagnosis there is a requirement of huge amount of data. Many efforts are still required to collect healthcare data from all kind of patients and from each corner of the world.

(vii) Integration and Privacy: Every medical institute has its own patient population and concern data. Integration of this diverse data to build a computational model based on ML or DL without affecting and leaking sensitive information of patients and institutes is a crucial problem. Considering security aspect in DL models are more complex due to the enormous parameters [110, 111]. Another privacy based issue is the deployment of computational models in cloud environment. Tramèr et al. [112] showed the issues in considering ML and DL networks as a service. Although some articles to answer such issue have been published. A differential private method to secure the parameters trained for the logistic regression model has been developed by Chaudhuri et al. [113]. Consequently to develop computational models for such federated environment in secure manner is also an important research topic, in which researchers have to incorporate other security domains like cryptography, digital signature etc. [114, 115].

(viii) Medical Vocabulary: To standardize the medical terms a standard controlled medical terminology (CMT) [116] has been created. But the adaptability of CMTs by some medical practitioner is still a big issue and leads the challenges like evaluation, understanding and collection of diverse medical terminologies.

(ix) Data Repository creation: Repositories are normally created by collecting data from variety of sources. Each source may have different database design and structure. Medical images normally have high resolution and for proper training we should have huge number of images. This may lead to challenge of feasibility of IT infrastructure in terms of storage, processing speed, security, reliability, maintainability, networking etc.

(x) Errors in medical data: The recorded medical data may have noise these noises may be due to some incident, errors in measurement, or may be due to the errors of recording tools [33]. A patient's electrocardiogram (ECG) signal may get modified by the patient's movements, breathing speed or some other incident [117]. The noise in signal suppresses its original characteristics. Although some researchers have worked on filtering techniques with biomedical signals For example, band pass filters are used to filter high frequency signals and low frequency noise from ECG signal by Balli et al. [118]. Band pass filter is also used on electronystagmography (EMG) signals and electroencephalography (EEG) signals [119] to filter different parameters, but still it is a research challenge to get right pattern of biomedical signals for the purpose of ML and DL.

(xi) Large feature size: Medical images normally have very high dimensions. Matrix size of radiology images range from 64 × 64 for some examination to 4000 × 5000 for mammogram images. Matrix size of MRI and CT scans are 256 × 256 and 512 × 512 respectively. For every examination there may be requirement of over 100 images, hence number of extracted features would also be very large. It raises the challenge of increasing complexity of computational model and requires new solutions.

(xii) Acceptability of models: The many computational tools designed in medical domain are facing the problem of lack of theoretical understanding of issues of this domain especially from the perspective of non physician. It results in use of many computational models ineffective. The physicians are susceptible to accept the recommendations of computational model designed with the help of ML and DL hence some solid validation and verification schemes are required for acceptability of developed computational models.

4 Research Directions

All above mentioned challenges introduce several opportunities and research directions to make computational models more accurate, precise, fast and adaptable. Therefore, by keeping all them in consideration, some research directions are mentioned below:

(i) Integration: For better results of detection, analysis and measurement of biomedical information diverse parameters from different aspects and sources should be integrated. Data from different sources would be able to describe different aspects of a problem. With the help of integrative analysis comprehensive insights of a particular disease can be obtained.

(ii) Security: The security and privacy of individual patient's data is very important and needs attention. Sun et al. [120] described that very slight modification in lab record value in a patient's EHR can completely change the prediction made by a predictor model. It is very important for researchers to develop defence mechanism by considering adversarial attacks.

(iii) Transparency: The functioning of computational models for some areas of medical domain is very important. Presently mostly presented models function as black box. These black boxes should be explored more to achieve better accuracy, precision and adaptability.

(iv) Database creation: The database for many diseases like dengue, aids, flu, heart diseases, mental disorders etc. is not sufficiently available or not considering all parameters and aspects, or for a specific region or medical institute. There is a still requirement of a lot of databases with different parameters.

5 Conclusion

The availability of huge EHRs, zeal and necessity of improving health of any individual or community, processing power of computers, tools and technologies, computational concepts like natural language processing, machine learning and deep learning are attracting the attention of many researchers and research communities. Many types of medical texts and images are available for getting inference, prediction and analysis. Research communities have published a lot of open source datasets and created repositories for analysis and getting inference. Lot of tools to assist physicians have been designed which are mainly based on text processing. Many Classification and regression models also have been designed for scanning MRI and CT images. But still there are a lot of challenges like data collection, heterogeneity of data, data integration, noise in data, large number of features, model selection, and feature selection, privacy of patients and medical institutions, IT infrastructure support, adaptability etc. All these challenges lead to many research directions and despite of these some tools have been adopted and deployed by some medical practitioners.

References

1. S.E. Dilsizian, E.L. Siegel, Artificial intelligence in medicine and cardiac imaging: harnessing big data and advanced computing to provide personalized medical diagnosis and treatment. Curr. Cardiol. Rep. **16**(1), 441–448 (2014)
2. F. Jiang, Y. Jiang, H. Zhi, Y. Dong, H. Li, S. Ma, Y. Wang, Q. Dong, H. Shen, Y. Wang, Artificial intelligence in healthcare: past, present and future. Stroke Vasc. Neurol. **2**(4), 230–243 (2017)
3. J. He, S.L. Baxter, J. Xu, J. Xu, X. Zhou, K. Zhang, The practical implementation of artificial intelligence technologies in medicine. Nat. Med. **25**(1), 30–36 (2019)
4. G. Quer, E.D. Muse, N. Nikzad, E.J. Topol, S.R. Steinhubl, Augmenting diagnostic vision with AI. Lancet **390**(10091), 221 (2017)
5. A. Esteva, A. Robicquet, B. Ramsundar, V. Kuleshov, M. DePristo, K. Chou, C. Cui, G. Corrado, S. Thrun, J. Dean, A guide to deep learning in healthcare. Nat. Med. **25**(1), 24–29 (2019)
6. T. Ching, D.S. Himmelstein, B.K. Beaulieu-Jones, A.A. Kalinin, B.T. Do, G.P. Way, E. Ferrero, P.M. Agapow, M. Zietz, M.M. Hoffman et al., Opportunities and obstacles for deep learning in biology and medicine. J. R. Soc. Interface **15**(141), 20170387 (2018)
7. D. Rav'i, C. Wong, F. Deligianni, M. Berthelot, J. Andreu-Perez, B. Lo, G.Z. Yang, Deep learning for health informatics. IEEE J. Biomed. Health Inform. **21**(1), 4–21 (2017)
8. J. Luo, M. Wu, D. Gopukumar, Y. Zhao, Big data application in biomedical research and health care: a literature review. Biomed. Inform. Insights 8:BII–S31559 (2016)

9. P.B. Jensen, L.J. Jensen, S. Brunak, Mining electronic health records: towards better research applications and clinical care. Nat. Rev. Genet. **13**(6), 395–405 (2012)
10. M.L. Littman, Reinforcement learning improves behaviour from evaluative feedback. Nature **521**(7553), 445–451 (2015)
11. O. Gottesman, F. Johansson, M. Komorowski, A. Faisal, D. Sontag, F. Doshi-Velez, L.A. Celi, Guidelines for reinforcement learning in healthcare. Nat. Med. **25**(1), 16–18 (2019)
12. Y. Bengio, A. Courville, P. Vincent, Representation learning: a review and new perspectives. IEEE Trans. Pattern Anal. Mach. Intell. **35**(8), 1798–1828 (2013)
13. Y. LeCun, Y. Bengio, G. Hinton, Deep learning. Nature **521**(7553), 436–444 (2015)
14. N. Ganapathy, R. Swaminathan, T.M. Deserno, Deep learning on 1-D biosignals: a taxonomy-based survey. Yearb. Med. Inform. **27**(01), 098–109 (2018)
15. S. Sonoda, N. Murata, Neural network with unbounded activation functions is universal approximator. Appl. Comput. Harmonic Anal. **43**(2), 233–268 (2017)
16. J.T. Springenberg, A. Dosovitskiy, T. Brox, M. Riedmiller, Striving for simplicity: the all convolutional net. arXiv preprint arXiv 1412(6806) (2014)
17. N. Srivastava, G. Hinton, A. Krizhevsky, I. Sutskever, R. Salakhutdinov, Dropout: a simple way to prevent neural networks from overfitting. J. Mach. Learn. Res. **15**(1), 1929–1958 (2014)
18. S. Ioffe C. Szegedy, Batch normalization: accelerating deep network training by reducing internal covariate shift, in *International Conference on Machine Learning* vol. 1502, no. 03167 (2015), pp. 448–456
19. A. Graves, M. Liwicki, S. Fernandez, R. Bertolami, H. Bunke, J. Schmidhuber, A novel connectionist system for improved unconstrained handwriting recognition. IEEE Trans. Pattern Anal. Mach. Intell. **31**(5), 855–868 (2009)
20. S. Hasim, S. Andrew, B. Francoise, Long short-term memory recurrent neural network architectures for large scale acoustic modeling. In: *Fifteenth Annual Conference of the International Speech Communication Association* (2014)
21. L. Xiangang, W. Xihong, Constructing long short-term memory based deep recurrent neural networks for large vocabulary speech recognition (2014). ArXiv 1410(4281)
22. R. Flynn, T.M. Macdonald, N. Schembri, G.D. Murray, A.S.F. Doney, Automated data capture from free-text radiology reports to enhance accuracy of hospital inpatient stroke codes. Pharmacoepidemiol. Drug Saf. **19**(8), 843–847 (2010)
23. L.L. Popejoy, M.A. Khalilia, M. Popescu, C. Galambos, V. Lyons, M. Rantz et al., Quantifying care coordination using natural language processing and domain-specific ontology. J. Am. Med. Inform. Assoc. **22**(e1), e93–e103 (2015)
24. H. Yang, I. Spasic, J.A. Keane, G. Nenadic, A text mining approach to the prediction of disease status from clinical discharge summaries. J. Am. Med. Inform. Assoc. **16**(4), 596–600 (2009)
25. J. Gao, Y. Yang, P. Lin, D.S. Park, Computer vision in healthcare applications. J. Healthc. Eng. **10**, 5157020 (2018)
26. Medical Imaging and Technology Alliance (2018) Medical Imaging Modalities, MITA, Arlington, VA. Accessed 12 Oct 2019
27. V.L. Clark, J.A. Kruse, Clinical methods: the history, physical, and laboratory examinations. JAMA **264**(21), 2808–2809 (1990)
28. I. Gur, D. Gur, J.A. Recabaren, The computerized synoptic operative report a novel tool in surgical residency education. Arch. Surg. **147**(1), 71–74 (2012)
29. Joint Commission on the Accreditation of Healthcare Organizations. Standard IM.6.10, EP 7 Website. http://www.jointcommission.org/NR/rdonlyres/A9E4F954-F6B5-4B2D-9ECF-C1E792BF390A/0/D_CurrenttoRevised_DC_HAP.pdf. Accessed 31 Aug 2019
30. D. Volkland, R.L. Iles, *Guidebook to Better Medical Writing* (Island Press, Washington, DC, 1997)
31. R.A. Rison, A guide to writing case reports. J. Med. Case Rep. **7**, 239 (2013)
32. Y. Luo, P. Szolovits, A.S. Dighe, J.M. Baron, Using machine learning to predict laboratory test results. Am. J. Clin. Pathol. **145**(6), 778–788 (2016)

33. NIH Clinic Center (2017) Chest X-ray images, meta data and diagnosis. https://nihcc.app.box.com/v/ChestXray-NIHCC. Accessed 31August 2019
34. The Cancer Imaging, Archive Medical images of cancer like PET/CT (2019). https://www.cancerimagingarchive.net. Accessed 3 Sept 2019
35. National Biomedical Imaging Archive, Image for development and validation of analytical software tools (2018). https://imaging.nci.nih.gov/ncia/login.jsf. Accessed 31 Aug 2019
36. The Open Access Series of Imaging Studies, Neuroimaging data sets of the brain (2010). http://www.oasis-brains.org/. Accessed 31 Aug 2019
37. The Federal Interagency Traumatic Brain Injury Research, MRI, PET, Contrast, and other data on a range of Traumatic brain injury (TBI) (2012). https://fitbir.nih.gov/. Accessed 31 Aug 2019
38. Clemson University, Structured Analysis of the Retina (STARE) (1975). http://cecas.clemson.edu/~ahoover/stare/. Accessed 1 Sept 2019
39. Alzheimer's Disease Neuroimaging, MRI and PET images, genetics, cognitive tests, CSF and blood biomarkers as predictors of the Alzheimer's disease Initiative (2014). http://adni.loni.usc.edu/. Accessed 1 Sept 2019
40. The Center for In Vivo Microscopy, Duke University, Medical Center Highest resolution images of MRI, CT, X-Ray, ultrasound, confocal, optical, and SPECT (2013). http://www.civm.duhs.duke.edu/SharedData/DataSupplements.htm. Accessed 1 Sept 2019
41. X. Luke, A.T. Layton, N. Wang, P.E.Z. Larson, J.L. Zhang, V.S. Lee, C. Liu, G.A. Johnson, Dynamic contrast-enhanced quantitative susceptibility mapping with ultrashort echo time MRI for evaluating renal function. Am. J. Physiol. Renal Physiol. **310**(2), F174–F182 (2015)
42. Midas Platform: Open-Source Toolkit, Flexible, intelligent data storage system (2010). https://www.insight-journal.org/midas/. Accessed 2 Sept 2019
43. Midas Platform: Open-Source Toolkit, National Alliance for Medical Image Computing (NAMIC) (2010). https://www.insight-journal.org/midas/community/view/17. Accessed 2 Sept 2019
44. Midas Platform: Open-Source Toolkit, Imaging Methods Assessment and Reporting (IMAR) (2010). https://www.insight-journal.org/midas/community/view/15. Accessed 2 Sept 2019
45. Digital Retinal Images for Vessel Extraction, Database for comparative studies on segmentation of blood vessels in retinal images (2012). https://drive.grand-challenge.org/. Accessed 3 Sept 2019
46. Digital Database for Screening Mammography, Mammographic images (2006). http://www.eng.usf.edu/cvprg/Mammography/Database.html. Accessed 3 Sept 2019
47. Public Lung Database to Address Drug Response, A public image database to support research in computer aided diagnosis (2009). http://www.via.cornell.edu/crpf.html. Accessed 3 Sept 2019
48. A.P. Reeves, A.M. Biancardi, D. Yankelevitz, S. Fotin, B.M. Keller, A. Jirapatnakul, J. Lee, A public image database to support research in computer aided diagnosis, in *31st Annual International Conference of the IEEE Engineering in Medicine and Biology Society* (2009), pp. 3715–3718
49. The Osteoarthritis Initiative, Data related to prevention and treat knee osteoarthritis (2013). https://oai.epi-ucsf.org/datarelease/. Accessed 3 Sept 2019
50. Image Sciences Institute, University Medical Center Utrecht, SCR database: Segmentation in Chest Radiographs (2018). http://www.isi.uu.nl/Research/Databases/SCR/. Accessed 3 Sept 2019
51. Japanese Society of Radiological Technology (2004) Digital Image Database. http://db.jsrt.or.jp/eng.php. Accessed 4 Sept 2019
52. J. Shiraishi, S. Katsuragawa, J. Ikezoe, T. Matsumoto, T. Kobayashi, K. Komatsu, M. Matsui, H. Fujita, Y. Kodera, K. Doi, Development of a digital image database for chest radiographs with and without a lung nodule: Receiver operating characteristic analysis of radiologists' detection of pulmonary nodules. Am. J. Roentgenol. **174**(1), 1–74 (2000)
53. LUT School of Business and Management, Standard Diabetic Retinopathy Database Calibration level 1 (2007). http://www2.it.lut.fi/project/imageret/diaretdb1/. Accessed 4 Sept 2019

54. T. Kauppi, V. Kalesnykiene, J.K. Kamarainen, L. Lensu, I. Sorri, A. Raninen, R. Voutilainen, H. Uusitalo, H. Kälviäinen, J. Pietilä, DIARETDB1 diabetic retinopathy database and evaluation protocol, in *Proceedings of the 11th Conference on Medical Image Understanding and Analysis* (Aberystwyth, Wales, 2007)
55. Cornell Visualization and Image Analysis Group, ECLAP public database of whole lung CT images (2019). http://www.via.cornell.edu/databases/. Accessed 4 Sept 2019
56. Omni Medical Search, Medical image and study databases (2008). http://www.omnimedicalsearch.com/image_databases.html. Accessed 4 Sept 2019
57. SpineWeb, A platform for getting spinal images for image analysis (2014). http://spineweb.digitalimaginggroup.ca/. Accessed 5 Sept 2019
58. Facebase, Repository for datasets by organisms, experiment type, age stage, mutation, genotype and more (2019). https://www.facebase.org/chaise/recordset/#1/isa:dataset. Accessed 5 Sept 2019
59. CLAMP, Natural Language Processing Software (2018). https://clamp.uth.edu/. Accessed 16 Oct 2019
60. A.R. Aronson, Effective mapping of biomedical text to the UMLS Metathesaurus: the MetaMap program, in *Proceedings of the AMIA Symposium* (2001), pp. 17–21
61. A.R. Aronson, F.M. Lang, An overview of MetaMap: historical perspective and recent advances. J. Am. Med. Inform. Assoc. **17**, 229–236 (2010)
62. A Medical Language Extraction and Encoding System http://www.medlingmap.org/taxonomy/term/80. Accessed 28 Oct 2019
63. C. Friedman, P.O. Alderson, J.H. Austin, J.J. Cimino, S.B. Johnson, A general natural-language text processor for clinical radiology. J. Am. Med. Inform. Assoc. JAMIA **1**, 161–174 (1994)
64. H. Xu, Z. Fu, A. Shah et al. Extracting and integrating data from entire electronic health records for detecting colorectal cancer cases, in *Proceedings of AMIA Symposium* (2011), pp. 1564–1572
65. J.H. Chiang, J.W. Lin, C.W. Yang, Automated evaluation of electronic discharge notes to assess quality of care for cardiovascular diseases using Medical Language Extraction and Encoding System (MedLEE). J. Am. Med. Inform. Assoc. JAMIA **17**, 245–252 (2010)
66. G.K. Savova, J. Fan, Z. Ye, S.P. Murphy, J. Zheng, C.G. Chute, I.J. Kullo, Discovering peripheral arterial disease cases from radiology notes using natural language processing, in *AMIA Annual Symposium Proceedings* (2010), pp. 722–726
67. T. Groza, S. Köhler, S. Doelken, N. Collier, A. Oellrich, D. Smedley et al., Automatic concept recognition using the human phenotype ontology reference and test suite corpora. Database **2015** (2015)
68. M. Lobo, A. Lamurias, F.M. Couto, Identifying human phenotype terms by combining machine learning and validation rules. BioMed Res. Int. **2017** (2017)
69. M.J. Sobrido Gómez, M. Pardo Pérez,. Automated semantic annotation of rare disease cases: a case study. Database J. Biol. Databases Curation (bau045) (2014)
70. J.B. Hawkins, J.S. Brownstein, G. Tuli, T. Runels, K. Broecker, E.O. Nsoesie, D.J. McIver, R. Rozenblum, A. Wright, F.T. Bourgeois, F. Greaves, Measuring patient-perceived quality of care in US hospitals using Twitter. BMJ Qual, Saf. **25**(6), 404–413 (2016)
71. Automated clinical record keeping. https://trykiroku.com/. Accessed 25 Oct 2019
72. Clinical documentation for iPhone. http://mdops.com/. Accessed 27 Oct 2019
73. D. Demner-Fushman, W.W. Chapman, C.J. McDonald, What can natural language processing do for clinical decision support. J. Biomed. Inform. **42**, 760–772 (2009)
74. V.M. Pai, M. Rodgers, R. Conroy, J. Luo, R. Zhou, B. Seto, Workshop on using natural language processing applications for enhancing clinical decision making: an executive summary. J. Am. Med. Inform. Assoc. **21**, e2–e5 (2014)
75. S. Pradhan, N. Elhadad, B.R. South, D. Martinez, L. Christensen, A. Vogel, H. Suominen, W.W. Chapman, G. Savova, Evaluating the state of the art in disorder recognition and normalization of the clinical narrative. J. Am. Med. Inform. Assoc. **22**, 143–154 (2014)

76. W.W. Chapman, M. Fiszman, J.N. Dowling, B.E. Chapman, T.C. Rindflesch, Identifying respiratory findings in emergency department reports for bio surveillance using MetaMap. Stud. Health Technol. Inform. **107**, 487–491 (2004)
77. L. Cui, S.S. Sahoo, S.D. Lhatoo, G. Garg, P. Rai, A. Bozorgi et al., Complex epilepsy phenotype extraction from narrative clinical discharge summaries. J. Biomed. Inform. **51**, 272–279 (2014)
78. C. Shivade, P. Raghavan, E. Fosler-Lussier, P.J. Embi, N. Elhadad, S.B. Johnson et al., A review of approaches to identifying patient phenotype cohorts using electronic health records. J. Am. Med. Inform. Assoc. **21**, 221–230 (2014)
79. S. Baker, A. Korhonen, S. Pyysalo, Cancer hallmark text classification using convolutional neural networks, in *The Workshop on Building and Evaluating Resources for Biomedical Text Mining* (2016), pp. 1–10
80. M. Asada, M. Miwa, Y. Sasaki, Extracting drug-drug interactions with attention CNNs, in *BioNLP* (2017), pp. 9–18
81. S. Mohan, N. Fiorini, S. Kim, Z. Lu, Deep learning for biomedical information retrieval: learning textual relevance from click logs, in *BioNLP* (2017), pp. 222–231
82. L. Sulieman, D. Gilmore, C. French, R.M. Cronin, G.P. Jackson, M. Russell, D. Fabbri, Classifying patient portal messages using Convolutional Neural Networks. J. Biomed. Inform. **74**, 59–70 (2017)
83. Y.H. Lee, Efficiency improvement in a busy radiology practice: determination of musculoskeletal magnetic resonance imaging protocol using deep-learning convolutional neural networks. J. Digit. Imaging **31**(5), 604–610 (2018)
84. W. Salloum, G. Finley, E. Edwards, M. Miller, D. Suendermann-Oeft, Deep learning for punctuation restoration in medical reports, in *BioNLP* (2017), pp. 159–164
85. H. He, K. Ganjam, N. Jain, J. Lundin, R. White, J. Lin, An insight extraction system on biomedical literature with deep neural networks, in *Proceedings of the 2017 Conference on Empirical Methods in Natural Language Processing* (2017), pp. 2691–2701
86. R. Chalapathy, E.Z. Borzeshi, M. Piccardi, Bidirectional LSTM-CRF for clinical concept extraction (2016). arXiv 7–12 (2016)
87. S.M. Shortreed, E. Laber, D.J. Lizotte, T.S. Stroup, J. Pineau, S.A. Murphy, Informing sequential clinical decision-making through reinforcement learning: an empirical study. Mach. Learn. **84**(1–2), 109–136 (2011)
88. T. Pham, T. Tran, D. Phung, S. Venkatesh, Deepcare: a deep dynamic memory model for predictive medicine, in *Pacific-Asia Conference on Knowledge Discovery and Data Mining* (Springer, Cham, 2016), pp. 30–41
89. E. Choi, M.T. Bahadori, J. Sun, J. Kulas, A. Schuetz, W. Stewart, Retain: an interpretable predictive model for healthcare using reverse time attention mechanism, in *Advances in Neural Information Processing Systems* (2016), pp. 3504–3512
90. C. Pettit et al., Developing a multi-scale visualization toolkit for use in climate change response. Landscape Ecol. **2012** (2012)
91. Y. Liu, A. Kot, F. Drakopoulos, C. Yao, A. Fedorov, A. Enquobahrie et al., An ITK implementation of a physics-based non-rigid registration method for brain deformation in image-guided neurosurgery. Front. Neuroinform. **8**(33) (2014)
92. I. Larrabid, P. Omedas, Y. Martelli, X. Planes, M. Nieber, J. Moya et al., GIMIAS: an open source framework for efficient development of research tools and clinical prototypes, in *International Conference on Functional Imaging and Modeling of the Heart* (Springer, Berlin, 2009), pp. 417–426
93. A. Esteva, B. Kuprel, R.A. Novoa, J. Ko, S.M. Swetter, H.M. Blau, S. Thrun, Dermatologist-level classification of skin cancer with deep neural networks. Nature **542**(7639), 115–118 (2017)
94. A. Esteva, B. Kuprel, R.A. Novoa, J. Ko, S.M. Swetter, H.M. Blau et al., Dermatologist-level classification of skin cancer with deep neural net-works. Nature **542**, 115–118 (2017)
95. R. Poplin, A.V. Varadarajan, K. Blumer, Y. Liu, M.V. McConnell, G.S. Corrado et al., Prediction of cardiovascular risk factors from retinal fundus photographs via deep learning. Nat. Biomed. Eng. **2**(3), 158 (2018)

96. S. Liu, W. Cai, S. Pujol, R. Kikinis, D. Feng, Early diagnosis of Alzheimer's disease with deep learning, in *2014 IEEE 11th International Symposium on Biomedical Imaging (ISBI)* (2014), pp. 1015–1018
97. A. Prasoon, K. Petersen, C. Igel, F. Lauze, E. Dam, M. Nielsen, Deep feature learning for knee cartilage segmentation using a triplanar convolutional neural network, in *International Conference on Medical Image Computing and Computer-Assisted Intervention* (Springer, Berlin, 2013), pp. 246–253
98. Y. Yoo, T. Brosch, A. Traboulsee, D.K. Li, R. Tam, Deep learning of image features from unlabeled data for multiple sclerosis lesion segmentation, in *International Workshop on Machine Learning in Medical Imaging* (Springer, Cham, 2014), pp. 117–124
99. J.Z. Cheng, D. Ni, Y.H. Chou, J. Qin, C.M. Tiu, Y.C. Chang et al., Computer-aided diagnosis with deep learning architecture: applications to breast lesions in US images and pulmonary nodules in CT scans. Sci. Rep. **6**, 24454 (2016)
100. V. Gulshan, L. Peng, M. Coram, M.C. Stumpe, D. Wu, Narayanaswamy et al., Development and validation of a deep learning algorithm for detection of diabetic retinopathy in retinal fundus photographs. JAMA **316**(22), 2402–2410 (2016)
101. Y. Cheng, F. Wang, P. Zhang, J. Hu, Risk prediction with electronic health records: a deep learning approach, in *Proceedings of the 2016 SIAM International Conference on Data Mining* (2016), pp. 432–440
102. F. Sahba, H.R. Tizhoosh, M.M. Salama, A reinforcement learning framework for medical image segmentation, in *The 2006 IEEE International Joint Conference on Neural Network Proceedings* (2006), pp. 511–517
103. F. Sahba, H.R. Tizhoosh, M.M. Salama, Application of opposition-based reinforcement learning in image segmentation, in *2007 IEEE Symposium on Computational Intelligence in Image and Signal Processing* (2007), pp. 246–251
104. D. Liu, T. Jiang, Deep reinforcement learning for surgical gesture segmentation and classification, in *International Conference on Medical Image Computing and Computer-Assisted Intervention* (Springer, Cham, 2018), pp. 247–255
105. A. Alansary, O. Oktay, Y. Li, L. Le Folgoc, B. Hou, G. Vaillant et al., Evaluating reinforcement learning agents for anatomical landmark detection. Med. Image Anal. **53**, 156–164 (2019)
106. S.R. Soroushmeh, K. Najarian, Transforming big data into computational models for personalized medicine and health care. Dialogues Clin. Neurosci. **18**(3), 339–343 (2016)
107. F. Cabitza, R. Rasoini, G.F. Gensini, Unintended consequences of machine learning in medicine. JAMA **318**(6), 517–518 (2017)
108. R.B. Correia, L. Li, L.M. Rocha, Monitoring potential drug interactions and reactions via network analysis of instagram user timelines, in *Biocomputing 2016: Proceedings of the Pacific Symposium* (2016), pp. 492–503
109. A. Nikfarjam, A. Sarker, K. O'Connor, R. Ginn, G. Gonzalez, Pharmacovigilance from social media: mining adverse drug reaction mentions using sequence labeling with word embedding cluster features. J. Am. Med. Inform. Assoc. **22**(3), 671–681 (2015)
110. M. Abadi, A. Chu, I. Goodfellow, H.B. McMahan, I. Mironov, K. Talwar, L. Zhang, Deep learning with differential privacy, in *Proceedings of the 2016 ACM SIGSAC Conference on Computer and Communications Security* (2016), pp. 308–318
111. R. Shokri, V. Shmatikov, Privacy-preserving deep learning, in *Proceedings of the 22nd ACM SIGSAC Conference on Computer and Communications Security* (2015), pp. 1310–1321
112. F. Tramèr, F. Zhang, A. Juels, M.K. Reiter, T. Ristenpart, Stealing machine learning models via prediction apis, in *25th USENIX Security Symposium* (2016), pp. 601–618
113. K. Chaudhuri, C. Monteleoni, A.D. Sarwate, Differentially private empirical risk minimization. J. Mach. Learn. Res. **12**, 1069–1109 (2011)
114. R. Gilad-Bachrach, N. Dowlin, K. Laine, K. Lauter, M. Naehrig, J. Wernsing, Cryptonets: applying neural networks to encrypted data with high throughput and accuracy, in *International Conference on Machine Learning* (2016), pp. 201–210
115. A.C. Yao, Protocols for secure computations, in *23rd Annual Symposium on Foundations of Computer Science* (1982), pp. 160–164

116. D.E. Oliver, Y. Shahar, E.H. Shortliffe, M.A. Musen, Representation of change in controlled medical terminologies. Artif. Intell. Med. **15**(1), 53–76 (1999)
117. K. Majumdar, Human scalp EEG processing: various soft computing approaches. Appl. Soft Comput. **11**(8), 4433–4447 (2011)
118. T. Balli, R. Palaniappan, Classification of biological signals using linear and nonlinear features. Physiol. Meas. **31**(7), 903 (2010)
119. P.S. La Rosa, A. Nehorai, H. Eswaran, C.L. Lowery, H. Preissl, Detection of uterine MMG contractions using a multiple change point estimator and the K-means cluster algorithm. IEEE Trans. Biomed. Eng. **55**(2), 453–467 (2008)
120. M. Sun, F. Tang, J. Yi, F. Wang, J. Zhou, Identify susceptible locations in medical records via adversarial attacks on deep predictive models, in *Proceedings of the 24th ACM SIGKDD International Conference on Knowledge Discovery & Data Mining* (2018), pp. 793–801

Dr. Shweta Agrawal has more than 15 years of teaching experience and 10 years of research experience. 8 Ph.D. scholars are doing their research work under her guidance. She has guided more than 15 PG scholars. She is coordinator of nationwide project of Bennett University Noida and Royal Academy London for AI skilling and research. She has worked in the areas of Distributed computing, orchestration, machine learning, deep learning, medical image processing. She has delivered sessions in FDPs and conference on machine learning and chaired session in International conference. She has received best teaching award, work of excellence award and aatammanthan award at University level. She has reviewed Elsevier and other reputed journal papers.

Mr. Sanjiv Kumar Jain has more than 16 years of teaching experience and 7 years of research experience in the field of electrical and electronics. He is M.E. & Ph.D. (Thesis submitted). Mr. Jain has worked in the areas of power system security, optimization, renewable energy sources, artificial intelligence, machine learning. He has received several accolades like best faculty and best project mentor at University level. He is also reviewer of reputed international peer-reviewed journals.

Machine Learning Methods for Managing Parkinson's Disease

Kunjan Vyas, Shubhendu Vyas and Nikunj Rajyaguru

Abstract A neurodegenerative disorder without permanent cure, Parkinson's Disease (PD) increasingly hinders motor and cognitive abilities. Timely intervention of neuroprotective therapies can help minimize the early impairments in PD. Early diagnosis would play major role in facilitating such proactive treatment plan. However, the conventional methods of PD diagnosis suffer from less accessibility, high costs, human bias and patient inconvenience. Moreover, there is a dearth of high-frequency monitoring systems to track the progression. Deficient monitoring and management of the progression diminishes both quality of life and life expectancy of the patient. The challenges and concerns in conventional methods of diagnosis and treatment of PD call for use of advanced technology like Machine Learning (ML) and Internet of Things (IoT). The proposed chapter is aimed at giving insights into robust and effective practical implementation of ML with IoT in PD care. Non-invasive biomarkers data from human voice and keystrokes (tapping) are demonstrated as promising base for early diagnosis. These illustrations focus on ease of building cost efficient and scalable PD prediction systems. In addition, a multitude of contemporary developments and inspiring future opportunities for managing PD with ML are highlighted.

Keywords Parkinson's disease · Machine learning · Internet of Things · Healthcare · Diagnosis · Monitoring

---

K. Vyas (✉) · S. Vyas
TotemXAI, Karlsruhe, Germany
e-mail: kv.totemxai@outlook.com

S. Vyas
e-mail: sv.totemxai@outlook.com

N. Rajyaguru
Robert Bosch GmBH, Karlsruhe, Germany
e-mail: nikunj.rajyaguru@de.bosch.com

V. Jain and J. M. Chatterjee (eds.), *Machine Learning with Health Care Perspective*, Learning and Analytics in Intelligent Systems 13,
https://doi.org/10.1007/978-3-030-40850-3_12

1 Introduction

Neurodegeneration leads to the continued deterioration in the function of neurons eventually leading to death of these essential cells. Among hundreds of neurodegenerative disorders, Alzheimer's disease and Parkinson's disease (PD). Increasing growth rate of aging people is one of the most significant factors for rise in the number of people with a neurodegenerative disorder.

Looking at developed countries, as per US Parkinson foundation, there will be over 1 million PD patients in US alone by 2020, a number larger than total patients with multiple sclerosis, muscular dystrophy and ALS. While the National Institute of health reports growth rate of 50,000 new patients every year. Worldwide nearly 10 million people are estimated to be affected by PD. PD prevalence is likely to be twofold by 2030 [1]. In these estimates, one should also factor in the lesser amount of awareness and diagnosis in developing countries.

With these numbers, the diagnosis and treatment for PD have much to achieve and improve. PD care suffers from lack of early diagnosis which could control the progression and stop irreversible damage. The patient treatment is also constrained by insufficient, infrequent symptom monitoring and lesser access to specialist care globally. The intermittent encounters with medical practitioners lead to intuitive decision-making and risk misdiagnosis. For the patient it results in hampered quality of life (QoL) and high-end expenditures of therapy, medications and surgery at times. Hence, timely prediction of patients at risk of PD can help with more befitting and personalized treatment planning.

The authors aim to show that Machine learning (ML) and Internet of Things (IoT) can help restrain the impact of this chronic disease with early diagnosis and continuous symptom monitoring. Use of widely available tools like computers, smartphones, digital assistants and wearable sensors facilitates high frequency data collection and management. Tests like walking, voice, tapping and memory can be devised as an ML application on smart devices to manage PD. In present chapter, current scenario in PD care is discussed and to improve it, practical ML implementations using voice and tapping data are demonstrated.

2 Current Scenario

PD is close second to Alzheimer in terms of prevalence [2] and predominantly affects the motor functions and cognitive abilities.

Traditional methods are still prevalent in medical science and PD care is not an exception to it. Below are the traditional methods used for PD patient care.

Table 1 Symptoms classification

Motor		Non-motor	
	Involuntary tremors Slow rhythmic rest tremors **Rigidity** Stiffness in Torso **Bradykinesia** Slow movement Difficult motor coordination Micrographia (small/mixed hand writing) Facial masking **Postural imbalance** **Walking disabilities or gait** **Dystonia** Involuntary movements in eyes, neck, trunk and limbs **Vocal changes** Softer lower voice Stuttering or rapid words		**Fluctuations in the olfactory sense** **Disturbed sleep cycles** **Depression and distress** **Pain, sweating and fatigue** **Psychosis** Hallucinations **Cognitive shifts** Slow thinking **Weight loss** **Digestive difficulties** **Dizziness** **Urination problems** **Sexual challenges** **Melanoma** Skin lesions Risk of skin cancer **Personality changes** **Eye and vision issues**

2.1 Diagnosis

2.1.1 Understanding the Symptoms

Here is a short compilation of symptoms explained by American Parkinson Disease Association (APDA) (Table 1).

Progression of PD is highly variable, meaning different patients exhibit different combinations of symptoms and severity of symptoms on an individual level [3].

A useful list of early warning signs of PD is compiled by Mayo Clinic is as shown in Fig. 1.

2.1.2 Tests

Unlike most diseases, PD cannot be detected by blood tests or regular lab tests. Standard method for diagnosing PD is performing a clinical assessment of motor symptoms. However, the probability of absence or varying severity of the symptoms of PD during the clinical assessment may lead to incorrect observations. For example, many cases motor impairment appears much later to cognitive symptoms causing much of neurological damage.

Second test is by administering Levodopa and check for a positive response which is again subjective to individual patient response and of course the unsolicited side effects.

Another level of confirmation is possible with brain scans like with imaging and tomography tests.

Fig. 1 Early signs of PD

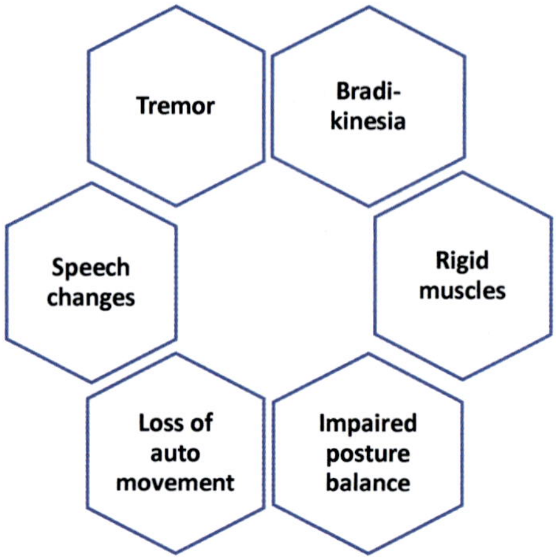

Early symptoms largely coincide the effects of aging. In addition, PD symptoms also resemble to other neurodegenerative or non-neurodegenerative conditions, making the exact identification difficult.

2.2 Treatment

It is needless to mention that the disease has no cure till date and unfortunately the treatment involves only controlling the symptoms and progression through medicines, therapies and surgery.

For more than 5 decades, the symptomatic treatment of PD focuses on controlling the Dopamine deficit with help of Levodopa (Brand name Sinemet). MAO-B inhibitors is another category to tackle Dopamine breakdown. Though the severe adverse effects on motor performance resulting from such medicine defeat the very purpose [4]. PD patients are prone to develop fluctuations in motor and/or cognitive functions as side effects, e.g. levodopa induced Dyskinesia, levodopa induced Psychosis, vomiting and so on. Deep brain simulations or surgery are the options depending on the stage of the patient when medications are not effective enough.

Identification of the stage and severity of the disease is carried out by the Movement Disorder Society-Unified Parkinson's Disease Rating Scale (MDS-UPDRS) and Hoehn and Yahr (HY) Scale [5].

3 Concerns and Challenges

3.1 Diagnosis

Main priority for any case is to halt the progression with minimal treatment, saving the neurophysiological integrity of the neurons—which can only be possible with early diagnosis.

Readers are encouraged to refer about the indicators of misdiagnosis as summarized with evident statistics in [6]. Here we will briefly evaluate the diagnosis modalities listed above.

The accuracy of PD tests in clinical setting has been varying based on the individual expertise level and human bias. Approx. 25% of PD diagnoses are incorrect when compared to post-mortem autopsy. Sadly, in certain cases the putamenal Dopamine is depleted to nearly 80%, and around 60% of SNpc dopaminergic neurons have already been lost, before physical symptoms are detectable (in terms of conventional diagnosis criterion) [7, 8].

While the Levodopa challenge predict 70–81% cases, researchers suggest it to be redundant and trust the physical-neurological tests more [9]. Not to mention the uncertain risk of subjecting the patient to unnecessary side effects.

Among the scan tests, MRI tests are next level recommendation for cases where the primary symptoms cannot be ascertained. PET requires to subject the patient to a radioactive substance injection to check the abnormalities of Dopamine transmission. CT scans create a 3D picture for the doctors with main aim to rule out other conditions and not direct diagnosis of PD. DaTscan, another radioactive IV scanning test can only confirm a prognosis and not make first-hand diagnosis of PD. In addition, it has limitations in distinguishing PD with other Dopamine depleting disorders. Given the cost to effectiveness trade-off combined with the administration complexity, these tests are more suitable for later stages and not for early diagnosis.

Thus, requirement of simple yet effective diagnostic tools to capture early signs as shown in Fig. 1, is evident. Under applications of ML, such practical implementation examples would be discussed.

3.2 Treatment

In the treatment area, the options are limited. The primary option is medication as explained above. The dosage is decided based on clinical assessment, HY scores and MDS-UPDRS scoring system. Correct identification of severity and stage (patient profiling) is crucial to ensure the patient is not under-medicated or over-medicated.

The HY process of stage identification is questionnaire-based while MDS-UPDRS is four-scale structured test, performed by a movement disorder specialist. Certainly, they are the gold standards to measure the progression of PD, the individual variance in symptoms leaves the judgement to human bias in absence of adequate statistical inferences. The patient stats/observations are "snapshots" in clinical setting. However, the ongoing impact of "On" and "Off" phases with Levodopa dosage daily life are continuously varying. Advance yet simple-to-use monitoring IoT tools and ML supported management system can help the doctors to capture nuanced data and deep implicit insights about treatment effectiveness for the specific patient, as well as ensure patient's adherence to the treatment plan.

4 Application of Machine Learning in Parkinson Disease

To address the concerns and challenges in diagnosis and treatment in PD, the authors propose leveraging ML and IoT.

It has been a popular belief that ML requires a large database, state-of-art computing facilities and knowledge of complex algorithms.

The authors have proposed easy-to-implement models. There is a vast variety of applications of ML in PD. In the proposal a few examples are discussed in order to open the discussion for larger arena of possibilities.

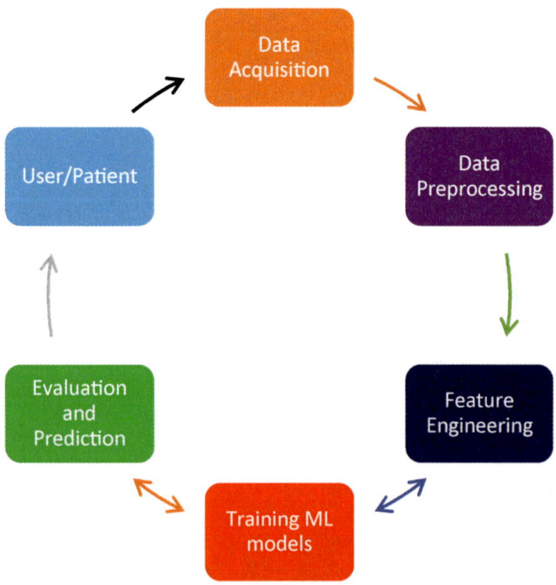

Fig. 2 Machine learning model workflow

4.1 General Workflow

Figure 2 represents the generic process flow of a typical Machine Learning model. It starts with raw data acquisition from user. To make data understandable, we need to extract features from the data which can be fed further to ML algorithm. So, in the next step the data is cleaned and preprocessed.

In feature engineering, we remove unnecessary features as well as create some new features which would help with training the ML model. The dataset with selected features is split in Train and Test dataset, where Test dataset is reserved for evaluation and prediction purpose. The Train dataset is further split for validation purpose. ML model is trained on Train dataset and validated with validation dataset. Once model is finalized the predictions are made for Test dataset, which are provided to user. From feedback of users and predictions the model is improvised with more data and/or better ML algorithms.

4.1.1 Data Acquisition

Biomarkers are essential for accurate and early diagnosis of any disease or disorder. Advancement of health-tech has made obtaining biomarkers less complicated nowadays. The collected biomarkers data have become tools to drive ML applications in healthcare. In this chapter our particular focus is non-invasive biomarkers data to develop ML model for diagnosis of PD. In case of PD some of the non-invasive biomarker resources are human voice, finger movements, imaging etc.

Two different PD diagnosis ML models are developed with following datasets:

1. Sound Recordings of PD patients and healthy controls
2. Keystrokes of PD patients and healthy controls

The selection of these datasets has been made based on factors such as

- Ease and cost of data acquisition,
- Type and size of data,
- Continuous data update,
- Amount of clinical attention required and
- Physical visit to clinic by patient.

Table 2 Biomarkers dataset selection factors

Dataset	Data acquisition		Data update	Clinical intervention	Physical visit	Analysis complexity
	Ease	Cost				
Voice	Easy	Low	Continuous	Partial	No	Low
Keystrokes	Easy	Low	Continuous	No	No	Moderate
Imaging	Moderate	High	Occasional	Must	Yes	High
Genetic	Hard	Very high	Rare	Must	Yes	High

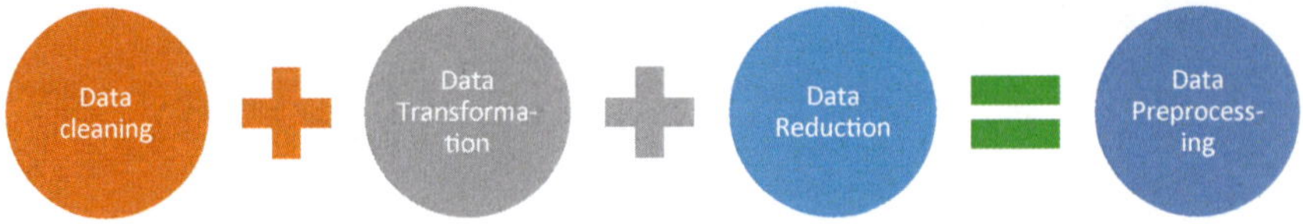

Fig. 3 Data preprocessing

As shown in Table 2, imaging datasets are expensive and not available easily. Patient needs to physically visit the lab to take the MRIs and the analysis or diagnosis through it requires an expert evaluation. Moreover, even with help of ML, analysis and prediction are computationally intensive.

Voice and Keystrokes datasets are simpler and cheaper to obtain. The features and amount of data can be continuous which can facilitate regular improvements to the ML model. Since human voice and typing keystrokes can be recorded from anywhere, no physical visit to clinic is required. Data interpretation and analysis can be accomplished with minimal expert assistance.

4.1.2 Data Preprocessing

The data recorded from any medium, e.g. sensors, needs to be converted in machine comprehensible form. Data cleaning is the first step carried out to remove unnecessary data and outliers. It is very important to identify significant data required for particular problem. For example, there is a possibility of corrupt or missing data fields which should be eliminated. Consequently, transformation of data i.e. attributes selection, scaling and normalization of the data is done. Sometimes in the case of large dataset the size of the data is reduced to save on computational efforts (Fig. 3).

4.1.3 Feature Engineering

Most essential and crucial step, feature engineering can make or break an ML model. This is the part where domain knowledge can be helpful (Fig. 4).

Feature engineering makes data interpretable for ML algorithms. Creating new features, removing unused features, checking correlation of features, one-hot encoding categorical features, detecting outliers, grouping operations, transformation of features are part of feature engineering process. Feature selection impacts the model performance significantly. The selection is made using different feature combinations, feature importance and model evaluation.

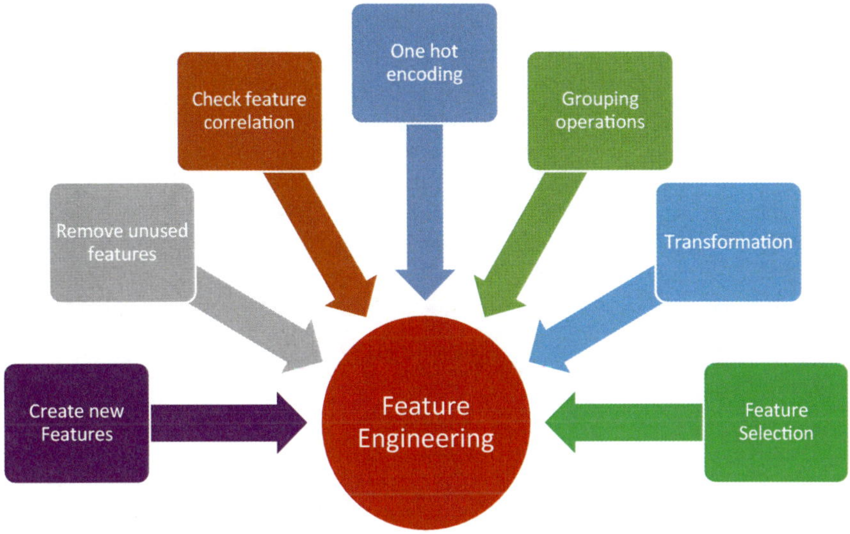

Fig. 4 Feature engineering

4.1.4 Train Machine Learning Models

Before training ML model, feature engineered data is split into Train and Test datasets to reserve a data subset for final evaluation of the model. Training dataset is further divided with validation set for Cross Validation (CV) purpose. Since it is not completely straightforward to choose an apt ML model, general practice should be

Fig. 5 Training cycle

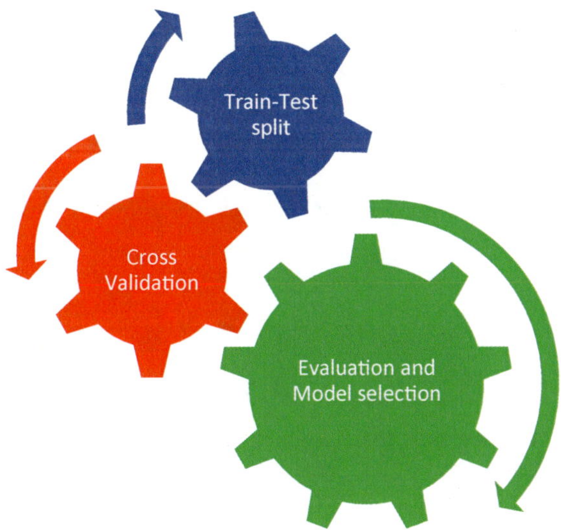

training dataset through multiple machine learning algorithms (Fig. 5). There are many impressive ML open source libraries consisting various algorithms available. To name a few popular libraries and algorithms are Support Vector Machine (SVM), Linear and Logistic Regression, K-Nearest Neighbors (KNN), Naïve Bayes, Tree Regressors, Boosted Tree Regressors and Neural Networks.

Next in the process a few selected algorithms are cross validated and compared using technique such as K-fold CV method. More details on K-fold CV and ML algorithms are covered in Example 1: Voice data based early diagnosis. In cases of insufficient accuracy or overfitting/underfitting, any/all the processes of more data acquisition, data preprocessing and feature engineering are repeated, and models are retrained.

4.1.5 Evaluation and Prediction

On the basis of the evaluation from cross validation and test dataset, we can employ an ideal ML candidate (algorithm) to predict on future input data. The evaluation metrics are picked according to problem definition and output requirements. Some of the popular metrics are Root Mean Squared Error (RMSE), Mean Absolute Error (MAE), Log Loss, Accuracy, F1-Score, Confusion Matrix, Area Under Curve (AUC), Matthews Correlation Coefficient (MCC) etc. The test datasets in present chapter are evaluated with accuracy, sensitivity, specificity and MCC.

$$accuracy = \frac{(TP + TN)}{(TP + FP + TN + FN)}$$

$$sensitivity = \frac{TP}{(TP + FN)}$$

$$specificity = \frac{TN}{(FP + TN)}$$

$$MCC = \frac{(TP \times TN) - (FP \times FN)}{\sqrt{((TP + FP)(TP + FN)(TN + FP)(TN + FN))}}$$

TP = number of true positives, TN = true negatives, FP = false positives, FN = false negatives

True Positive (TP) True label: PD ML model predicted: PD	**False Positive (FP)** Reality: Healthy ML model predicted: PD
False Negative (FN) Reality: PD ML model predicted: Healthy	**True Negative (TN)** Reality: Healthy ML model predicted: Healthy

Accuracy accounts for the quantity of correct predictions (both TP and TN) from the total number of examples. Accuracy is not considered a decisive evaluation to check performance of a model especially if the data is unbalanced (more samples of particular class than the other). For example, if there are 5 PD patients and 45 healthy controls in the data, an ML classifier might classify all the subjects as healthy. The accuracy for this classifier would be 90% but it only classifies healthy controls, and it doesn't recognize PD patients.
Sensitivity (true positive rate) is the ratio of the rightly predicted PD subjects by classifier to all who are actually PD patient. That means with higher sensitivity, fewer actual PD patient would go undiagnosed.
Specificity (true negative rate) is the ratio of correctly predicted healthy controls by classifier to all who are actually healthy controls. So higher specificity means lesser subjects would be labeled PD patient.
MCC shows the quality of binary classification model. It helps one to gauge how balanced the binary classifier is. The value of MCC ranges between of -1 to 1. The lower bound value (-1) is sign of highly incorrect binary classifier while an upper bound value (1) indicates a completely correct binary classifier.

4.1.6 Continuous Improvement

There is no perfect or 100% accurate ML model. Continuous improvements can be rendered by recurrent cycle fetching more data, creating better features, employing advanced ML algorithms and training more robust ML models.

5 Implementation: Machine Learning in Parkinson Disease

With minor adaptions, the biomarkers serve useful data points for both diagnosis or treatment monitoring purposes. In the present context, authors would discuss practical examples of two diagnosis examples on two different open source datasets.

Given the data security compliances, on-going treatment data is usually not made public. Open secured data sharing would help improving the treatment globally. However, such policy issues are already under discussion in Healthcare arena, hence are not made part of this chapter.

With two detailed diverse diagnosis examples on open datasets, the authors provide a practical beginner's guide of ML in PD care for healthcare professionals. Although, a conceptual summary of existing implementations in treatment would also be provided to motivate the readers in that direction.

5.1 Example 1: Voice Data Based Early Diagnosis

Vocal disorders are exhibited by 90% of PD patients in the earlier stages of the disease [10]. To detect and diagnose PD, voice can be used as it can indicate decrease in motor control which is the hallmark of PD [11]. Low cost of gathering voice samples and doing signal processing on them is the primary reason behind the popularity of PD diagnosis using speech impairments [12, 13]. The interest in diagnosis of PD through vocal tests has grown over the years. One can refer [10] to know almost all the relevant research done so far.

5.1.1 Objective

To show that how simple it is to develop a predictive binary classification ML model to distinguish PD patients from healthy controls using only human voice.

5.1.2 Solution

To build voice-based ML classification model and explain the general ML workflow, outcome and future possible improvements. Two datasets, different ML classifier algorithms and corresponding evaluation metrics are discussed.

5.1.3 Data Acquisition

In this example we have acquired open source voice datasets available on UC Irvine Machine Learning Repository. These datasets were donated by [10, 14]. The datasets are:

I. Parkinson Speech Dataset with Multiple Types of Sound Recordings [14]
II. Parkinson's Disease Classification Data Set [10]

I. Parkinson Speech Dataset with Multiple Types of Sound Recordings

20 patients (6 female, 14 male) with Parkinson's Disease and 20 healthy controls (10 female, 10 male) are involved in building this dataset by Department of Neurology in Cerrahpasa Faculty of Medicine, Istanbul University. From all subjects, 26 voice samples are recorded. From each voice sample 26 linear as well as time frequency-based features are extracted (Fig. 6).

One can use this dataset for regression too by using UPDRS (Unified Parkinson's Disease Rating Scale) score of each patient. All the data features are organized in columns and the voice samples are structures in rows. Feature extraction of features from voice samples is carried out through Praat [15] acoustic analysis software.

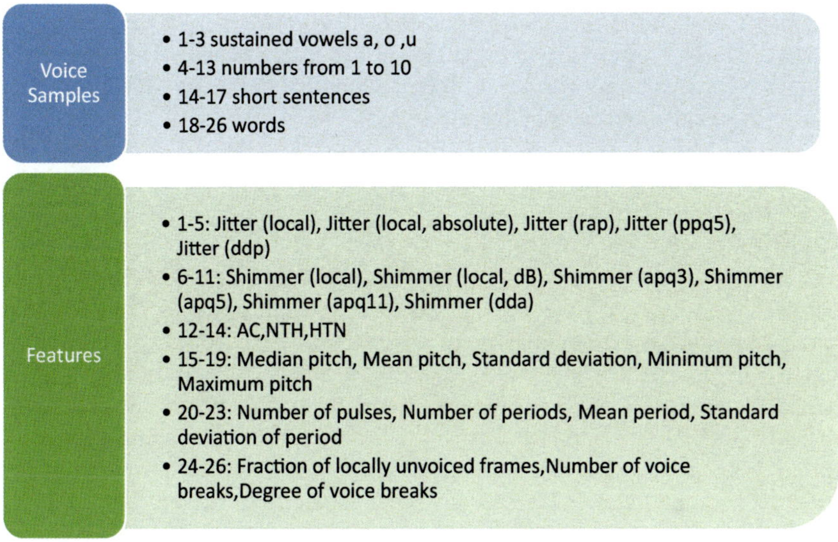

Fig. 6 Dataset I, voice samples and features

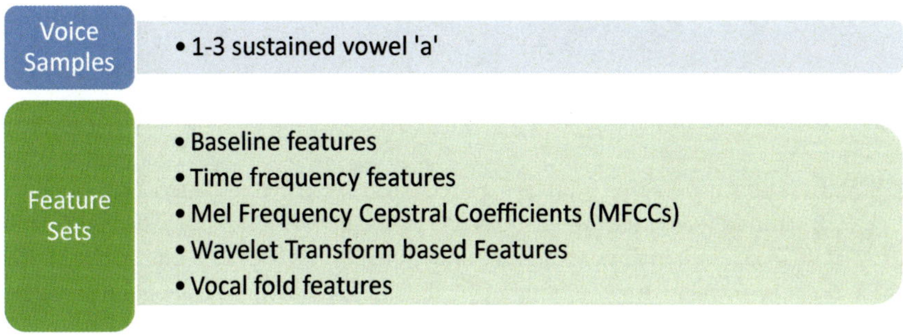

Fig. 7 Dataset II, voice samples and features

II. Parkinson's Disease Classification Data Set

188 patients with PD (107 men and 81 women) of age between 33 and 87 (65.1 ± 10.9) and the control group of 64 healthy controls (23 men and 41 women) with ages ranging from 41 to 82 (61.1 ± 8.9) are considered for this dataset. This dataset is also compiled at Department of Neurology in Cerrahpasa Faculty of Medicine, Istanbul University. The sustained phonation of the vowel 'a' was collected from each subject using the microphone set to 44.1 kHz and the physician's examination. The same vowel is recorded by every subject three times.

There are 5 feature sets having multiple number of features in each feature-set. Here only feature-set names are mentioned (Fig. 7). For full list of features, please refer paper [10]. Baseline features are extracted with the Praat acoustic analysis software. Moreover, from the spectrograms of the speech signals, Time frequency features too are extracted with Praat. Refer [10] for the details on MFCC, Wavelet Transform based and Vocal fold features.

5.1.4 Data Preprocessing

Both the datasets are carefully collected and structured by the researchers. Hence no data cleaning is required. Moreover, datasets are considerably small, so data reduction is necessary either. As a part of data preprocessing, first step is to identify the data into feature columns as X and labels (also called target variable) as y. Feature columns X are already defined in data acquisition section. Here in both the datasets, y consists labels '0' for healthy controls and '1' for patients with PD. X contains all type of features where we may have numeric values in different units or scale. For example, in dataset (I) some Jitter features have values ranging from 10^{-1} to 10^{-5} whereas some pitch features are of 10^{2}. Hence scaling and normalization of X is performed. Also, for some of the ML algorithms to produce predictions correctly, normalization and scaling of the data is advised. We have labeled 'Class' data column as y in both the datasets. All the X feature columns are scaled/normalized using a function from open source library called scikit-learn.

5.1.5 Feature Engineering

Both the datasets are created with the help of domain experts and therefore there are very well curated features in the dataset. As a result, we will not have to deal with many of the processes in feature engineering. Some of the features such as patient ID which do not contribute to prediction of PD/healthy label and so have been dropped. Since there are no categorical variables/features, one-hot-encoding is considered.

For dataset (II), there are 750+ features, out of which we will select only certain number of features. We will start by taking only Baseline features for initial model development. We will include a greater number of features and observe the change in output. Our aim is to demonstrate the simplest way of building a PD classification model with ML, complex feature selection technique details are avoided. We will also present results with all the features. Although it is usually recommended to perform feature selection and include only most important features in the final model. Significant amount of Feature Engineering shall be covered in Sect. 5.2.

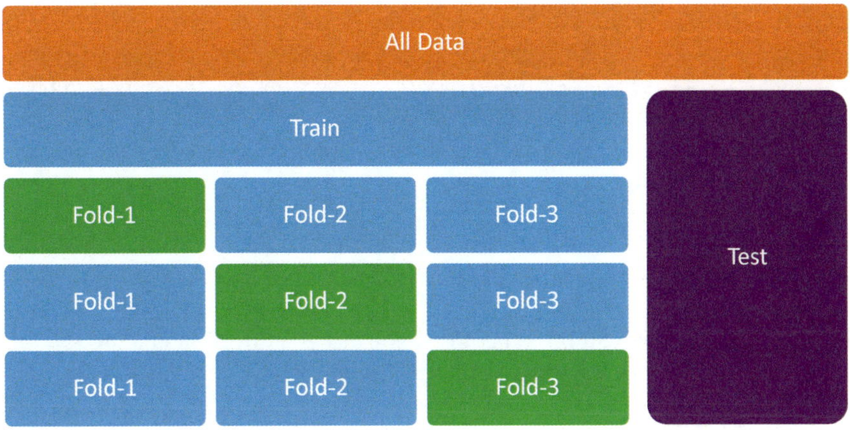

Fig. 8 K-Fold CV with k = 3

5.1.6 Train Machine Learning Models

Train-Test Split and Cross Validating Model

It would be a methodological mistake if the model is trained to learn the prediction function parameters and testing the model on the same data. Such model leads to a situation called overfitting. In this situation model would not predict correctly on yet unseen data. To avoid it, firstly we will split our feature engineered data into two datasets. Namely Train dataset and Test Dataset. Test dataset is the unseen dataset on which we verify our model performance before deploying the model into system. In current example the 80% dataset is split for Train with 20% reserved for Test.

To build a robust ML model it is important to make predictions on data not used during the training of the model. In machine learning cross-validation (CV) is a popular technique which is applied to estimate how the model will perform in general on new data. For this we can hold out some more part of the training dataset in terms of validation dataset. However, this may cause available training samples reduced to the amount which is not sufficient for learning the model. To address this issue, a method called *K-fold Cross Validation* can be implemented. In this method the training data is split by k number of folds. This method is very useful for dataset with small number of samples.

The procedure for k-fold CV is as follows

- Randomly split your train dataset into k folds
- Building a model on k − 1 folds of the dataset for each k-fold and validating model performance on the k-th fold
- The process is repeated until every k-folds is used as the validation set.

As shown in Fig. 8, for k = 3 the train dataset is divided in 3 folds, where the dataset will be iterated 3 times. In every iteration 1 fold (in green) is held out for validation purpose and other 2-folds will be used to train the model. In our example for both the datasets we use k = 10. In present example we have implemented a variation of K-fold CV called *StratifiedKfold* from scikit-learn library. The only difference is, in this method the class/label information is taken into consideration. This helps with imbalanced class datasets. In other words, the folds are created with retaining percentage of each class in the samples.

Machine Learning Classification Algorithms

As there is no all-powerful medicine for all the disease, there is no one-for-all ML classifier algorithm for every classification problem. We have implemented and compared some of the most effective and industry standard ML classifiers. We will give brief introduction of the algorithms we have used here. Theoretical and practical literature of all these algorithms is openly available.

I. Support Vector Machine (SVM): SVM is a well-known supervised ML algorithm. In practice it can be used for both classification or regression problems. Usually it works very good for classification of small dataset. SVM produces an optimal hyperplane which classifies new examples for given labeled training dataset. This hyperplane would be a line dividing a plane in two parts if we are dealing with a binary classification problem. Here each class lays in either side of the line.
II. Logistic Regression (LR): LR is one of the most reliable method for binary classification problems. Logistic regression uses a function called sigmoid to return a probability value of output. LR predicts the probability of the default class, i.e. the first class. Let's say we want to predict if person suffers with PD or not. In this scenario PD patient could be the first class. Given a person's age, the logistic regression model is written as the probability of PD patient. So, it will imply that the older person will have higher possibility of having PD than younger person.
III. K-Nearest Neighbors (KNN): This algorithm assumes that similar things are near to each other. The standard steps of KNN are

   i. Load the data
   ii. Calculate distance (Euclidian) from the new data to already classified data
   iii. Based on distance values do the sorting
   iv. Pick k top sorted values (value of k is defined by user)
   v. Count the frequency of each class that appears and select the class which appeared the most
   vi. Return the value of selected class

IV. Multi-layer Perceptron (MLP) Classifier: MLP uses neural-network (NN) to perform classification. A simple NN is made of at least 3 layers, namely an input layer, a hidden layer and an output layer. MLPs execute in two motions, forward pass and backward pass. The input is fed from the input layer through the hidden layers to the output layer, and the prediction is measured against the true labels. This is called forward pass. For the backward pass, backpropagation is used to move the MLP one step closer to the error minimum. In general, to perform this operation, an algorithm called stochastic gradient descent is employed. The procedure is repeated with an aim is to attain convergence, i.e. achieve lowest error possible.

V. Random Forest (RF): RF is probably the most popular classification algorithm. The underlying concept for RF is based on decision tree classifier. In other words, decision trees are building blocks of RF. RF being an ensemble learning method, it works quite differently than above discussed algorithms. Ensemble of individual decision trees creates Random Forrest. Usually a greater number of trees leads to higher accuracy. Simple example: let's say we have features [x1, x2, x3, x4] and corresponding targets as [y1, y2, y3, y4]. From the input features RF would generate three decision trees [x1, x2, x3], [x1, x2, x4], [x2, x3, x4] and the majority of votes from each of the created decision trees predicts the outcome.

VI. Extreme Gradient Boosting (XGBoost): Similar to RF, XGBoost is also an ensemble learning method. It has recently been dominating applied machine learning due to it speed, performance and scalability. XGBoost is the gradient boosting decision tree implementation. XGBoost algorithm is parallelizable so it can harness the power of multi-core computers. It is parallelizable onto Graphics Processing Unit (GPUs) too which enables it to train on very large datasets as well. XGBoost provides inbuilt algorithmic advancements like regularization for avoiding overfitting, efficient handling of missing data and cross validation capability. Reader may refer the XGBoost library documentation for more details and practical implementation.

Except XGBoost, all the above algorithms are implemented from scikit-learn library. In machine learning, a hyperparameter is a parameter whose value is initialized before starting the training of model. Since hyperparameters govern the training process, all of the mentioned algorithms accomplish better results with hyperparameters tuning. We are going to implement all the classifiers with default hyperparameter values only.

5.1.7 Evaluation and Prediction

The ultimate goal of ML classifier is to classify new and unseen data with maximum number of correct predictions. As explained earlier, to get robust prediction model we execute K-fold CV first on training data. The metrics we use for CV are accuracy, precision, recall and Matthews Correlation Coefficient (MCC). With k = 10 value in K-fold CV, all the metrics for 10 folds are calculated. From all these folds, for each metric, mean value is obtained. We run K-fold CV for all the above referred ML classifiers. The mean value of each metric is computed for every classifier. An

Table 3 Evaluation of different classifiers for PD dataset (I)

Classifier	Accuracy %	Sensitivity	Specificity	MCC
SVM	69	0.72	0.66	0.39
LR	65	0.64	0.65	0.30
KNN	68	0.70	0.66	0.37
NN	65	0.60	0.67	0.31
RF	66	0.59	0.73	0.32
XGBoost	68	0.69	0.67	0.38

apt ML classifier can be selected based on the mean values of the metrics and can be applied on the Test dataset to verify the model learning.

In this example we have applied all the described ML algorithms on the Test dataset to show the comparison of their performance with different data size, quality and number of features, number of samples etc.

I. Parkinson Speech Dataset with Multiple Types of Sound Recordings

Table 3 shows the evaluation of metrics on the Test data for each ML classifier. As described earlier, accuracy is not completely reliable metric to select an apt algorithm for classification. We can see that NN and RF have significant accuracy. However, their sensitivity and specificity values indicate that these classifiers are not able to predict enough number of positive PD patients. Moreover, if we observe MCC values, classifiers LR, NN and RF are not as balanced as other classifiers. SVM, KNN and XGBoost have somewhat performed equally. Note that here we have not used any kind of hyperparameter tuning for any of these algorithms. To select only one of the classifiers we can tune the hyperparameters and compare the results. At the end the most ideal candidate can be chosen to put into real life predictor system. The metrics exhibit the fact that the outcome of the predictor system would not be sufficient for diagnosing majority of the PD patients. Which leaves us with the questions if more data, feature, samples can help build better diagnosis ML aided system. We will find that out with dataset (II).

II. Parkinson's Disease Classification Data Set

In his dataset we start with evaluation classifiers by using only 21 baseline features (refer Data Acquisition of this example), which are similar to dataset (I) features (Table 4).

Here we are getting very interesting results. Accuracy of all the classifiers has increased. One of the primary reasons for this the data samples. The data samples consist only of vowel 'a'. It is observed from research that sustained vowels such as 'a' are very effective in PD diagnosis. The data creators have chosen to take 3

Table 4 Evaluation of different classifiers for PD dataset (II) with basic features

Classifier	Accuracy %	Sensitivity	Specificity	MCC
SVM	76	0.96	0.17	0.30
LR	78	0.93	0.36	0.36
KNN	77	0.88	0.31	0.23
NN	74	0.75	0.69	0.40
RF	75	0.84	0.49	0.33
XGBoost	77	0.87	0.49	0.37

Table 5 Evaluation of different classifiers for PD dataset (II) with basic + time frequency + vocal fold features

Classifier	Accuracy %	Sensitivity	Specificity	MCC
SVM	78	0.95	0.28	0.32
LR	75	0.88	0.29	0.36
KNN	78	0.94	0.33	0.35
NN	82	0.89	0.62	0.52
RF	82	0.89	0.59	0.50
XGBoost	80	0.89	0.54	0.46

Table 6 Evaluation of different Classifiers for PD dataset (II) with all the features

Classifier	Accuracy %	Sensitivity	Specificity	MCC
NN	87.5	0.94	0.69	0.66
RF	84	0.94	0.56	0.56
XGBoost	85	0.95	0.58	0.58

repetition of the same vowel to create the voice samples. Notably, in the outcome of all the classifiers specificity is low. Only NN performs better with specificity and MCC comparatively. One of the causes for this is imbalanced dataset. The dataset contains higher number of PD subjects than healthy controls. We would see all the algorithms performing very good in predicting if subject is PD patient due to high sensitivity. However, they would fail to predict if subject does not have PD. Let's take a greater number of features.

With increased number of features, in Table 5 we see some better performance of classifiers on the test data. SVM, LR and KNN are still suffering with low specificity and MCC. NN, RF and XGBoost have performed quite better with increased number of features (56 features). Finally, taking all the feature sets (750+) and evaluating NN, RF and XGBoost.

From Table 6 metric values, we would have NN as preferred choice of classification algorithm to learn and deploy the model on this dataset.

A few points to consider developing binary classifier for PD patient diagnosis:

- More and effective data samples can help to improve performance of classifier such as sustained vowels such as ‘a’.
- Metrics sensitivity, specificity and MCC are good indicators for selecting a right classifier.
- More features certainly help the performance of some ML classifiers like NN, Random Forest and XGBoost. Although trade-off like computation cost and overfitting should be considered.
- For scalable prediction system algorithms such as NN and XGBoost are good choices.
- Employing IoT devices such as mobile or digital assistants (Google assistant, Amazon Alexa) can help collecting data as well as diagnosing PD using ML/AI powered app.

5.2 Example 2: Tappy Keystroke

The base of this example is ongoing research project—PDLab initiative at Charles Sturt University, Australia by Warwick Adams and team.

The focus is on characterizing the difference in the typing patterns of users. By capturing the Bradykinesia symptoms, slower hand movements and diminishing cognition, this research identifies users that might be at risk of PD much before they are clinically diagnosed.

There are two main papers [16, 17] discussing the research design and the accuracy levels achieved by ensemble of ML algorithms in distinguishing healthy controls and PD patients.

In current discussion, authors use publicly available version of the user's key stroke dataset and implement a different set of ML algorithms to show the implementation.

5.2.1 Objective

To develop a predictive classification model to distinguish PD patients from healthy controls (HC) based on daily keystrokes data. With key strokes the following patient specific signs can be observed:

- Time to react: from decision making by the brain and completing the task—key press
- Decreasing movement speed (e.g. Latency)
- Difference in response in left side and right-side keys
- Decline in speed for repetitive movements (similar to 10 times finger-thumb tap test)
- Inconsistency in similar keypresses due to tremor

- Shuddering movements—sudden strike or pauses due to diminishing cognition
- Tiredness or decline within the day with timestamps
- Overall behavior with multiple data points over time

5.2.2 Solution

The data is collected with help of software called Tappy. The installation guidelines are availed on PDlab website. The software can be used on standard windows desktops/laptops by the users—volunteers in their daily routine. The privacy policies clarify that the app does not collect any information about the text typed or personal user identification. Permissions from respective Human research ethics committee can also be observed. Users can access their data and contribute their data by uploading. This data can be tested in ML algorithms and an early diagnosis is facilitated. In present chapter we create a miniature version of this project covering all the ML workflow steps.

5.2.3 Data Acquisition

The open source data of Tappy keystrokes as availed in [18]. More than 500 users participated from 4 different countries. Two types of data are collected for each user as shown in Fig. 9.

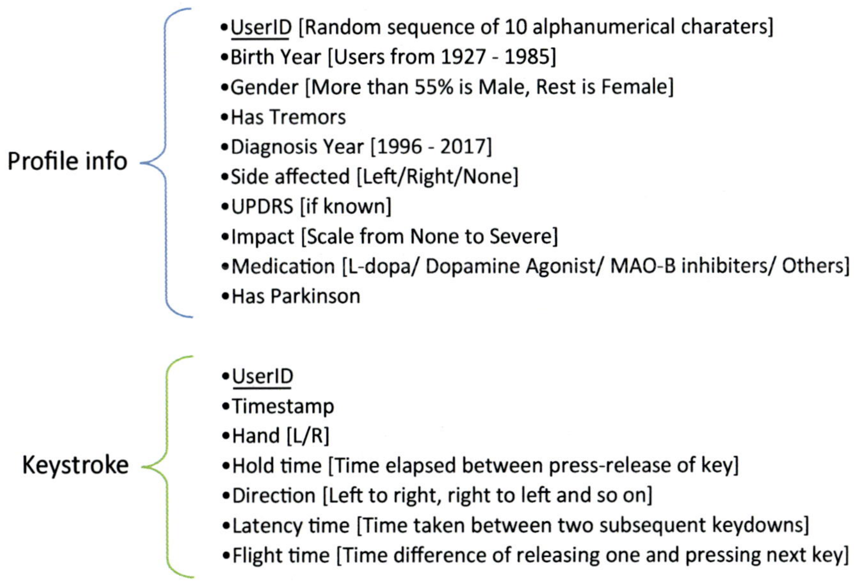

Fig. 9 Attributes understanding for given datasets in Tappy keystroke problem

Attribute name	Original label	Numeric label
Parkinsons	False	0
	True	1
Gender	Female	0
	Male	1

Fig. 10 Numeric encoding of labels

For new diagnosis, one can request the patients to start using Tappy and acquire above mentioned datapoints. In the current data source, there are 646 user profiles.

5.2.4 Data Preprocessing

The data is contained in form of space delimited text files. Single line command would read this data and organize it into 2 different data frames, one for user profile info and another for user keystrokes.

Data Cleaning, Filling and Formatting

1. It involves removing incorrect mixed or empty entries. For example, patient records with missing Birth year can be removed. In case of statistical measures null values handled with standard imputation techniques like mean, median, mode or zero values replacements depending on specific problem. These strategies are problem specific. In present case, mean values will be used.
2. Most of the variables are categorical, e.g. Gender [Male/Female], Impact [None, mild, medium, severe] or Parkinson's condition [True/False] and so on. These categories can simply be converted into numeric labels like True: 1, False: 0. The encoding is as shown in Fig. 10.
3. The birth year and year of observation are used to create new numeric feature of Age (Fig. 11).
4. After cleaning for empty or mixed entries, 500+ user profiles remain (Fig. 12).

Filtering

Listed below are few of the selection criteria.

1. Out of 646 user profiles, 523 users remain when filter for incomplete Birth year, Gender or Parkinson condition information is applied.

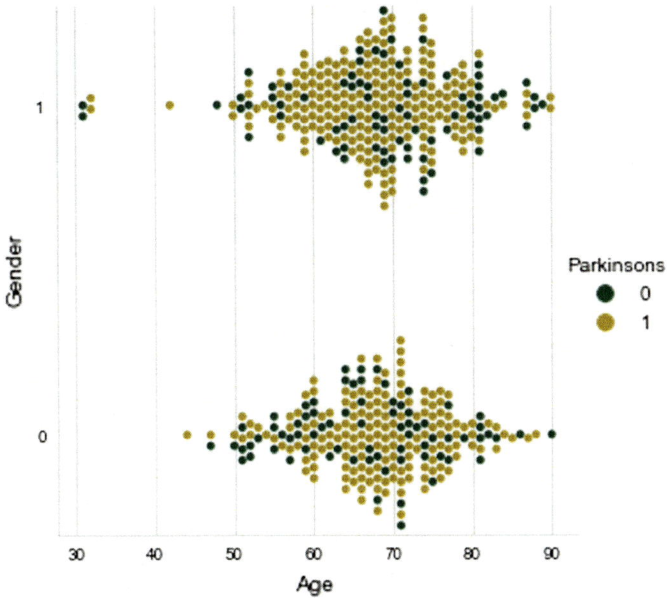

Fig. 11 Subjects demographics (PD patients + HC)

Fig. 12 Medication filtering

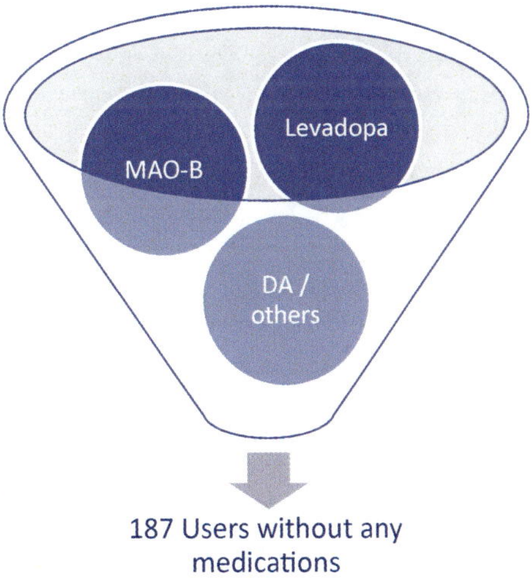

Users diagnosed with PD	Healthy Controls
371	152

2. It is important to avoid any influence of medications on patient's behavior. So, the dataset will be further filtered and patients' profiles with on-going medication (Levadopa, DA, MAO-B or others) are not included. Total 187 users clear the criteria in which 35 users are diagnosed with PD but are not under any medications or treatment. Data of these patients serves as an uninfluenced training data for the ML model.

Users diagnosed with PD	Healthy controls
35	152

3. There are two dataframes, user profile info and user keystrokes. The keystroke dataframe has relatively large amount of data. Hence, a list of coinciding userIDs is prepared. The program should only read keystroke data for the shortlisted users.
4. Next step is to read the keystroke data of shortlisted users and apply the minimum keystroke criteria—only users with 500 or more keystrokes are included.

With reference to the original paper [16], authors have adopted to slightly different selection approach. Keystroke criteria for training data has been kept at 500 instead of 2000. The authors agree to the point in original paper that more keystrokes would result into better accuracy. So, as an experiment and a challenge for the model, minimum number of keystrokes is set at smaller number to review the performance. This criterion further shortlists the subset of users in present example.

Furthermore, in original investigation, only the users with mild or none impact are considered. In current example, users with medium or severe impact are also included as long as they are not being treated with any medications.

5.2.5 Feature Engineering

The model is expected to distinguish between a healthy control and a PD patient with minimum possible information. Logically, attributes like Tremor presence, Diagnosis year, affected side, UPDRS, Impact etc. can give a direct hint to the model. Hence these attributes must be hidden from the model. So, in user info dataframe, only 3 features—Age, Gender and Parkinsons are selected. Parkinsons will later be the target variable that the model is expected to predict.

1. For shortlisted users, more than 10Mn keystrokes are retrieved. In user keystroke dataframe, statistical features are engineered in order to aggregate user specific datapoints.

   Figures 13 and 14 are few possible aggregations.

   For simplification purpose, only basic statistical parameters like mean and standard deviation (std) are used unlike the original paper [16]. If both hold time

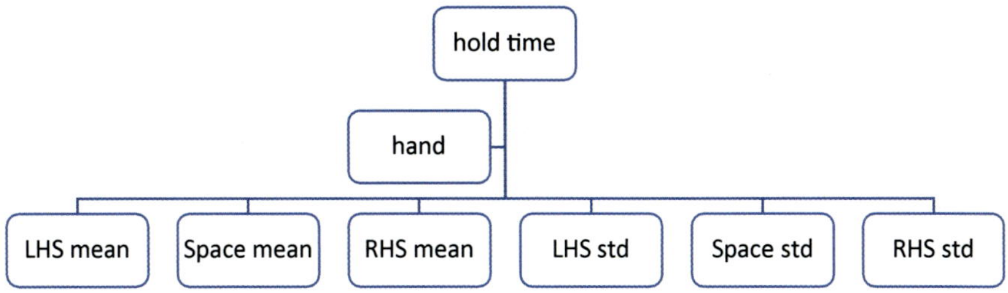

Fig. 13 Hold time features

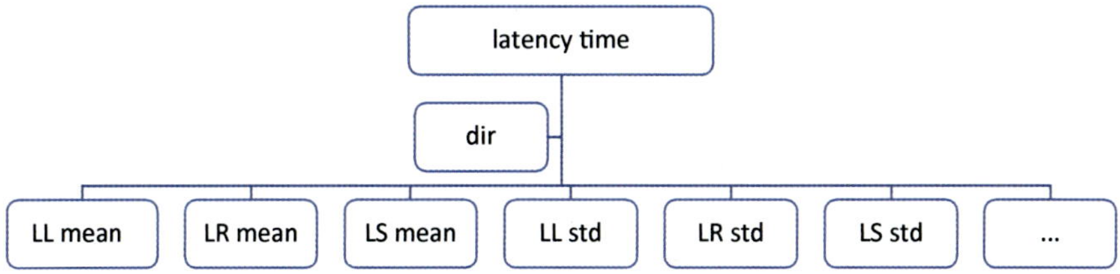

Fig. 14 Latency time features

and latency features are considered, 27 new attributes would be added. For least complexity, only 6 hold time features are calculated—as shown in Fig. 13. For each user the hold time is aggregated (mean and std) for left hand side(LHS), space and right hand side (RHS).

After initial understanding, readers can create latency time features and advance their implementation.

2. User info and user keystroke are joined on basis of common attribute i.e. userID. So, the combined dataframe contains user specific profile info and aggregated keystroke data (Fig. 15).
3. As result of medication filtering, number of PD patients are underrepresented in the data. While most classification algorithms risk focusing on majority class, so balanced binary class training data is essential for a robust and unbiased prediction model. Undersampling HC is suboptimal solution. Widely recommended techniques to tackle such cases are SMOTE (Synthetic Minority Oversampling) or ADASYN (Adaptive Synthetic Sampling) [19]. In current scenario, use of SMOTE is demonstrated (Figs. 16 and 17).

5.2.6 Training ML Models

In voice dataset example, the algorithms are already thoroughly explained. Following algorithms are used for Tappy keystroke example:

1. SVM: Support vector machine classification

User info dataframe

UID	Age	Gender	Parkinsons
XXXXXXXXXX	...	...	...
XXXXXXXXXX	...	...	...
XXXXXXXXXX	...	...	...
XXXXXXXXXX	...	...	...
XXXXXXXXXX	...	...	...

User keystroke dataframe

hold_time			
		Mean	Std
UID	Hand	...	...
XXXXXXXXXX	L	...	...
	R	...	...
	S	...	...

Final dataframe

UID	Age	Gender	Parkinsons	hold_time_mean_L	...
XXXXXXXXXX	...	...	...	...	...
XXXXXXXXXX	...	...	...	...	...

Fig. 15 Joining user info and keystroke dataframes

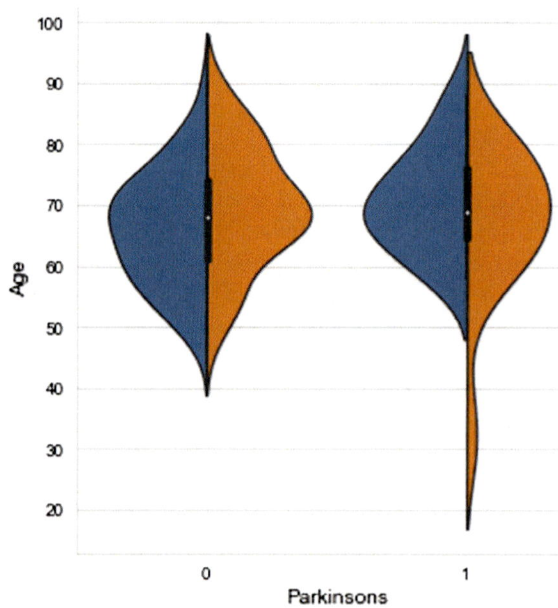

Fig. 16 Original distribution of dataframe

2. KNN: K-nearest neighbors
3. Random forest
4. XGboost

There are various cross validation techniques. In voice dataset the K-fold cross validation has been discussed. Same is followed in this example.

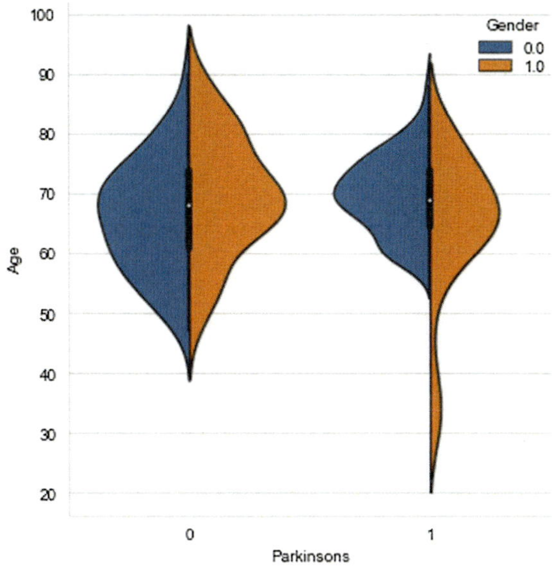

Fig. 17 Similar distribution achieved after SMOTE

5.2.7 Evaluation and Prediction

Multiple algorithms are evaluated on this dataset. Evaluation parameters for few algorithms are listed in Table 7.

As discussed earlier, accuracy is important but cannot be the sole judgement factor in algorithm selection. The trade-off of sensitivity and specificity is an important consideration. While specificity demonstrates the percentage of HC correctly predicted as normal, sensitivity is the percentage of PD patients correctly identified as having that condition. Hence in present context, sensitivity is more important than specificity. The results of SVM show specificity of 1.0 which is the best possible measure, however low sensitivity clearly explains that it is a biased model. Moving further, the MCC value ensures a balanced classifier. A value nearer to -1 means a highly imbalanced classifier while value nearer to 1 suggests it to be a neutral classifier. Based on these evaluations, XGBoost outperforms other algorithms used in this example. The selection and performance of algorithm heavily depends on problem

Table 7 Evaluation of classification algorithms

Classifier	Accuracy	Sensitivity	Specificity	MCC
SVM	61.4	0.31	1.0	0.41
KNN	80.70	0.78	0.84	0.61
Random forest	80.70	0.84	0.76	0.61
XGBoost	84.21	0.94	0.72	0.68

Table 8 Confusion matrix for XGBoost

True label	Predicted label: 0	Predicted label: 1
0	0.72	0.28
1	0.06	0.94

definition and approach. For further understanding, let us look at confusion matrix of XGBoost (Table 8).

In 94% cases, the model makes correct prediction if the user is a PD patient. The success in predicting an HC as normal is 72%. The promising observation is that in only 6% cases the model has been wrong where it has confused a PD patient as normal. For any preventive diagnosis model, this error i.e. false negatives must be as minimal as possible.

5.2.8 Review and Refinement

The model presented here is a start point with openly available source code with multiple possibilities of scaling up. Here are few quick suggestions:

1. Increase number of features i.e. include latency and flight time features.
2. Ingest more data in training and evaluate the performance.
3. Experiment with other algorithms variants such as Neural Networks as discussed in Example 1.

With these two informative examples in diagnosis, the readers should be able to explore further the applications in treatments.

5.3 Treatment Monitoring

The variation in symptom's presence and severity complicates stage identification for the patient. If the patient profiling is done with sufficient accuracy, certain treatment course can be applied on the patient with reasonable confidence that the patient's system will respond well to it.

ML similarity algorithms can find similar patients through deep clustering process, exploring every permutation combination. This can help segment patients and ML recommender algorithms can provide list of treatment options with estimated success probabilities for each patient group.

As mentioned earlier, monitoring the treatment effectiveness requires attention of movement disorder specialist and patient's physical presence in the clinical setup, constraining the frequency as well as subjective approach and bias. With help of easy-to-use sensors data is acquired at much higher sampling rates yielding timely therapeutic interventions.

A combination of ML algorithms can detect unusual movement fluctuations and classify the patterns specifically for each patient. This can help design personalized treatment for each patient with the goal of uplifting the overall quality of life (QoL).

Here are few contemporary development possibilities in treatment arena-

1. Published in Sept 2019, Medical professionals of University hospital of Italy have implemented simple smartphone and sensor-based home monitoring system to assess motor fluctuations. In FoG (Freezing of Gait—for 38 patients) and LA (Leg agility—for 93 patients) experiments, smartphone is tied to patient's waist and thigh respectively. The data relayed from mobile sensors, when fed in ML algorithms has produced 80–90% accurate results. The next steps are to make observations and data processing on a small IoT module combining numerous sensors. It is supported with a basic processor, bluetooth interface and a memory card. Eventually it becomes an electronic diary capturing everyday motor fluctuations, posture and dyskinesia for the patient. Please refer [20] for more details.
2. A considerable scale of research has been ongoing under the mPower project. This clinical observational study uses iPhone app interface. The research evaluated the facets PD with surveys and recurrent sensor measurements from a large group of 9000 users including both PD patients and healthy controls. Wide acceptance and high sampling rates availed sufficient quality and quantity of data was promising in clearly identifying distinct symptom progression patterns. To invite a collective community effort, the authors have been courteous to provide open source code for the app and data collection modules [21].
3. Since PD medication is yet the least explored field, Pfizer and IBM have announced a collaborative IoT project named Bluesky—to improve the system of collecting data for clinical trials for PD drug development. The details of this project can be found on IBM research blog.

6 Futuristic Developments

In addition to above applications, here are some more interesting prospective applications one can explore:

EEG as non-invasive biomarker for PD: Researchers from university of Oregon have suggested on online journal *eNeuro*, that noninvasive EEG readings can act as easily visible electrophysiological biomarkers for diagnosis and to refine the treatments for the disease [22].

Predict severity of PD based on patient's pose data: On Kaggle—an open data science community platform has the dataset [23] of 2D human pose estimates of Parkinson's patients as they carried out various activities. At Toronto Western Hospital, the patients were administered with 2 h of Levadopa infusion before the 2 h examination sessions. 120–130 video recordings per task were produced for later evaluation by neurologists. The videos are parsed and converted into line pose estimates that can be used by ML algorithms to assess severity. Researchers can access the data and develop useful PD applications around it.
Discriminate HC and PD patient, Distinguish the "On" and "Off" states of a PD patient: Quite comprehensive app—iMotor, performs 3 different tests: Finger tapping, Pronation-supination and reaction time check. The sourced data is processed for ML algorithms that classify PD versus HC with more than 90% accuracy and "On"/"Off" state classification with more than 75% accuracy. Such applications can be further enhanced with domain specific inputs and supplement the clinical care of this movement disorder. Researchers can observe and adept from such apps [24].

Another such app is i-Prognosis that focuses on voice distortions as early sign of PD and aims in capturing it earlier than the clinical diagnosis in order to avail neuroprotective therapies for the patient.

7 Conclusion

With concrete examples stated, it is evident that ML and IoT can help build effective early diagnosis systems and data-driven well-informed treatment monitoring, empowering the health professionals to provide better care to the patients regardless to resource constraints.

The chapter also highlights the ease-of-implementation, encouraging more and more health care institutions/practitioners to leverage ML and IoT and to achieve improved quality of life for their patients as the output.

References

1. E.R. Dorsey, R. Constantinescu, J.P. Thompson, K.M. Biglan, R.G. Holloway, K. Kieburtz et al., Projected number of people with Parkinson disease in the most populous nations, 2005 through 2030. Neurology **68**, 384–386 (2007). https://doi.org/10.1212/01.wnl.0000247740.47667.03
2. A. Elbaz, L. Carcaillon, S. Kab, F. Moisan, Epidemiology of Parkinson's disease. Rev Neurol (Paris) **172**(1), 14–26 (2016)
3. Michael J. Fox Foundation for Parkinson's Research. Accessed online on 23/10/19 at https://www.michaeljfox.org/understanding-parkinsons/living-with-pd/topic.php?symptoms
4. R.D. Sweet, F.H. McDowell, Five years' treatment of Parkinson's disease with levodopa. Therapeutic results and survival of 100 patients. Ann. Intern. Med. **83**, 456–463 (1975). https://doi.org/10.7326/0003-4819-83-4-456

5. C.G. Goetz, W. Poewe, O. Rascol, C. Sampaio, G.T. Stebbins, C. Counsell, N. Giladi, R.G. Holloway, C.G. Moore, G.K. Wenning, M.D. Yahr, L. Seidl, Movement Disorder Society Task Force on Rating Scales for Parkinson's Disease, Movement Disorder Society Task Force report on the Hoehn and Yahr staging scale: status and recommendations. Mov. Disord. (Official Journal of the Movement Disorder Society) **19**, 1020–1028 (2004)
6. F. Pagan, Improving outcomes through early diagnosis of Parkinson's disease. Am. J. Manag. Care **18**, 176–182 (2012)
7. R. Pahwa, K.E. Lyons, Early diagnosis of Parkinson's disease: recommendations from diagnostic clinical guidelines. Am. J. Manag. Care **16**(4), 94–99 (2010)
8. W. Dauer, S. Przedborski, Parkinson's disease: mechanisms and models. Neuron **39**, 889–909 (2003)
9. C.E. Clarke, P. Davies, Systematic review of acute levodopa and apomorphine challenge tests in the diagnosis of idiopathic Parkinson's disease. J. Neurol. Neurosurg. Psychiatry **69**, 590–594 (2000). https://doi.org/10.1136/jnnp.69.5.590
10. C.O. Sakar, G. Serbes, A. Gunduz, H. Tunc, H. Nizam, B. Sakar, M. Tutuncu, T. Aydin, M. Isenkul, H. Apaydin, A comparative analysis of speech signal processing algorithms for Parkinson's disease classification and the use of the tunable Q-factor wavelet transform. Appl. Soft Comput. **74**, 255–263 (2019 Jan)
11. T.J. Wroge, Y. Özkanca, C. Demiroglu, D. Si, D.C. Atkins, R.H. Ghomi, Parkinson's disease diagnosis using machine learning and voice, in *2018 IEEE Signal Processing in Medicine and Biology Symposium (SPMB), Philadelphia, PA*, pp. 1–7 (2018). https://doi.org/10.1109/spmb.2018.8615607
12. M.A. Little, P.E. McSharry, S.J. Roberts, D.A.E. Costello, I.M. Moroz, Exploiting nonlinear recurrence and fractal scaling properties for voice disorder detection. BioMed. Eng. OnLine **6**, article 23 (2007)
13. M.A. Little, P.E. McSharry, E.J. Hunter, J. Spielman, L.O. Ramig, Suitability of dysphonia measurements for telemonitoring of Parkinson's disease. IEEE Trans. Biomed. Eng. **56**(4), 1015–1022 (2009)
14. B. Erdogdu Sakar, M. Isenkul, C.O. Sakar, A. Sertbas, F. Gurgen, S. Delil, H. Apaydin, O. Kursun, Collection and analysis of a Parkinson speech dataset with multiple types of sound recordings. IEEE J. Biomed. Health Inf. **17**(4), 828–834 (2013)
15. P. Boersma, Praat: doing phonetics by computer. Ear Hear. **32**(2), 266 (2011)
16. W.R. Adams, High-accuracy detection of early Parkinson's Disease using multiple characteristics of finger movement while typing. PLoS ONE **12**(11), e0188226 (2017). https://doi.org/10.1371/journal.pone.0188226
17. W. Adams, The detection of hand tremor through the characteristics of finger movement while typing (2018). https://doi.org/10.1101/385286
18. W. Adams, Keystroke dataset for 'The detection of hand tremor through the characteristics of finger movement while typing', Mendeley Data, v2 (2018)
19. H. He, Y. Bai, E. Garcia, S. Li, ADASYN: Adaptive Synthetic Sampling Approach for Imbalanced Learning in *Proceedings of the International Joint Conference on Neural Networks*, 1322–1328 (2008). https://doi.org/10.1109/ijcnn.2008.4633969
20. L. Borzì, M. Varrecchia, G. Olmo et al., J. Reliab. Intell. Environ. **5**, 145 (2019). https://doi.org/10.1007/s40860-019-00086-x
21. B. Bot, C. Suver, E. Neto et al., The mPower study, Parkinson disease mobile data collected using ResearchKit. Sci Data **3**, 160011 (2016). https://doi.org/10.1038/sdata.2016.1
22. https://medicalxpress.com/news/2019-05-noninvasive-biomarker-parkinson-disease-possibly.html
23. M.H. Li, T.A. Mestre, S.H. Fox, B. Taati, Vision-based assessment of parkinsonism and levodopa-induced dyskinesia with pose estimation. J. NeuroEng. Rehabil. **15**(1), 97 (2018). https://doi.org/10.1186/s12984-018-0446-z
24. G. Tsoulos Ioannis, G. Mitsi, A. Stavrakoudis, S. Papapetropoulos, Application of machine learning in a Parkinson's disease digital biomarker dataset using neural network construction (NNC) methodology discriminates patient motor status. Front. ICT **6** (2019). https://doi.org/10.3389/fict.2019.00010

Kunjan Vyas With more than 6+ years of experience in advance technology and management, currently Kunjan works with startups in healthcare and education, helping them build cost-efficient scalable systems. As a co-founder of TotemXAI—a small team of likeminded data scientists and ML engineers, she is responsible for the operations and delivery. Her passion is to promote AI as the assistive intelligence to resolve socio-economic challenges in developing and developed countries.

Shubhendu Vyas is a freelancing Machine Learning (ML) and data science consultant at TotemXAI. His current work involves applications of Deep Learning in healthcare. Shubhendu has knack for transforming real world problems into mathematical simulations. Being a mechatronics engineer and masters in advance computation, he is uniquely skillful in sourcing IoT data, combine with apt ML algorithms and distill deeply buried insights.

Nikunj Rajyaguru is an Industry professional working for than 10 years in a renowned multinational company based in Germany. Coming from Industrial Engineering background, while working in industry he has developed keen interest in areas of IoT and ML. Following his personal interest and enthusiasm, he is actively exploring different applications of IoT and ML.

An Efficient Method for Computer-Aided Diagnosis of Cardiac Arrhythmias

Sandeep Raj

Abstract In this chapter, an efficient features representation and machine learning methods are combined and developed to process the ECG signals. Initially, the raw heartbeats are pre-processed for eliminating various kinds of noises inherited within them. Consequently, the QRS-wave is located by applying Pan-Tompkins (PT) technique within the signals. Following the QRS-wave localization, a rectangular window of fixed size is selected for segmenting the heartbeats. Then, the empirical mode decomposition (EMD) algorithm is utilized for extracting the time domain information from heartbeats as features. Few coefficients are selected for an efficient representation of heartbeats using principal component analysis (PCA) which further reduces the complexity during processing using classifier. These output coefficients represent the characteristics of individual heartbeats and supports in distinguishing between them based on their morphology. Further, the R-peak to R-peak information between heartbeats are captured and concatenated with the output time-frequency coefficients. As a result, this final feature vector represents each heartbeat that are applied to support vector machine (SVM) model for recognizing these feature representations into corresponding classes of heartbeats. The classifier performance is also enhanced as its parameters are employed by employing the particle swarm optimization (PSO) algorithm under patient specific scheme. The proposed methodology is validated over Physionet database and the output of classifier model are compared to the labels of corresponding heartbeats of the database to formulate the results. The experiments conducted reported a higher overall accuracy of 95.86% over existing state-of-art methods.

Keywords Electrocardiogram (ECG) · Arrhythmias · Empirical mode decomposition · R to R wave · Support vector machines

S. Raj (✉)
Indian Institute of Information Technology Bhagalpur, Bhagalpur 813210, India
e-mail: sraj.ece@iiitbh.ac.in

V. Jain and J. M. Chatterjee (eds.), *Machine Learning with Health Care Perspective*, Learning and Analytics in Intelligent Systems 13,
https://doi.org/10.1007/978-3-030-40850-3_13

1 Introduction

The World Health Organization (WHO) statistics indicate that there has been a significant growth in count of global mortalities due to cardiac abnormalities. The mortalities due to cardiac disease shall prevail up to the next decade. In 2008, an approximate number of around 17.3 million people died due of CVDs which is estimated to increase up to 23.3 million by 2030 [1, 2]. Approximately 80% of mortalities has been occurred in low or moderate per capita income countries. Maximum of these deaths are caused due to misdiagnosis or late diagnosis of CVDs [1, 2]. It is a low-cost, accurate and non-invasive tool widely employed for diagnosis of cardiovascular diseases (CVDs), which is commonly used to analyze the behavior of cardiac processes [1, 2]. In ECG, the electric potentials are recorded and displayed in graphical manner that are produced due to the pumping action of the heart. It is the depolarization and repolarization of the heart tissue, which generates a potential difference to examine. This mechanism starts by triggering of the SA node, the pacemaker for cardiac muscle generating an electrical impulse that triggers a series of electrical occurrences in the heart. Such natural events are mentioned by the electrodes and fluctuations are represented in the wave component of the ECG signal heartbeat signals usually consisting a P-wave component which is an indicative of depolarization of atrias, a QRS complex wave component which is an indicative of depolarization in ventricles, ST slope component represents the proper of blood flow throughout the human body and T wave is an indicative of repolarization in the ventricles. The detection of cardiac diseases from the ECG recordings is performed using rhythm, heart rate variability measures and QRS duration [1]. The R-wave peak classification is much essential in automatic signal classification, especially in critical conditions and cardiac abnormalities. Today ECG is the most promising cardiac diagnostic approach around the world because it reveals essential clinical information [1]. It diagnoses the rhythmic episodes of the ECG and further arrhythmias are classified if the patient were having any cardiac disease. Arrhythmia comes due to damage of heart muscles, diabetes, habit of eating tobacco, low and high respiration, blood pressure, etc. [1]. These arrhythmias come under life-threatening (critical) and non-life-threatening (non-critical) [1]. Severe cardiac abnormalities do not allow any time to undergo treatment whereas noncritical requires special treatment for saving life. Simple diagnosis using naked eye may mislead the detection. Therefore, the cardiac healthcare demands an automated computer-aided diagnosis of longer duration electrocardiogram (ECG) signals or heartbeats [1].

In the last few decades, several methodologies are reported to overcome the limitations with classical heartbeat recognition methods to have a useful diagnosis. More often, four steps [1–29] are integrated together to perform automated diagnosis of heartbeat. They include, the filtering, R wave localization, extraction of features extraction and features recognition stages. Various robust filters are designed to eliminate the noises associated within the heartbeats. However, the main aim of the current work comprises of the signal processing and machine learning algorithms. The feature extraction step extracts the information and represents the heartbeats while they are being categorized into different classes using the machine learning tools [2–29]. Several feature extraction algorithms have been reported in literature. In most techniques, features of time [11–14], frequency [13], and non-linear dynamism [15, 16] are extracted and passed through the classification stage to achieve more accuracy. Sedjelmaci et al. gathered the varying nature of R wave to R wave and ST segment information as from every heartbeat as feature considering the variable heart rate of a subject. This time-domain information is applied to detect four kinds of cardiac abnormality i.e. arrhythmia. Rai et al. employed discrete wavelet transform (DWT) for representing the signals along with morphological features and artificial neural network for recognition into one of the sixteen different categories of heartbeats. George et al. filtered the heartbeat recordings and computed the fractal dimension of ECG using Hurst power to identify four kinds of arrhythmias. Martis et al. [17] employed the classical PT algorithm [30] for determining the position of R-peak within the ECG signals. Thenafter, the classical wavelet transform (WT) is employed to filter the input heartbeats. The corresponding feature set is determined for representing a particular heartbeat and are identified by employing support vector machines (SVMs) into three classes of arrhythmias and reported a performance accuracy of more than 99%. Vafaie et al. employed the features using artificial neural networks (ANNs) whose parameters are optimized using genetic algorithm (GA) and reported an accuracy of 98.67% in detecting the potential arrhythmias. Mary et al. employed the DFA technique to compute the fractal dimensions which is further employed for distinguishing between a normal and an arrhythmia beat. Mhetre et al. gathered the morphological information of the heartbeats as features and applied them for identification using an expert system into ventricular natural heartbeat. Raj et al. [20] proposed a DCT based DOST method for extracting significant characteristics from the heartbeats and further classification these features are performed by employing support vector machine (SVM) whose best performance parameters are determined by employing particle swarm optimization (PSO) algorithm.

The current study aims to develop a method which combines the feature extraction and machine learning algorithms to analyze the heartbeats efficiently in a real scenario. Here, the empirical mode decomposition (EMD) is utilized for extracting the non-linear and non-stationary information from the subsequent heartbeats. As a result, the IMFs extracted for each heartbeat are brought down in lower dimensions using principal component analysis (PCA). In addition to the morphological features, R wave to R wave information of the heartbeats are combined together to form a feature vector representing a particular heartbeat. These features represent each of the ECG signals and are further identified support vector machines (SVMs) into different categories of heartbeats. The optimal classifier parameters are chosen by employing the particle swarm optimization (PSO) algorithm. The developed methodology is validated over the Physionet data and evaluated under subject specific scheme to estimate its accuracy. The experimental results reported higher recognition accuracy that the previous studies.

The later body of the chapter is organized as follows: the methods are summarized in Sect. 2 while the proposed methodology involved in the automated diagnosis of heartbeat is explained in Sect. 3. The simulation performance of the proposed method is highlighted in Sect. 4 and lastly, the conclusion of the chapter is presented in Sect. 5.

2 Methods

This section highlights the methods such as empirical mode decomposition (EMD) for input representation, support vector machines (SVM) for recognition and artificial bee colony for optimizing the SVM parameters used in this study.

2.1 Feature Extraction Using EMD

EMD [31] is commonly employed for random and irregular time series signals like a statistics-analysis method and has two types: HSA and EMD respectively [31]. EMD disintegrates a sophisticated time series into several IMFs has decreasing order of higher frequency elements to elements with lower frequency order [31]. An IMF functions as a sequel to the SHM. The following properties should be met by any IMF:

(I) the number of extrema and zero crossings must differ by maximum of nil or unity, and
(II) the upper envelope mean value described by the lower envelope and local maxima designed by local minima that use that interjecting signal, like cubic spline feature should be zero.

EMD method alters the following adaptive method for separating the input signal and its IMFs:

i. The possible number of local maxima and minima within an input ECG data a(t) are determined marked with $F_i, i \in \{1, 2, \ldots\}$ and $f_j, j \in \{1, 2, \ldots\}$, individually.
ii. Let F(t) be the upper envelope as well as the lower envelope is denoted by f(t), interpolating signal will be calculated.
iii. Thus estimate envelope's local mean as $P(t) = (F(t) + f(t))/2$ using such interpolating signals.
iv. Deduct the mean from initial signal like $m(t) = a(t) - P(t)$. When the IMF characteristics are not satisfied by m(t), avoid steps (v) and (vi) and jump to step (i) Repeat the method with new input h(t) by substituting x(t).
v. Until m(t) satisfies the IMF properties, it is stored as IMF, for example $d_i(t) = m(t)$ where the initial raw signal is subtracted, for example, $d(t) = e(t) - f_i(t)$, here 'j' alludes to the jth IMF.
vi. Restart from (i) as novel contribution d(t), and $f_i(t)$ is stored as an IMF.

The adaptive method is generally prevented if a monotonic IMF is acquired, however in this report the sifting method stop criterion in the EMD algorithm is one mentioned in [31]. Therefore, the EMD method produces a sequence of IMFs added to a last leftover element, e(t).

Second stage includes rebuilding that initial signal by summarizing all IMFs and the remaining element. When the analyzed signal is received, the Hilbert transform estimation is performed in order to generate an analytical sign. The instant amplitude and frequency may be calculated as follows, as per the nature of an analytical signal, in order to create the combined allocation of amplitude, frequency and energy gradually [31]. The initial value of function p(t) can be determined using Hilbert transform as

$$q(t) = H[Q(t)] = \frac{1}{\pi} PV \int_{-\infty}^{+\infty} \frac{p(\tau)}{(t-\tau)} d\tau \quad (1)$$

where the value of the singular integral is indicated by PV. The analytical signal is therefore now described below:

$$w(t) = x(t) + iy(t) = a(t)e^{i\theta(t)} \quad (2)$$

And,

$$a(t) = \sqrt{(p^2 + q^2)}, \text{ and } \varphi(t) = arctan\left(\frac{q}{p}\right) \quad (3)$$

In this case, *a(t)* is instant amplitude as well as φ is the phase [31], whereas instant frequency can be represented as

$$\omega = \frac{d\varphi}{dt} \quad (4)$$

Like this, the instant amplitude and frequency for every IMF will be calculated utilizing the HHT method as well as the essential characteristics ingrained in ECG beats can be obtained further.

2.2 Support Vector Machine (SVM)

Support vector machines (SVMs) [32, 33] are machine learning tools employed for classifying an input data. In the beginning, it was used for solving binary classification problem. It generates a hyperplane which is surrounded by the feature points of the different classes and gets clustered. Every plane in the higher dimensional space divdies the data of the two input classes and makes them apart as far as possible. In SVMs [32, 33] all data points are distributed. Here, the objective function resembles a particular category of any input data. On the other hand, the constraints are computed by patterns of the another category [32, 33] of input signal. The hyperplane is obtained

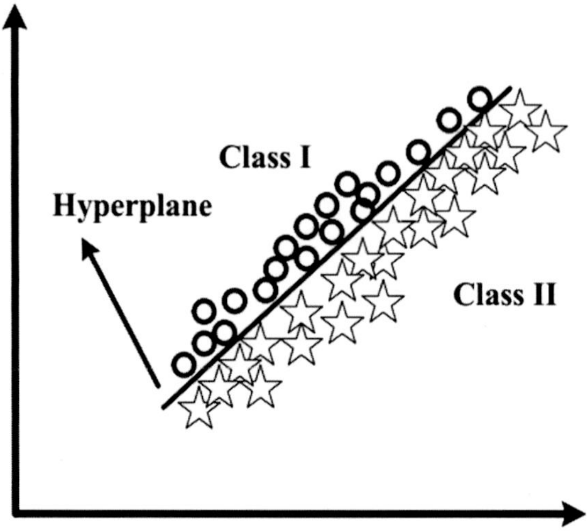

Fig. 1 Hyperplane separating binary classes

by solving a single quadratic programming problems (QPPs) one for each class. The classification of the new point is done as per a hyperplane for a given point of any input data is closest to. Figure 2 shows a hyperplane separating the two classes of data problem in the higher dimensional space. In d-dimensional real space R_d, the matrix X_1 represents the positive class data samples while X_2 represents the negative class data samples. Further, a hyperplane [32, 33] in the real space can be represented as:

$$x^T w + z = 0$$

Here, symbols z represents the bias term of the hyperplane whereas w indicate vector normalized to hyperplane. The SVM developed for performing linear and non-linear classification are reported in [32, 33]. Therefore, the SVM method is chosen as an alternative for analysis of heartbeats (Fig. 1).

2.3 *Particle Swarm Optimization (PSO)*

Particle swarm optimization (PSO) [34] technique is based on the social behavior of bird and fish, to attend the minimum of quadratic programming problem (QPP). It is a stochastic optimization technique developed by Kennedy and Eberhart [34]. Many particles (variable) participate in finding the global minimum value for QPP by communicating with each other. Position vector and velocity vector are two components associated with each particle. After every iteration, ith particle share its personal minimum value and updates the global minimum if this value is less than the existing global minimum [34]. All particles update its velocity vector using its

Fig. 2 Flowchart of PSO method

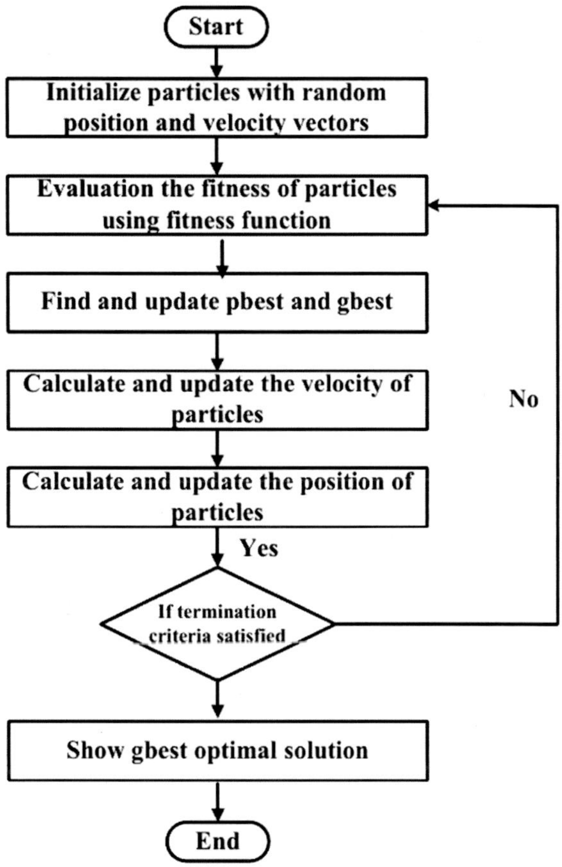

previous velocity vector (inertia term), the personal minimum value (cognitive component), the global minimum value (social component), and user-defined coefficients [34]. Once the velocity vector updated, each particle computes new position vector and examine its new position for the global minimum. This process repeats until all particle attends the global minimum. PSO is comparatively faster than other optimization techniques [34]. The step-by-step procedure used in the implementation of PSO algorithm is depicted in Fig. 2.

3 Proposed Method

The proposed methodology comprises of combination of four significant stages such as preprocessing, R-wave localization, feature extraction, and feature recognition stages depicted in Fig. 3 to classify different categories of heartbeats. Initially, the raw ECG signals extracted from the human body are preprocessed to improve their quality

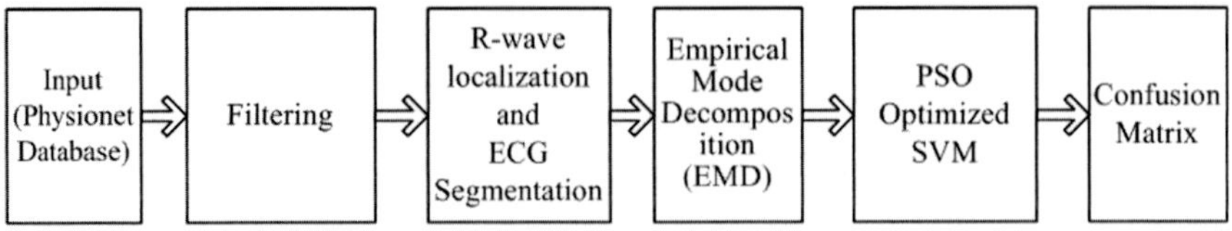

Fig. 3 Steps involved in automated classification of cardiac arrhythmias

by removing various noises associated within them during data acquisition. The preprocessing step is followed by the feature extraction approach where significant characteristics from the heartbeats are extracted to represent them into lower dimensions. Finally, these features are applied to learning algorithms for classification.

3.1 ECG Data

The proposed methodology is validated on a well-known database i.e., Massachusetts Institute of Technology-Beth Israel Hospital (MIT-BIH) arrhythmia data [35]. The database contains the records of forty-seven subjects comprising forty-eight files or records having a sum of 110,109 heartbeat labels [35]. The data contain the annotations of the signals which is finally used to formulate the results in the supervised learning classifier mechanism. A total of sixteen classes of heartbeats including normal is available in the database for experimental purpose. The input heartbeat data are sampled at 360 samples per second while the signals are digitized using an ADC having 11-bit resolution within 10 mV range [35]. For performing the experiments all the records comprising different classes of signals are used which is filtered using a band-pass filter (BPF) with a cut-off frequency of 0.1–100 Hz [35]. The training and testing dataset are determined by choosing a certain fraction from all the sixteen categories of the heartbeats are selected and mapped as per AAMI recommendations under subject-specific scheme. Under this scheme, recordings, i.e., 102, 104, 107, and 217 are not considered into any of the training and testing datasets. Hence, only 44 records are considered to conduct the experiments. Here, two different types of analysis are performed. The first one in which the records are divided (i.e. 22 records each) equally to constitute the datasets. The mapping of ECG beat categories from the Physionet database to AAMI recommendation is presented in [35]. The overfitting of the classifier model is avoided by conducting 22-fold cross validation [36] over the entire dataset.

3.2 Preprocessing

The performance of any diagnosis system is greatly affected by the quality of heartbeats. The noises associated with a heartbeat may contain baseline drift, power line interference, muscle artifacts, contact noise, electrosurgical noise, and quantization noise. It is necessary to eliminate these different kinds of noises failing which can lead to false alarms. Further, this step enhances the signal-to-noise (SNR) ratio which helps in accurate detection of the fiducial points within the of the heartbeats. In order to remove noise, different filters are employed to remove different kinds of noises.

A set of two median filters are employed for eliminating the baseline wander [24] within the heartbeats. A 200 ms primary median filter is used to demarcate the QRS wave [24] and P wave. Whereas, a 600 ms secondary filter demarcates the T wave within the heartbeat. Finally, the baseline wander is removed by subtracting the output of the secondary filter from the raw ECG data [24]. Thenafter, the power-line interference and high-frequency noises are removed by using a 12-tapped low-pass filter (LPF) [24]. This LPF has a cut-off frequency of 35 Hz. The output of this filter is considered as pre-processed heartbeat which is allowed to pass through the R wave localization and segmentation steps for automated recognition of ECG signals [24]. Figures 4 and 5 shows the raw heartbeat and the filtered ECG output from the record #222 of the database respectively.

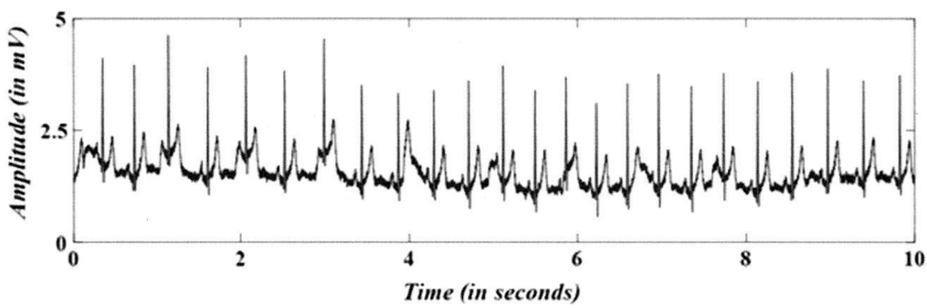

Fig. 4 Raw ECG signal corrupted with noise (record #222)

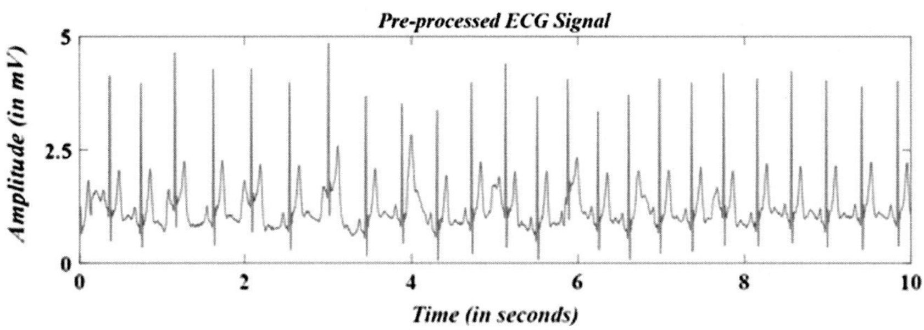

Fig. 5 Pre-processed ECG signal

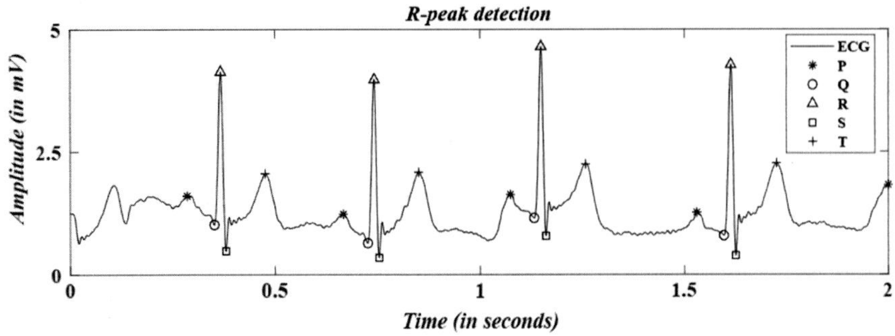

Fig. 6 Fiducial point detection

3.3 Localization of R-Wave and Heartbeat Segmentation

This study classifies the various types of cardiac abnormalities depending upon the localization of R-waves within the ECG signals. Prior to segment the ECG signals before feature extraction, it is necessary to determine the locations of R-waves. A lot of research works have been reported in literature for detecting the R-peaks [30] among which this study employs a well-established Pan-Tompkins (PT) algorithm [30]. It is chosen due to its proven lower computational burden, higher performance under noisy environments. The detected R-waves are verified with the positions of annotations of R-peaks provided in the database. Figure 6 depicts the R wave localization within the heartbeats of record #222 of the database.

In this study, the segmentation step is a bit different from almost all the works reported. They use a rectangular window of constant time or samples. A new segmentation step is proposed as proportional segmentation around the QRS wave is considered. The samples are taken as 65% of posterior R peak in the right and from the left, 35% of anterior R peak are selected for estimating the length of every heartbeat. It is ensured that every information of the ECG from starting P wave and ending T wave is preserved and no information regarding any wave is lost (Fig. 7).

3.4 Input Representation in Lower Dimensions

For any pattern recognition system, a significant role is played by the feature extraction stage. *These features represent the characteristics of an input heartbeat. In literature several types of features are extracted from the subsequent heartbeats, such as fiducial points within ECG signals, wavelet features, RR interval features, and high order cumulants. Many of the works concatenated different features together which results in achieving a higher accuracy. On contrary, concatenation of different types of features results in increased computational complexity of the classification system.*

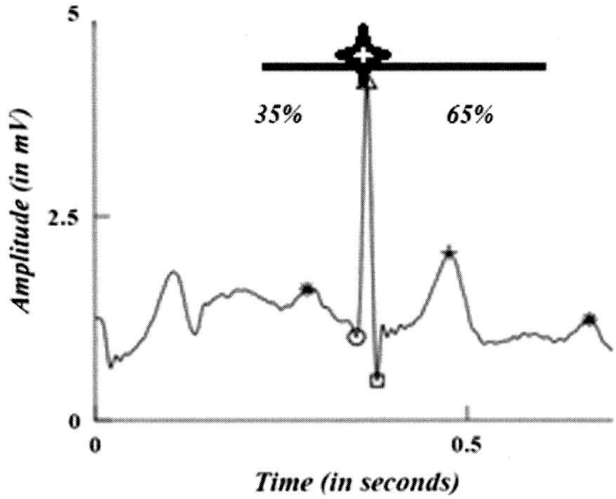

Fig. 7 Heartbeat segmentation

Therefore, an efficient representation of an input is very essential for any classification system. This chapter employs the empirical mode decomposition (EMD) to capture the significant time-domain (TF) information as features of the subsequent heartbeats to detect potential arrhythmias. The EMD considers an input signal as non-linear and a non-stationary which is applied to input heartbeat to provide intrinsic mode functions (IMFs) i.e. a heartbeat of length *N* is transformed into *M X N* length. *The EMD method decomposes an input signal into functions which form a complete and nearly orthogonal basis for the original signal. These functions are named as Intrinsic Mode Functions (IMFs). These IMFs are quite capable to distinguish between different patterns of a particular category. Hence, they can be considered as sufficient to completely describe the signal, even though they are not necessarily orthogonal. Figure* 8 *depicts the decomposition of the normal ECG signal into its corresponding IMFs. It can be observed from Fig.* 8 *that nine IMFs are obtained as a result of application of EMD on the normal ECG signal.*

Every heartbeat data is decomposed into IMFs. It is observed that different number of IMFs are obtained for different types of heartbeats. As such, only five IMFs have been extracted from these heartbeats by restricting the number. Out of the five IMFs extracted it has been found that the second IMF has the highest frequency content of the heartbeat. In other words, the second IMF contain the maximum amount of information than the rest of the IMFs. Therefore, only second IMF is chosen to represent the heartbeats. This second IMF can be termed as feature vector corresponding morphological characteristics of the heartbeats.

These IMFs are applied as input to the principal component analysis (PCA) [37]. *The use of the PCA allows to reduce the feature vector in reduced dimensions which further reduces the computational complexity of machine learning tools. Also, the training time of the classifier is also reduced and therefore, the use of PCA can serve as an important tool in feature reduction.*

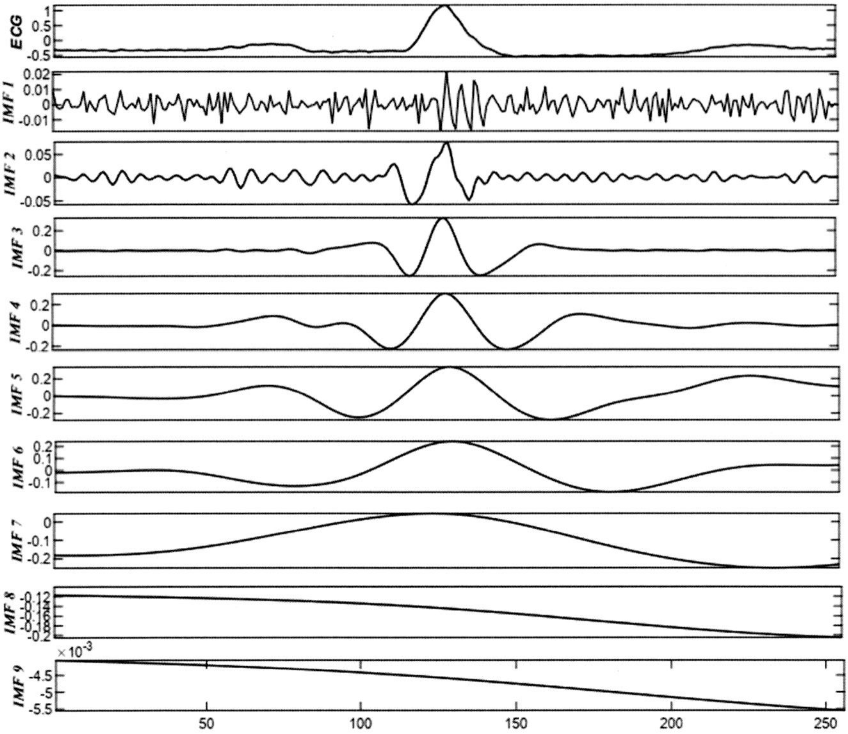

Fig. 8 EMD decomposition of the normal signal

Figure 9 depicts the three-dimensional feature vector plot of normal heartbeat and left bundle branch block heartbeat signals. From Fig. 9, it can be concluded that the features are very closely located to each other and overlaps with each other. Therefore, it becomes necessary to adopt a non-linear classification classifier model that can efficiently handle these non-linear features.

3.5 Heartbeat Variability Features

In addition, the R wave to R wave information between the consecutive heartbeats are measured to estimate the heart rate variability by heartbeats. Hence, four types of R to R wave information as features are calculated that resembles the pattern of heartbeat, namely previous, post, local, and average R peak to R peak information. Here, the period between a previous and present R-wave is estimated to compute the previous R to R information. The period between the present R-wave and upcoming R-wave is estimated to compute the post R to R information. The combination of the previous and post R to R wave period as characteristic of the heartbeat can be termed as an instantaneous rhythm feature. The average R wave to R wave characteristic is

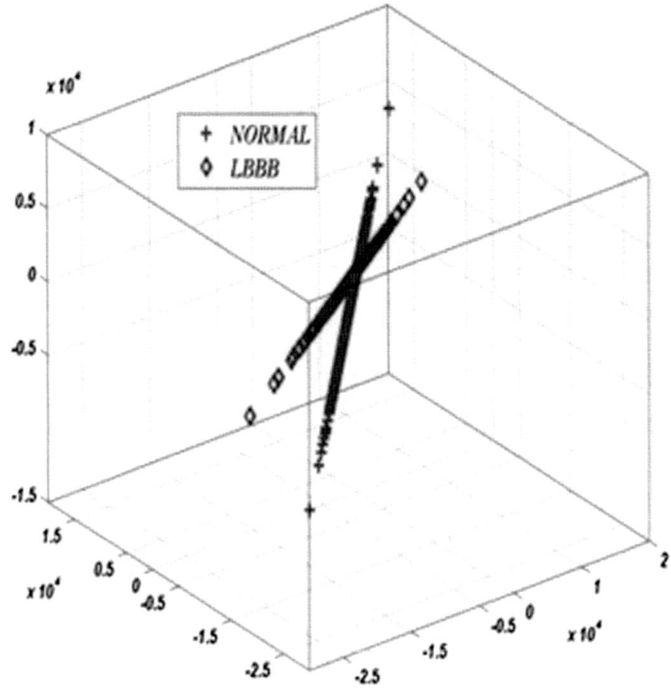

Fig. 9 Three dimensional representation of features

estimated by taking average of R wave to R wave interval of previous three minutes episode of each and every heartbeat. Similarly, the local R wave information is estimated by taking average of all the R wave to R wave interval of the previous eight seconds events of a current heartbeat. Both these features i.e. average and local correspond to average features for a number of heartbeats.

Finally, these heartbeat variability features are concatenated to the output coefficients obtained as a result of principal components for a particular heartbeat.

3.6 Feature Classification

In this work, a support vector machine (SVM) classifier model [32, 33] is employed for classification of extracted PCs and heart rate variability features representing each heartbeat into sixteen classes. Initially, the SVMs were designed to solve a binary category recognition problem [32, 33]. This study addresses the multi-category recognition problem by using the one-against-one (OAO) SVM model [32, 33]. Under this scheme, the selection of the kernel argument and the cost function parameters plays a significant role in reporting higher classification performance. The use of kernel functions enables non-linear classification of features. In other words, the use

Table 1 Accuracy reported using different kernel functions

Kernels	Training accuracy (%)	Testing accuracy (%)
Linear	97.4	91.36
Radial basis function $K(x, x') = \exp\left(-\frac{\|x-x'\|^2}{2\sigma^2}\right)$	99.53	93.64
Multi-layer perceptron $K(x, x') = \tanh(\rho\langle x, x'\rangle + \varrho)$	99.02	92.71
Polynomial $K(x, x') = \langle x, x'\rangle^d$	99.39	93.45

The performance is computed in terms of accuracy

of kernel helps in achieving better classification performance on non-linear data or overlapping features in the high-dimension space. As such, all the kernel functions such as linear, radial basis function (RBF), and polynomial functions are employed to determine the accuracy. A summary of the classification accuracies reported by these different kernel functions is presented in Table 1. Table 1 concludes that the RBF kernel achieved the highest accuracy among all the kernels utilized for classification purpose. It is noted that the training and testing datasets assumed to measure the performance is same as presented in Sect. 3.1. Therefore, the classification performance of RBF kernel is only reported and studied in detail.

Further, the kernel (γ) and cost function (C) metrics are gradually optimized using PSO algorithm for enhancing the accuracy of the classifier model from the results reported in Table 1. The step-by-step procedure involved in the implementation of the PSO technique is presented in [34]. The PSO technique aims to determine the optimal classifier model by evaluating the fitness of each particle for a specific set of particles at every iteration and aims to determine the classifier model optimally. Since this current study involves the classification of five categories of heartbeats using OAO-SVM model, the performance metrics of 10 binary SVM classifiers are tuned by employing the PSO algorithm. Here, simple SV count technique is chosen as a criterion of fitness in order to optimize the PSO framework to restrict the error bound condition resulting in the unbiased performance of the recognition model. During training stage, the learning metrics of the SVM model i.e., C and γ, are optimized by employing the PSO technique. These parameters are selected based on m-fold cross-validation technique [36] in the training stage. The biasing of the classifier model is avoided by using 22-fold cross-validation (CV) is conducted over the testing and training data sets, in order to estimate the better parameters estimation of C and γ parameters which results in best classification performance of the developed model.

Figure 10 depicts the hyperplane plot the features of the normal and lbbb signals. It can be concluded from this figure that the RBF kernel used for non-linear classification of ECG signals is quite significant in making these features apart which can help to easily recognize the features of a particular category of ECG signal.

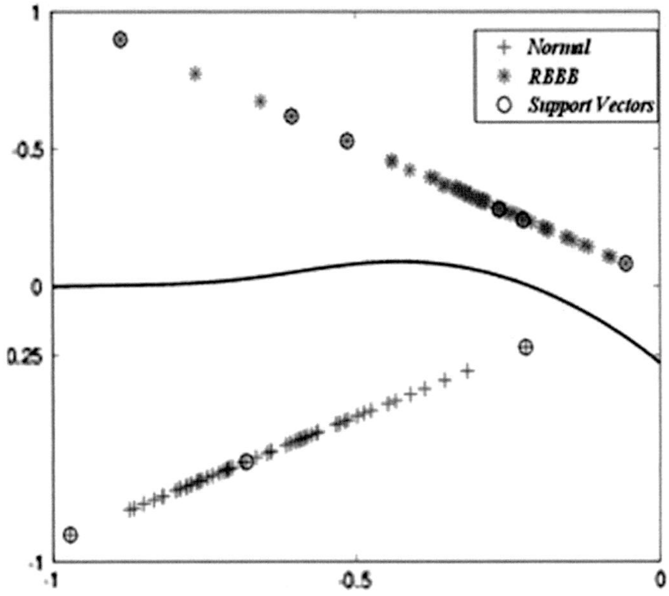

Fig. 10 Hyperplane plot separating normal and lbbb features

3.7 *Performance Metrics*

After the confusion matrix is estimated, the performance metrics for every category of ECG beat are determined in terms of five performance metrics, namely, accuracy (A_C), sensitivity (S_E), F-score (Fs), error (E_R), and positive predictivity (P_P). The S_E is defined as the ratio of correctly recognized instances over total number of instances as $S_E = T_P/(T_P + F_N)$ [21]. The P_P is determined as the ratio of correctly recognized ECG beats over the total number of detected ECG beats, $P_P = T_P/(T_P + F_P)$ [21]. The A_C is defined as fraction of total figure of correctly identified instances and the number of instances classified, $A_C = (T_P + T_N)/(T_P + T_N + F_P + F_N)$ and F-score (Fs) is termed as $(2T_P/2T_P + F_N + F_P)$. These discussed performance metrics are calculated on the benchmark Physionet data for the developed method which is analyzed in patient-specific scheme [21].

4 Simulation Results

The developed method is implemented on personal computer (windows 10 platform, Intel core i5 (CPU), 2.5 GHz, 8 GB RAM) in MATLAB (version 8.2, R2018b) simulation environment whose performance is estimated by identification of various classes of heartbeats evaluated under patient-specific assessment scheme. The recognition performance report of proposed method is demonstrated by performing the experiment on benchmark MIT-BIH arrhythmia data described in Sect. 3.1. The performance of the trained classifier has been presented in this section in form of a confusion matrix as presented in Tables 2 and 3 for every category of heartbeat in patient-specific strategy. The confusion matrix is formulated by mapping the correctly identified instances and misidentified instances into their corresponding categories identified by the developed methodology. The column of the confusion matrix denotes the total number of cardiac events identified by the proposed method. Whereas the row of confusion matrix denotes the ground truth or annotations used for reference provided in the benchmark database [35]. In total of 43,112 and 89,414 testing ECG instances, 85,712 instances are correctly identified by the developed methodology and reported more accuracy of 95.18% and 95.86% respectively in the equally split dataset and 22-fold CV strategy along with an error rate of 4.82% an 4.14%. In Tables 2 and 3, it is observed that the accuracy of classes 'e' and 'q' is quite less when compared to other categories of heartbeats which is due to lesser number of ECG instances considered for training in these two classes. It must be noted that experiments are conducted for all the data available for these two classes in the benchmark Physionet data.

After the confusion matrix is computed, the performance measures such as sensitivity (S_E), F-score (Fs) and positive predictivity (P_P) metrics are calculated for every category of heartbeats. Before determining these performance metrics, it is

Table 2 Accuracy reported using equally split dataset

	Classified results						
	Category	n	s	v	f	q	Total
Ground truth	N	39,188	917	1270	2822	41	44,238
	S	491	1210	239	23	9	1972
	V	269	160	2631	143	17	3220
	F	188	1	116	81	0	386
	Q	2	1	2	0	2	7
	Total	40,138	2289	4258	3069	69	49,823

Table 3 Accuracy reported using 22-fold CV

	Classified results						
	Category	n	s	v	f	q	Total
Ground truth	N	81,072	1051	3664	4199	97	90,083
	S	738	1681	531	17	5	2972
	V	441	261	6361	398	19	7480
	F	361	1	139	298	3	802
	Q	4	1	5	3	2	15
	Total	82,616	2995	10,700	4915	126	101,352

important to calculate the other parameters like true positive (T_P), false positive (F_P) and false negative (F_N) metrics for a particular category of heartbeat and shown in Tables 4 and 5. Under the patient-specific scheme, Tables 4 and 5 presents the Fs, P_P, S_E parameters computed for all five categories of heartbeats is reported to be overall 86.53% each in equally split dataset and 88.22% each in 22-fold cross-validation strategy respectively.

In Figs. 11 and 12, the three bars together resemble the value of sensitivity, positive predicitivty and F-score respectively for class 1. Here, class 1 indicate the normal

Table 4 Metrics under equally split dataset

TP	FP	FN	S_E	P_P	F_s
40,197	950	5050	88.58	97.63	92.88
1210	1079	762	61.35	52.86	56.79
2631	1627	589	81.70	61.78	70.36
81	2988	305	20.98	2.63	4.68
2	67	5	28.57	2.89	5.26
43,112	6711	6711	95.18	95.18	95.18

Table 5 Metrics under 22 fold CV

TP	FP	FN	S_E	P_P	F_s
81,072	1544	9011	96.67	97.66	92.95
1681	1314	1291	68.75	60.15	56.12
6361	4339	1119	90.11	70.00	70.65
298	4617	504	29.76	12.77	4.91
2	123	12	36.57	15.03	5.47
89,414	11,937	11,937	95.86	95.86	95.86

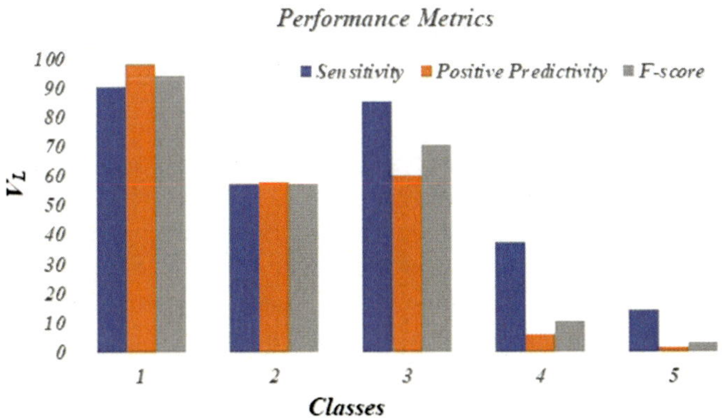

Fig. 11 Metrics plot under equally split datasets

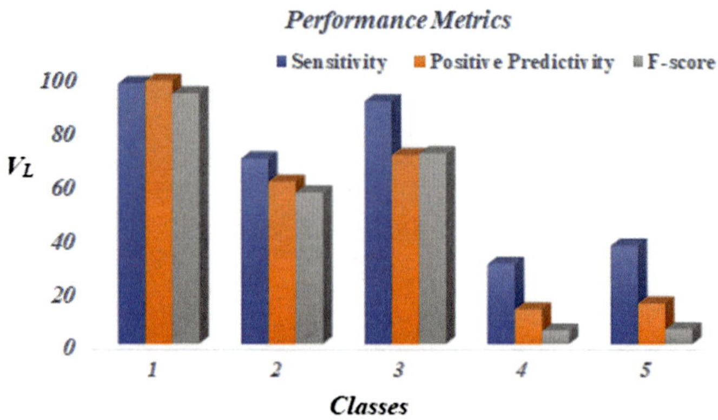

Fig. 12 Metrics plot under 22-fold cross validation

category of heartbeat. And similarly, 2, 3, 4 and 5 indicates the s, f, v, q categories of heartbeats along x-axis. In addition, V_L is values of the parameters out of 100 along y-axis.

4.1 Comparative Study

A brief comparison is presented between the accuracy achieved by the developed method and the previous methods available in the literature under patient-specific assessment scheme. A fair comparison is quite tedious to make which is due to the fact that different works have been evaluated on a different database and different numbers and classes of heartbeats classified. A few of the works have validated their methods over the datasets considered from the healthcare units or patients. In account of these factors, a fair comparative study is presented over the existing methodologies evaluated on the benchmark Physionet arrhythmia data and reported in Table 6.

Table 6 concludes that the proposed methodology has achieved more accuracy under patient-specific scheme when compared with the available methodology provided along the literature. In comparison with some of the works, the current study identifies more number of cardiac events. As such, the quantity of heartbeats varies greatly among the different classes of heartbeats in the datasets, the metrics parameters reported for a particular class can be considered as more reliable and significant. As the current work achieves a higher accuracy than the existing works [37–40], it is implicit to conclude that features extracted in time-frequency space using empirical mode decomposition (EMD) technique are efficient along with heart rate variability features and significant in discriminating between various categories of heartbeats when combined with the PSO optimized support vector machine to efficiently detect and recognize heartbeats.

Table 6 Comparative study with existing works

Works	Feature extraction	Classifier	Classes	Accuracy (%)
Oresko [28]	RR-interval	NN	5	90
Jeon et al. [41]	WT	SVM	3	95.1
Melgani and Bazi [12]	Morphology + PCA	SVM	6	91.67
Lagerholm et al. [15]	Block processing	OSEA	2	92.36
Nambhash et al. [27]	Wavelet	Fuzzy	3	85 (avg.)
Martis et al. [17]	Geometrical	OSEA	2	92.59
Ince et al. [11]	PCA	ANN + PSO	2	95.58
Martis et al. [17]	PCA	LSSVM	5	93.48
Proposed	EMD	PSO + SVM	5	95.86

WT Wavelet transform; *PCA* Principal Component Analysis; *NN* Neural Networks; *LSSVM* Least square support vector machines; *HOS* Higher order statistics; *SVM* Support vector machines; *TSVM* Twin support vector machines

5 Conclusion

This chapter reported a new method by combining the empirical mode decomposition (EMD) as morphological features and heart rate variability as dynamic features. This final feature vector is applied to principal component analysis (PCA) for representing the input signals in reduced dimensions. The final feature vector applied as input to support vector machines (SVMs) for automated recognition of heartbeats into five classes as per AAMI recommendation. The parameters of SVM model developed are chosen using particle swarm optimization (PSO) algorithm. The developed method can be utilized to monitor long-term heartbeat recordings and analyzing the non-stationary behavior of heartbeats. The validation of the developed method is performed on the Physionet data while its evaluation is done under patient-specific scheme. An improved accuracy of 95.86% is achieved by the methodology under patient specific scheme. In future, this current work is subjected to include more number of heart rhythms for monitoring and analysis, to develop more efficient algorithms and their implementation of mobile platforms. This developed methodology can be considered as an efficient technique and employed in the computer-aided diagnosis allowing subjects to lead a healthy lifestyle for cardiovascular diseases.

References

1. S. Raj, Development and hardware prototype of an efficient method for handheld arrhythmia monitoring device. Ph.D. Thesis IIT Patna (2018), pp. 1–181
2. G.K. Garge, C. Balakrishna, S.K. Datta, Consumer health care: current trends in consumer health monitoring. IEEE Consum. Electron. Mag. **7**(1), 38–46 (2018)

3. R.V. Andreao, B. Dorizzi, J. Boudy, ECG signal analysis through hidden Markov models. IEEE Trans. Biomed. Eng. **53**(8), 1541–1549 (2006)
4. P. de Chazal, R.B. Reilly, A patient-adapting heartbeat classifier using ECG morphology and heartbeat interval features. IEEE Trans. Biomed. Eng. **53**(12), 2535–2543 (2006)
5. K.T. Chui, K.F. Tsang, H.R. Chi, B.W.K. Ling, C.K. Wu, An accurate ECG-based transportation safety drowsiness detection scheme. IEEE Trans. Ind. Inform. **12**(4), 1438–1452 (2016)
6. M. Faezipour, A. Saeed, S. Bulusu, M. Nourani, H. Minn, L. Tamil, A patient-adaptive profiling scheme for ECG beat classification. IEEE Trans. Inf. Technol. Biomed. **14**(5), 1153–1165 (2010)
7. G.D. Fraser, A.D. Chan, J.R. Green, J.R. Macisaac, Automated biosignal quality analysis for electromyography using a one class support vector machine. IEEE Trans. Instrum. Meas. **63**(12), 2919–2930 (2014)
8. C.W. Hsu, C.J. Lin, A comparison of methods for multiclass support vector machines. IEEE Trans. Neural Netw. **13**(2), 415–425 (2002)
9. Y.H. Hu, S. Palreddy, W.J. Tompkins, A patient-adaptable ECG beat classifier using a mixture of experts approach. IEEE Trans. Biomed. Eng. **44**(9), 891–900 (1997)
10. Y.H. Hu, W.J. Tompkins, J.L. Urrusti, V.X. Afonso, Applications of artificial neural networks for ECG signal detection and classification. J. Electrocardiol. **26**, 66–73 (1993)
11. T. Ince, S. Kiranyaz, M. Gabbouj, A generic and robust system for automated patient-specific classification of ECG signals. IEEE Trans. Biomed. Eng. **56**(5), 1415–1426 (2009)
12. F. Melgani, Y. Bazi, Classification of electrocardiogram signals with support vector machines and particle swarm optimization. IEEE Trans. Inf. Technol. Biomed. **12**(5), 667–677 (2008)
13. K. Minami, H. Nakajima, T. Toyoshima, Real-time discrimination of ventricular tachyarrhythmia with Fourier-transform neural network. IEEE Trans. Biomed. Eng. **46**(2), 179–185 (1999)
14. P. de Chazal, M. O' Dwyer, R.B. Reilly, Automatic classification of heartbeats using ECG morphology and heartbeat interval features. IEEE Trans. Biomed. Eng. **51**(7), 1196–1206 (2004)
15. M. Lagerholm, C. Peterson, C. Braccini, L. Edenbrandt, L. Sornmo, Clustering ECG complexes using Hermite functions and self-organizing maps. IEEE Trans. Biomed. Eng. **47**(7), 838–848 (2000)
16. T.H. Linh, S. Osowski, M. Stodoloski, On-line heart beat recognition using Hermite polynomials and neuro-fuzzy network. IEEE Trans. Instrum. Meas. **52**(4), 1224–1231 (2003)
17. R. Martis, U. Acharya, K. Mandana, A. Ray, C. Chakraborty, Cardiac decision making using high order spectra. Biomed. Signal Process. Control **8**(2), 193–203 (2013)
18. S. Raj, G.S.S. Chand, K.C. Ray, Arm-based arrhythmia beat monitoring system. Microprocess. Microsyst. **39**(7), 504–511 (2015)
19. S. Raj, S. Luthra, K.C. Ray, Development of handheld cardiac event monitoring system. IFAC Papers-On-Line **48**(4), 71–76 (2015)
20. S. Raj, K.C. Ray, ECG signal analysis using DCT-based dost and PSO optimized SVM. IEEE Trans. Instrum. Meas. **66**(3), 470–478 (2017)
21. S. Raj, K.C. Ray, A personalized arrhythmia monitoring platform. Sci. Rep. **8**(11395), 1–11 (2018)
22. S. Raj, K.C. Ray, Automated recognition of cardiac arrhythmias using sparse decomposition over composite dictionary. Comput. Methods Programs Biomed. **165**, 175–186 (2018)
23. S. Raj, K.C. Ray, Sparse representation of ECG signals for automated recognition of cardiac arrhythmias. Expert Syst. Appl. **105**, 49–64 (2018)
24. S. Raj, K.C. Ray, A personalized point-of-care platform for real-time ECG monitoring. IEEE Trans. Consum. Electron. **66**(4), 1–9 (2018)
25. S. Raj, K.C. Ray, O. Shankar, Development of robust, fast and efficient QRS complex detector: a methodological review. Aust. Phys. Eng. Sci. Med. **41**(3), 581–600 (2018)
26. S. Raj, A real-time ECG processing platform for telemedicine applications, in *Advances in Telemedicine for Health Monitoring: Technologies, Design, and Applications* (IET, 2019), (Accepted)

27. M.S. Nambakhsh, V. Tavakoli, N. Sahba, FPGA-core defibrillator using wavelet-fuzzy ECG arrhythmia classification, *in 2008 30th Annual International Conference of the IEEE Engineering in Medicine and Biology Society* (2008), pp. 2673–2676
28. J.J. Oresko, A wearable smartphone-based platform for real-time cardiovascular disease detection via electrocardiogram processing. IEEE Trans. Inf. Technol. Biomed. **14**(3), 734–740 (2010)
29. B. Pourbabaee, M.J. Roshtkhari, K. Khorasani, Deep convolutional neural networks and learning ECG features for screening paroxysmal atrial fibrillation patients. IEEE Trans. Syst., Man, Cybern: Syst. **48**(12), 2095–2104 (2018)
30. J. Pan, W.J. Tompkins, A real-time QRS detection algorithm. IEEE Trans. Biomed. Eng. **32**(3), 230–236 (1985)
31. N.E. Huang, Z. Shen, S.R. Long, M.L. Wu, H.H. Shih, Q. Zheng, N.C. Yen, C.C. Tung, H.H. Liu, The empirical mode decomposition and Hilbert spectrum for nonlinear and non-stationary time series analysis. Proc. Roy. Soc. London A **454**, 903–995 (1998)
32. V. Vapnik, *The Nature of Statistical Learning Theory* (New York, 1995)
33. B.E. Boser, I.M. Guyon, V.N. Vapnik, A training algorithm for optimal margin classifiers, in *Proceedings of the Fifth Annual Workshop on Computational Learning Theory* (1992), pp. 144–152
34. J. Kennedy, R.C. Eberhart, *Swarm Intelligence* (Morgan Kaufmann, San Mateo, CA, USA, 2001)
35. G.B. Moody, R.G. Mark, The impact of the MIT-BIH arrhythmia database. IEEE Eng. Med. Biol. Mag. **20**(3), 45–50 (2001)
36. M. Stone, Cross-validatory choice and assessment of statistical predictions. J. Roy. Stat. Soc. B (Methodol.) **36**(2), 111–147 (1974)
37. S. Raj, K.C. Ray, A comparative study of multivariate approach with neural networks and support vector machines for arrhythmia classification, in *2015 International Conference on Energy, Power and Environment: Towards Sustainable Growth (ICEPE)* (2015), pp. 1–6
38. S. Raj, K.C. Ray, O. Shankar, Cardiac arrhythmia beat classification using DOST and PSO tuned SVM. Comput. Methods Programs Biomed. **136**, 163–177 (2016)
39. S. Raj, K.C. Ray, Application of variational mode decomposition and ABC optimized DAG-SVM in arrhythmia analysis, in *2017 7th International Symposium on Embedded Computing and System Design (ISED)* (2017), pp. 1–5
40. R.K. Tripathy, M. Paternina, J. Arrieta, A. Mendez, G. Naik, Automated detection of congestive heart failure from electrocardiogram signal using Stockwell transform and hybrid classification scheme. Comput. Methods Programs Biomed. **173**, 53–65 (2019)
41. T. Jeon, B. Kim, M. Jeon, B. Lee, Implementation of a portable device for real-time ECG signal analysis. BioMed. Eng. Online **13**, 160 (2014)

Sandeep Raj received the B. Tech. degree in Electrical and Electronics Engineering from Allahabad Agricultural Institute-Deemed University, Allahabad, India, in 2009 and the M. Tech. (Gold Medalist) degree in Electrical Engineering in 2011. He was a Visiting Faculty with the National Institute of Technology Patna, India, in 2012. He received DST-INSPIRE fellowship to pursue Ph.D. in the department of Electrical Engineering from Indian Institute of Technology Patna, Bihta, India and completed his Ph.D. in 2018.

He is currently serving as an Assistant Professor in the department of Electronics and Communication Engineering at Indian Institute of Information Technology Bhagalpur, Bhagalpur 813210, India. His current research interests include embedded systems, biomedical signal processing, machine learning, Internet-of-Things.

Clinical Decision Support Systems and Predictive Analytics

Ravi Lourdusamy and Xavierlal J. Mattam

Abstract The chapter introduces the history of the clinical decision support system beginning with the history of the system of decision making. It is an overview of how the clinical decision support systems developed through the years. The current technology used in decision making are also discussed. With the use of artificial intelligence, the clinical decision support systems have moved to the realm of predictive analysis to find out the possibilities of diseases rather than just the diagnosis and treatment. The chapter also elaborates the various types of clinical decision supports systems. Although the decision support systems are widely regard as an important and integral part of healthcare there has been a notable reluctance in the use of clinical decision support systems. The chapter also discusses the practical challenges in the implementation of the clinical decision support systems in healthcare organisations. Each of the topics in the chapter is dealt with summarily and the reference to a detailed study is provide. The idea is to provide a clear understanding of the system rather than to fully elaborate the system.

Keywords Clinical decision support systems · Predictive analytics · Decision making · Artificial intelligence · Medical diagnosis and treatment

1 Introduction

Clinical Decision Support Systems (CDSS) are Decision Support Systems (DSS) used in Healthcare. Digitalisation of medical records or the Electronic Health Record (EHR) was the precursors of CDSS. When DSS, which was primarily implemented for business, was adapted in healthcare, the EHR was used as the knowledge base in many CDSS. CDSS evolved with the development and incorporation of advanced technologies. Accordingly, the functions and use of CDSS also varied. While in

R. Lourdusamy (✉) · X. J. Mattam
Department of Computer Science, Sacred Heart College, Tirupattur, Tamil Nadu, India
e-mail: ravi@shctpt.edu

X. J. Mattam
e-mail: xaviermattam@gmail.com

V. Jain and J. M. Chatterjee (eds.), *Machine Learning with Health Care Perspective*, Learning and Analytics in Intelligent Systems 13,
https://doi.org/10.1007/978-3-030-40850-3_14

some CDSS, these functions are components of the CDSS, in others the CDSS performed only a few functions. This gave rise to different forms of CDSS. Research and literature on CDSS are also mainly into integrating newer technologies into the different components of the CDSS to make CDSS more efficient of user-friendly. While advances in technology are established as the primary driving force in the evolution of CDSS, there are goals of CDSS that are to be met. The greatest challenge of an efficient CDSS is to achieve a perfect balance between technological adaptations in CDSS while not compromising the primary task of CDSS. Commercialisation of CDSS has led to its popularity and has slowly turned it into an ERP-like solution in many cases. Therefore, there is need to define CDSS more accurately for all emerging CDSS.

Recent advances in medical knowledge and health care systems combined with the advances in the use of huge knowledge bases has led to the creation of whole new CDSS that are more efficient and less time consuming. CDSS has moved from a system for data presentation to predictive analytics. With the use of newer technologies such as the Internet of Things (IoT), big data analytics, and machine learning components, the emerging CDSS are capable of functions such as predictive diagnosis, advanced automated knowledge retrieval and newer delivery models for the ease of the user. The CDSS will then be services running in the cloud leading to uniformity in the knowledge and structure. The emerging CDSS will redefine the functions of those involved in the healthcare system. While the ease of use of the CDSS will be much greater and the CDSS will be more efficient, the development and functioning of the system will be more complex. Moreover, the CDSS will be dynamic.

Since evolution of CDSS is linked to the development of technologies, the biggest challenge in the development and implementation of CDSS is the instability in the processes. Unlike some add-on tools in some other systems, the newer technologies replace the old ones resulting in newer forms of CDSS. So, the process of development and implementation of CDSS becomes a continuous procedure. In order to gain stability in the processes, the CDSS is structured in the form of independent components, each of which can be individually developed or modified without affecting the other parts. Other challenges to the successful development and implementation of CDSS include human factors, financial constraints and technical implementation problems. In trying to get out of problems like alert fatigue, newer problems arose with the use of expert systems. The recent AI based CDSS addresses the problems of the expert systems but it is still at its beta testing stages. Although some components of CDSS are proven to be successful, CDSS as whole, in many cases, is not a part of the clinician's workflow. The CDSS is still a separate system outside the clinician's routine and therefore is not considered for optional utilisation. Issues such as availability and accessibility of hardware and training in the use of systems have to be addressed to overcome the problem of integrating CDSS in clinicians' workflow.

The objective of the chapter is to present CDSS as an evolving system which has become effective and reliable with the use of predictive analytics. Basing on the evolving process of the CDSS, the chapter concludes with a prediction of the future CDSS and barriers that has to be overcome in order to achieve it. The chapter consist of the following sections: (1) History and Development of CDSS: this section will give a glimpse of how the CDSS developed through the years. (2) Emerging trends in CDSS: here the recent evolution of CDSS is discussed. (3) Problems in the implementation of CDSS: will be a summary of key disruptions, barriers, accelerators in the CDSS development. (4) Types of CDSS: will speak of different categories of CDSS that have evolved. (5) Artificial Intelligence (AI) and CDSS: is about the recent developments in CDSS with AI algorithms. (6) Predictive analytics in CDSS: this again is about the more recent developments that makes CDSS indispensable in healthcare.

2 History and Development of CDSS

Decision making has always been an important task in any process but in the case of any clinical process it is not only important but could be crucial as well. That is why there has been so much of interest in creating a system of decision support for clinicians. The research literature on CDSS has been steadily on the rise [1, 2]. There also has been studies on the origin and evolution of CDSS to understand the background of the development and to find the future possibilities [3–6]. To trace the origins of CDSS will also involve the tracing of the origins of a decision making system since the formalisation of decision making is the origin of a decision making system [7].

2.1 History of Decision Making

The origins of CDSS should be traced back to the origins of decision making. In ancient times, people depended on oracles and fortune tellers to find the outcome of their choices. The first use of permutations and combinations actually traces back to the study of cryptology by the Arabs between the 8th and 13th centuries [8]. The mathematics of decision making appeared first with the problem of points (also called the problem of division of the stakes) in the book 'Summa de arithmetica, geometrica, proportioni et proportionalità' written by Luca Pacioli in 1494 [9]. The problem of points is a classical probability theory problem that describes the expected value. The second published attempt to the problem of points is the 'La Prima parte del General Trattato di numeri et misure' by Niccolo Tartaglia in 1556. The next major contribution is that of Hieronymus Cardanus (Girolamo Cardano) in his book 'Practica arithmetice et mensurandi singularis' which was probably composed in 1539 but was published in 1663. A finally accepted solution is probably the one given by Blaise Pascal. His discussion with Pierre de Fermat was published by the title 'Traite du Triangle arithm ´ etique, avec quelques autres petits trait ´ es´ sur la meme mati ˆ ere' by Guillaume Desprez in 1665 [10].

The other problem of finding an expected value is the Game of Chance problem. It is also a classical probability theory problem that has been studied deeply over the years, much more than the problem of points. The problem was described by Cardano in his book 'Liber de Ludo Alea' (Book on Games of Chance) in 1526. The game of chance was next discussed in the book written in 1656 by Christian Huygens and published in 1657 with the title, 'De ratiociniis in aleae ludo' (On the Calculations in Games of Chance). The book remained as the book on mathematical probability for nearly half a century until the book of Jakob Bernoulli, Ars conjectandi (Art of Conjecturing) was posthumously published in 1713. Although his contribution to the game of chance with which he began his treatise as a discussion of Huygens' problems was minimal, the treatise is acclaimed as the founding of Mathematical Probability. A more elaborate work on the theory of probability was by Abraham De Moivre in his book 'The Doctrine of Chances' that was published in 1718. De Moivre tackles the problem of the game of chance just the way Bernoulli did, as a part of elaborate problem of expected outcome. A very convincing formula to the problem of probability came from Reverend Thomas Bayes in his book 'An Essay towards solving a Problem in the Doctrine of Chances' that was published in 1763, three years after his death. The Bayes' Theorem which speaks of the probability of an event E happening if a prior event F has happened and is given by the formula $P(E|F) = \frac{P(F|E)P(E)}{P(F)}$. It is the formula to calculate conditional probability commonly used in all fields even today. Pierre-Simon de Laplace is said to have laid the foundation of the classical theory of probability. His book 'Théorie Analytique des Probabilités' (Analytical Theory of Probability), was first published in 1812 and there were later editions to it in 1814, 1820, and 1825 [10, 11].

The next greatest contribution to the theory of probability was a hundred years later by Andrei Markov. Markov extended the results of Bernoulli and developed the calculations for linked probabilities. He came out with the idea of Markov process in 1913. The stochastic matrix or a Markov matrix that describes the Markov chain were also his great contributions [10].

2.2 *The Problem of Points*

It is the first classical uncertainty problem recorded. The problem is recorded as how will the reward be shared between two players if the game was interrupted after sometime while one person was leading in points. A similar problem can be presented in medical diagnosis. If a patient has a symptom s1, which is indicative of a disease d1 in 60% of the cases or a disease d2 which also exhibits the same symptom s1 in 40% of the cases. The pre-test probability (prior-probability) is 60% for d1 and 40% for d2. So, what should the physician suggest. The answer attempted by Luca Pacioli is to divide the points already got by a player by the number of games that have to be played to win the game. That will give the share of the reward. For example, if player A has 60 points and player B has 40 points, A gets A/(A + B) = 60/100 or 60% of the reward. Similarly, player B will get 40% of the reward. If we are to apply the solution to the medical diagnosis problem, then we have to divide the pre-test probability with the sum of the pre-test probabilities. Although this does not change the probabilities of the diseases, it was an attempt in finding the expected value [11, 12].

Tartaglia's solution is what each player should receive has to be in proportion to the difference in their points. Applying the solution to the medical diagnosis problem stated earlier, if there is a 50% chance of the patient having any one of the disease without the symptoms, then P(d1|s1) = 60% ((50) + (((60 – 40)/(60 + 40) * (50))) and P(d2|s1) = 40% ((50) + (((40–60)/(40 + 60) * (50))). Tartaglia's calculation goes a step further to increase or decrease from the prior contribution although in the given example, there is no change in the overall probability of d1 and d2 [11, 13].

Cardano makes use of progression that complicates the calculation of the expected outcome. According to Cardano's reasoning P(d1|s1) = 69% ((1 + 2 + ··· +40)/((1 + 2 + ··· + 60) + (1 + 2+ ··· +40)) and P(d2|s1) = 31% ((1 + 2 + ··· + 60)/((1 + 2 + ··· + 60) + (1 + 2 + ··· + 40)). Here there is some changes in the percentage of probability but it only increases the higher probability and decreases the lower probability. It does not actually make decision making changes [11, 14].

Although Pascal explained the solution using his arithmetic triangle, it could be calculated using the formula $\sum_{k=0}^{a-1}\binom{a+b-1}{k} \div 2^{a+b-1}$ where a and b are the known probabilities. Using this formula, the P(d1|s1) = 75% and P(d2|s2) = 25%. This gives a more consistent probability percentage unlike those calculated earlier [14].

2.3 The Game of Chance Problem

The problem was to find the minimum of throws of two dices to produce two sixes. There are many varieties of the problem of the game of chance but it is all about the probability of an event happening. Cardano's solution of eighteen was based on the miscalculation that it requires three throws of a single dice to get an even or an odd number. This problem was posed by Chevelier de Mere to Blaise Pascal after he realised that his calculation was inaccurate because although he rightly calculated that the chance of getting a double six in one roll of dice is 1/36, he erred in thinking that the chance of getting two sixes in 24 throws will be 24/36 or 52% [15].

Pascal solved the problem with inputs from Fermet by finding the probability of not getting a double six which 35/36. So, their solution is that in 24 throws, the probability of getting a double six is 1 – $(35/36)^{24}$ which is 1 – 0.5086 or 49%.

If the probability solution from the game of chance is applied to the probability problem of the medical diagnosis, it would mean that if the physician was to wager that the patient has d1, then the physician has only 40% (1–0.6) chance of being right. While the physician has a 60% (1–0.4) chance of being right about d2 [10].

Huygens' solution increases the chances of the patient with d2 since he uses a recursive formula in finding the minimum probability of winning. By using his method, if the first probability of d2 is 40% or 2 in 5 case, then the second chance is calculated using the formula $\frac{pd2+P(d2)q}{p+q}$ where p is number of cases with d2 = 2 and q is the number of case without d2 = 3 and P(d2) is (2/5)d2. By Huygens' formula, the probability of the patient to have d2 is 16–9 while the probability of the patient with d1 keeps decreasing.

Bernoulli analysis of the problem came to the same conclusion as Huygen but the formula Bernoulli used was $\frac{\log 2}{\log c - \log b}$ where c is the total number of chances (cases) and b is the possible number of failed chances or cases. De Moivre uses the formula $\frac{(probability\ of\ \mathrm{d2})^x}{(probability\ of\ \mathrm{d1} + probability\ of\ \mathrm{d2})^x}$ where x is the number of trials. Using logarithms De Movire formulated it as $\frac{\log 2}{\log(d1+d2) - \log d2}$. Taking the initial chances of d1 and d2 the probability of d1 is about 75% as also concluded by Huygens and Bernoulli. Laplace used modified version of the game of chance. The example Laplace used was a game of picking a white ticket from a bag of p white tickets and q black tickets. The question can be adapted to the medical diagnosis question mentioned earlier and question will be if there are 6 cases of d1 and 4 cases of d2 what will be the probability that the case in hand is d1. Laplace's solution is that if there exits x ratio of d1 and d2 cases then the probability of the present case being d1 is $\frac{x^p\,(1-x)^q\,dx}{\int_0^1 x^p\,(1-x)^q\,dx}$ [10].

According to Markov, the probability of d1 in n cases is 1/n then the probability of d2 will be (n − 1)/n [10].

2.4 Growth of CDSS Literature

The paper published by F. A. Nash in 1954 could probably be the first attempt to develop an automated CDSS [16]. From then there has been a constant growth in the literature regarding CDSS. Lipkln, M. and Hardy, J. D., described how eighty cases of hematologic disease taken from the files of a university college hospital could be classified and coded. They concluded by stating that the tabulation of their findings could be done using electronic computers available in 1957 [17]. A very significant contribution to the development of CDSS will probably be the articles of Lee and Lusted in which they analysed the decision-making process of physicians and described how the mathematical models of the same could be done using symbolic logic, probabilistic theory and value theory. They also presented a computerised card system to help diagnosis [18]. Crumb and Rupe also had proposed a somewhat similar technique but for a diagnostic procedure. Their proposed technique makes use of the computer capabilities to store and analyse data [19]. The article of Martin Lipkln et al. presents a CDSS system for hematologic disease with verifiable results. They conclude that the use of digital computers in tabulating the results of the various test led to a more efficient and accurate diagnosis and therefore the CDSS is very useful tool in medical diagnosis [20]. Warner et al. in the same year had presented an article on the mathematical approach to medical diagnosis where they used the probabilistic approach of Lee and Lusted for a set of mutually exclusive diseases with symptoms. They test their hypothesis using an application to Congenital Heart Disease and conclude that the use of computers will drastically reduce the time required for the huge number calculations required and so will speed up the critical medical diagnosis. The human-computer interface that they suggest for the use in the description of the CDSS is quite rudimentary because of the primitiveness of the technology of the early 1960s [6, 21].

Slack et al. describes a computer system that takes the inputs from the patient to record the history of a disease. It helped the patient better answer questions about the patient's disease and the system produces a clear printout of the history of the disease which was much clearer than that of the physician's recording [22]. A complete description of a CDSS with a flowchart of its various components was given by G. A. Gorry in 1968 [23]. It is probably the first programmed CDSS that proved to be a success in its effectiveness and speed. Bellman and Zudeh were the first to introduce the idea of fuzzy sets in decision making. Considering the real world scenario in medical diagnosis, many constraints and goals are not clearly defined. This can be solved using fuzzy sets in decision making [24].

The 1970s and 1980s saw a number of functional CDSS being implemented in specific categories. It was either a CDSS for diagnosis and treatment or a CDSS for cost effectiveness and resource management or a large scale CDSS for patient care with capacity to automatically solve issues relating to healthcare. Some of the known CDSS of the time were, CASNET, MYCIN, PIP, and INTERNIST-1. The CDSS that covers the whole medical organisation, from the entry of the details of the patient to the final discharge recommendations, where the prevailing features of the CDSS [6].

The advancements in CDSS of the 1990s was purely in the implementation of various research and technological developments. The adaptation of artificial intelligence in CDSS with user friendly interfaces made CDSS popular. Neural networks, fuzzy logic and machine learning algorithms used in pattern recognition and data analytics helped directly in prognosis, control and treatment of diseases. The efficiency and reliability of the CDSS led to the acceptance and popularity of CDSS. It also led to the standardisation of terminologies paving the way for global communication and consultation of findings and results [6].

The later developments in CDSS were all related to the evolution of technology and the understanding of the decision-making process.

3 Emerging Trends in CDSS

The growing popularity of CDSS due to its efficiency and reliability has led to the development of the state of art CDSS. Although a total decision making system is next to impossible, there has been marked improvements in the acceptance of CDSS in medical organisation and is considered a part of medical education. Some of the current trends in the CDSS relates to the use of current technology in CDSS and in the usefulness of the CDSS in a healthcare system.

3.1 CDSS for Cost Reduction and Risk of Readmission

It has been studied that there is a reduction in cost of healthcare and the length of hospital stay by adhering to CDSS alerts. It has also been studied that the odds of readmission has reduced when the CDSS alerts have been adhered to [25]. There are also other benefits of the CDSS [26].

One major factor that plays on any healthcare organisation is cost reduction. The important issue that plays on the organisation preventing reduction of cost is the fear of readmission. If a patient has to be readmitted, there is a possibility that the patient was not treated fully in the first instance. A lack of criteria to discharge a patient will lead to the discharge of the patient being done earlier or not being discharged. The decision needs to follow a certain set of rules that make the CDSS efficient in recommending a discharge. The dilemma of a discharge physician is to lower medical cost by decreasing the hospital stay while at the same time reducing the likelihood of readmission. The decision is usually made using the evidence available on the electronic medical record of the patient. By operationalising the use of the data in the electronic medical record and combining it with the criteria of discharge in the CDSS, the CDSS will be able to give a good option for discharge decision. An efficacious CDSS model will have to analyse all possible criteria [27]. An artificial intelligence algorithm can help make recommendation based on training sets that take into consideration the results of earlier recommendations.

The possibility of readmission can also be highly reduced if the CDSS has real-time patient monitoring at the time of discharge decision. The continuous real-time monitoring of the in-patient will also provide the CDSS with pharmacogenetics disorders that could alert doctors of possible readmission in case of adverse drug events. In many cases, especially among older people who need to use multiple medication, the adverse drug reaction is common leading to their readmission. The knowledge of the possibility of adverse drug reaction at the time of discharge could be a criterion for discharge decision to reduce readmission [28].

The Center for Medicare & Medicaid Services, CMS, which is part of the Department of Health and Human Services (HHS) of the American government has introduced the Hospital Readmissions Reduction Program (HRRP) in 2010. It has given cost formulas and penalties for readmission based on the rates of readmission. This could be coded into the CDSS to find the probability of readmission and the potential cost of readmission. This allows the hospitals to allocate resources according to the predicted expenditure. The CDSS helps to optimally select patients having a greater need for interventions in order to reduce cost and readmission rates [29].

3.2 Use of Big Data

Big data or extremely large sets of data that is available on the electronic medical records can be utilised to find trends or associations between the patient and

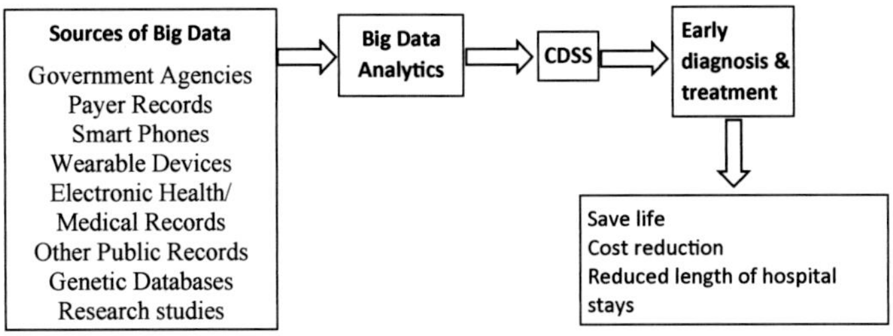

Fig. 1 Big Data in CDSS

the health condition. Collecting, analysing and using patient's physical, and clinical data that is too vast or complex cannot be easily understood by traditional means of data processing. The big data is processed by machine learning algorithms. With the digitalisation of health information, there has been a rise of healthcare big data. The volume, velocity, variety and veracity of the data has posed the challenges of healthcare data. As a result, the health systems have to find ways to use relevant technology capable of collecting, storing, and analysing the information to produce insights on which action can be taken. The sources for the Big Data include government agencies such as driving licence, public distribution systems, public accounts etc., payer records such as insurance, medical care etc., smart phone, wearable devices, electronic heath records and medical records in hospitals, other public records, genetic databases and research studies. The data from all the sources are analysed and the information is utilised in a CDSS to help decisions as in Fig. 1.

Big data analytics help in complex and challenging diagnosis uncertainty and in cases with unpredictable treatment. The data gathered and analysed leads to trajectory tracking in which, the general direction in which the disease will progress can be explained. It helps to take necessary precautions and plan ahead of the consequences. Big data analysis also helps in perspective taking whereby the perspective of a group is used for the decision making process. The metacognition process of focussing one's thinking is also helped by the analysis of the data by the system. Differential diagnosis procedure is accessed in the metacognition procedure and the clinical reasoning procedure is monitored [30].

In spite of the many advantages in the use of Big Data which have been exploited in many fields, the use of Big Data in healthcare has been comparatively very low. There are many reasons for the lesser use of Big Data in healthcare. Primary causes are the resistance of the stakeholders to change and the big investment needed for the change. Concerns of privacy and the lack of procedures are added reasons for the resistance to use Big Data [31]. The use of Big Data in the field of healthcare has the potential to improve care and reduce costs. But the data has to be analysed to find associations, patterns and trends that make meaningful information for decision making. The decision support using big data can potentially lead to the early diagnosis and treatment of diseases thus saving the organisation and consumer lot of money [32].

Mining the Big Data and utilising its analysis can go a long way in identifying high risk and high cost patients. The high cost or high risk patients can be categorised using their data available in other repositories other than the electronic health record of the hospital. These data are automatically collected and analysed to predict the cost or risk. Accordingly, decision of diagnosis and treatment are supported by the CDSS. It also gives a value for the readmission possibility by matching the data of the patient with the previous datasets of similar patients. The continuous monitoring of the patient's data will also assist in the creation of the readmission value. Once the high risk or high cost patients are identified, the decision regarding the allocation of resource can be formed in the CDSS. The CDSS keeps a complete monitoring of all resources to be able to help alert the medical practitioners at right moments thus saving lives and cutting costs [33].

3.3 Fuzzy Logic and Artificial Intelligence (AI)

Fuzzy logic is used in the area of soft computing where the is no clear demarcation between the true and false. There can only be degrees of truth and degrees of falsehood similar to the probability of the expectant value. Fuzzy logic for decision using uncertainties was used in the evolutionary stages of CDSS but it is still an emerging trend in combination with Bayesian network, artificial neural network, fuzzy inference system, genetic algorithms, swarm intelligence, and fuzzy cognitive maps [34].

Probability and Fuzzy Logic are different. Both are related to uncertainty. In probability the uncertainty is about some definite value but in fuzzy logic the uncertainty is quantified as degrees of the value. For example, in the case of probability the uncertainty is whether the expected value could be either true or false but in fuzzy logic, the expected value is quantified as amount of truth or amount of falsehood. While probability of A, P(A) is a value that maps to a set U, the fuzzy set F is a member function of U that gives the grade of membership of an element u in F. In a way, while probability defines a clear expected value, fuzzy logic defines the amount of belongingness of an event in a collection of vague events. But there are ways to use both the fuzzy logic and the probability theory together [35–38].

There are many ways to apply fuzzy logic to medical diagnosis [39–46]. In CDSS, fuzzy logic is also used in case of uncertainty in the clinical practice guidelines. Uncertainty in clinical practice guidelines arises due to a lack of information, non-specificity with measurements, the probabilistic nature of the result data, uncertainty in recommendations, conflicts in the various alternatives suggested and vagueness in the symptoms. In categorising the various uncertainties and classifying the vagueness in different fuzzy sets, the CDSS is able to recommend decisions that are effective and quick [47].

The method of applying fuzzy logic (Fuzzification) in a CDSS is by grouping the symptoms or test results in fuzzy sets which could have values like Low, Moderate, High. Each of these are given a fuzzy value of say, yes, maybe or no. These values are

Table 1 Fuzzy table for d1

Symptoms	Fuzzy value of the symptom		
	Low	Moderate	High
s_1	No	May be	Yes
s_2	May be	Yes	Yes
s_3	May be	Yes	Yes
s_4	May be	Yes	Yes

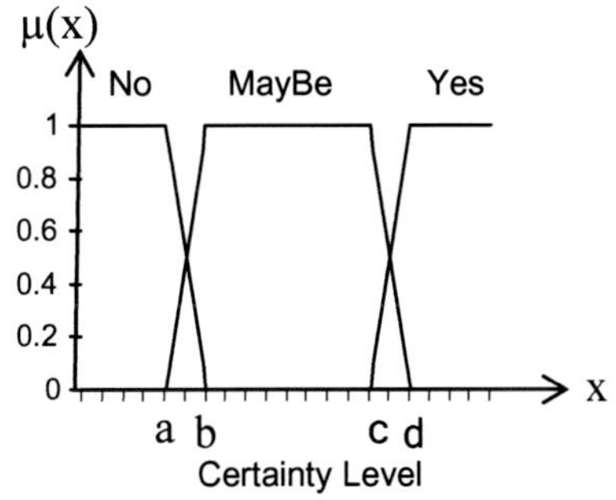

Fig. 2 Membership function

entered into a fuzzy table along the relevant feature. For example, the for a disease d1 with symptoms S = {s1, s2, s3, s4}, the fuzzy table will be as shown in Table 1.

The membership function μ(x) where x is mapped to a value between 0 and 1, is the fuzzy value assigned using triangular function [$\mu(x) = (x - a)/(x - m)$ or $(b - x)/(b - m)$ where a and b are the lower and upper limits and m lies between a and b], trapezoidal function ($\mu(x) = (x - a)/(b - a)$ or $(d - x)/(d - c)$ where a is lower limit, d is upper limit, b is a lower support limit, and c is an upper support limit, $a < b < c < d$) or gaussian function ($\mu(x) = e^{-\frac{(x-m)^2}{2k^2}}$, where m is a central value and k a standard deviation). The membership function of the above example using trapezoidal function will result in a map as shown in Fig. 2.

There are method used for Defuzzification, after the fuzzy sets are evaluated by the inference engine using rules, like the Max-Membership Method which uses highest output values, Centroid Method that choose the central value, Weighted Average Method which assigns weights according to the maximum membership function and the Mean-Max Membership Method where the central maximum membership function is chosen [48–50]. The overall flow of the CDSS using fuzzy logic is as shown in Fig. 3.

The fuzzy system in the CDSS using neural networks make use of artificial intelligence algorithms in the CDSS. A combination of fuzzy logic and artificial neural networks are found to be very effective in a CDSS. The neuro-fuzzy hybridisation is in general done using the neural network algorithms being applied to the fuzzy inference engine and the feedforward learning algorithms use the output of the inference engine to modified the neural network as shown in Fig. 4 [51].

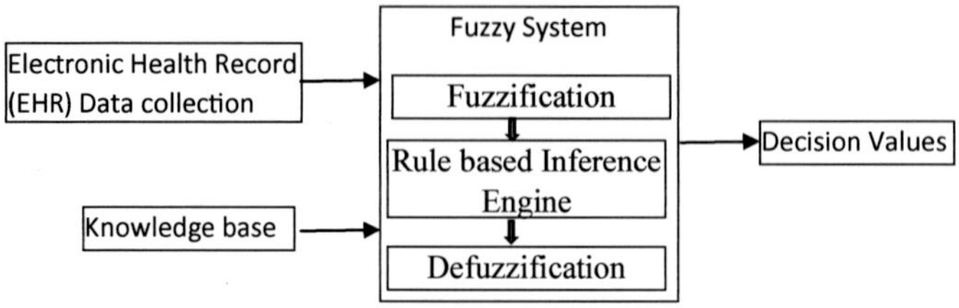

Fig. 3 CDSS using Fuzzy logic

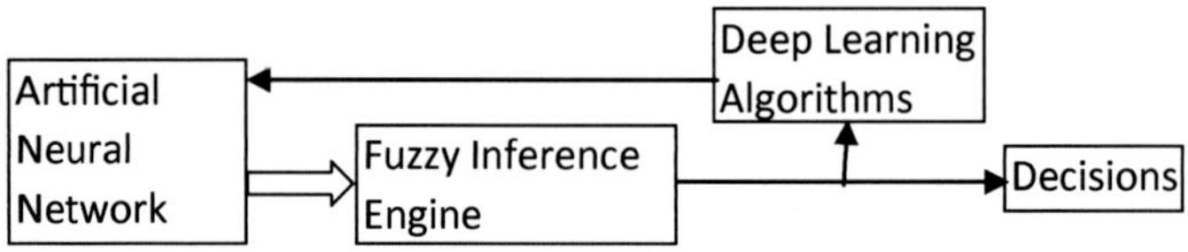

Fig. 4 Neuro-Fuzzy hybridisation in CDSS

4 Major Issues in the Implementation of CDSS

CDSS has evolved with the development of techniques and technology. Yet there has not been the desired CDSS that can give efficient, reliable and opportune decision options. The CDSS has to give accurate and comprehensive assessment. There has been a number of reasons why a fully optimal and functional CDSS is not being implemented. A few common issues in the implementation of CDSS are.

4.1 Conversion of Research into Practice

The greatest hurdle in any implementation is the slow rate of conversion of research findings. CDSS is a complex system involving all the departments of a healthcare organisation. Therefore, there are many factors that come to play when a research finding in CDSS is to be implemented in the organisation. The implementing issues can be classified as personal issues, team issue, organisational issue and CDSS issue. The personal issue for the implementation can be lack of skill, the lack motivation to change, the lack of time for training or updating and a general lack of necessity for change. The team issues could be the lack of coordination, the lack of communication infrastructure, the lack of management and lack of relationship between team members. The organisational issues include lack of time or resources, lack of infrastructural change possibilities, lack of understanding of the benefits and lack of leadership or policy. The CDSS issues for implementation could include the lack of proper documentation and training possibilities, the lack of feedback acceptance mechanism, the lack of cost effectiveness and rapid changes in research and development. All these issues are interconnected as shown in Fig. 5.

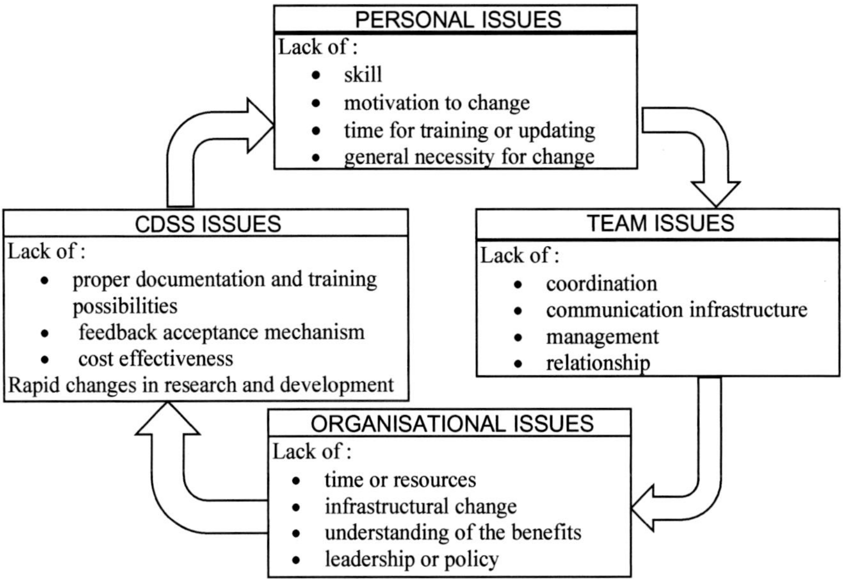

Fig. 5 Issues in implementation of CDSS research findings

Issues in implementation of research will be slower in the case of CDSS also due to the risk involved which is quantified more by lives saved rather that time or money [52, 53].

4.2 *Information Flow*

The foundation of any efficient CDSS is reliable information. The hospital has to have all the relevant information regarding the patients in its patient information database and the information regarding its resources in the management information database. Both these databases have to be linked in order to keep track of the allocation of the resources according to the need of the patients. So, there is already a constant flow of information in a hospital to which the CDSS has to be linked [54]. Moreover, the CDSS has to have its knowledge base and its knowledge acquisition process. A basic flow of information in a CDSS is as shown in Fig. 6.

In the hypothetico-deductive decision making model which most doctors use to diagnose and treat a patient, the flow of information is crucial. The patient's symptoms are compared with the medical knowledge and the resources available to make decision with regards to the likelihood or treatability of the cause using one or two hypotheses of the cause [55]. It is apparent that any break of right information could lead to a falsified hypothesis leading to more serious complications. Even in

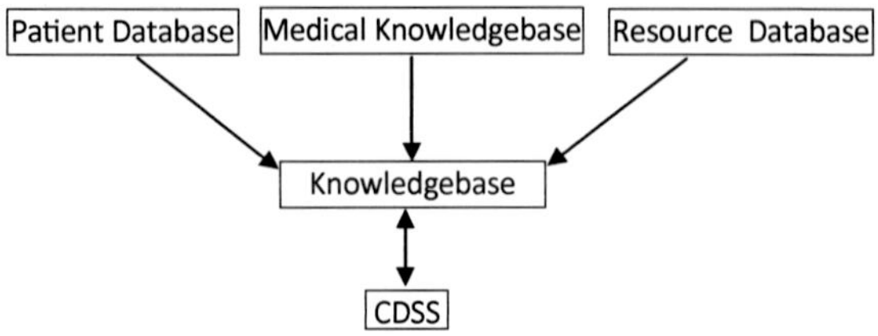

Fig. 6 Basic flow of information in a CDSS

the case of an inductive pattern-matching process of decision making, the flow of information cannot be ruled out.

According to a study published in 2006 [56] CDSS can be classified according to the flow of information. The study has found that those CDSS integrated well with the workflow were more accepted and there is the need for transparency in the creation of knowledge base. The use of nationalised guidelines and randomised trials helped in the creation of the knowledgebase in most CDSS but a more reliable creation will make the CDSS more effective.

There had been a debate on the methodology in the use of information resources [57–59]. The initial idea that the raw information resources is all that is needed to create knowledgebase was debunked by others who felt that unless a scientific process is added to the information resource there can be no meaningful knowledge. It is therefore, meaningful to have an unbridled information flow in order to create and access the knowledgebase in a CDSS. There should be relevant information to create the hypothesis, information to back the decision and more importantly information that makes decisions realistic. CDSS needs a flow of relevant information from the relevant sources.

The transparency in flow of information is part of the system design and it help in presenting the reason for the decision recommendation. The overall strategy for the information flow helps in better user interaction with the CDSS resulting in an efficient CDSS. The information from the clinical documentation is combined with the information that flows from the interaction with the patient. It is then assessed with an inference engine along with the doctor's hypotheses and a recommendation is formed. The transparency in this flow of data will not only build confidence but also serve for a better implementation and acceptance of the CDSS [60]. Apart from the visibility of the information to the stakeholders, the flow of information will also help, among other things, in the presentation of a well ordered list of things-to-do at each stage of the workflow. Another important aspect in the proper implementation of the flow of information is the grouping of information. Information for a particular stage in the workflow are grouped so that the relevant data can be entered or taken out of the system without exhibiting the whole lot of data available [61].

4.3 *Reluctance of the Stakeholders*

There are three teams involved in the successful implementation of a CDSS—the organisation, the users and the developers. As in the case of any implementation, the organisation is the primary stakeholder with regards to the implementation of the CDSS. The leadership of organisation matters for the acceptance and implementation of CDSS. The user's keenness in the CDSS is the next most important factor in the implementation of CDSS. The users include the clinicians and the patients. The developers of the CDSS will have to take into consideration the organisational structure and environment and the end user's requirements and limitations in the development [62]. All the three are important but most literature evaluating the implementation of CDSS focus on the human factor. In fact, there is very little or no literature on the study of the structure and environment of the organisation in the implementation of the CDSS although it is of primary importance [63].

4.3.1 Organisational Reluctance

Studies in organisational innovativeness often fail to find the relative cause of the failure to change. The studies often link the structural dimensions of an organisation like centralisation, formalisation, hierarchy and others. with the changes that happen to the organisation but secondary aspects like the resources, size and so on are not given due consideration. So, there is often an imbalanced understanding of the innovation processes in an organisation. Moreover, the data collected from top executives of an organisation is often found to be quite inadequate to judge the innovativeness of the organisation. Often that is the only data available. Surveys relate size of the organisation to its innovativeness because size is an accurately quantifiable variable. The size of an organisation may also represent other dimensions like resources, structure and the like [64]. There could be three reasons for the reluctance to implement CDSS from the part of the organisation. Resources are one important consideration that comes into play when a change in an organisation is mooted. The other important reasons are the lack of enough convincing features in the CDSS that allow the organisation to positively accept the CDSS and the disruption that the CDSS could cause in the workflow.

According to a literature review and Delphi study on the determinants of innovation within health care organizations, there are fifty determinants to innovation, of which twelve are directly related to the organisation. Decision-making process, hierarchical structure, formal reinforcement, organizational size, functional structure, relationship with departments, nature of the collaboration, staff turnover, degree of staff capacity, available expertise, logistical procedures and number of potential users are the determinants of innovation and any one of them could be a barrier to the implementation of CDSS by the organisation [65].

Organisational failure to implement a CDSS may also stem from the fact that the organisation has a wrong idea of the CDSS or it has priority of strategies that are not clear. It could be also that the CDSS itself presents a low motivation for the organisation to change as it may be doing well without it. Another factor that prevent the organisation to adapt CDSS, as the case of any other innovation changes, is its failure to make a proper evaluation of the situation resulting in a lack of commitment in the top management team [66]. Any of these possibilities in the organisational management could lead to the ambivalence that could result in the stalling of the implementation of the CDSS or and improper implementation that would render the CDSS ineffective. The organisational barrier to implementation is more complex as it has many levels of decision making with a hierarchy of people with different influential capacities involved. Different studies cite different reason that relate the organisational structure with the failure to implement CDSS. It has to be studied separately in order to find the underlying factor for the failure to successfully implement CDSS [67]. The conviction of the organisation is the first stage to the successful implementation of CDSS.

4.3.2 User's Reluctance

There is always a need to effectively communicate changes to various stakeholders. Those impacted by the change have to be convinced about the usefulness of the change in order for the change to be accepted. The success of the implementation lies in the acceptance by the users. According to the DeLone and McLean model for the successful implementation of information systems, use and user satisfaction play a key role in the acceptance of a system and the impact the system has on the organisation [68]. The are many other models to study user acceptance of information technology and these models can be compared and contrasted. The underlying factors to user acceptance may be defined by performance expectancy, effort expectancy, social influence, and facilitating condition of the user [69].

One of the key factors that determines a user's acceptance or rejection of the CDSS is the user's perceived ability to use the system. Although training can help increase the skill to use the system, the willingness to undergo the training or to equip oneself with the needed skill depends on one's self-assessment of the ability. If a person has very little knowledge or skill in computer's, the person can outrightly deny the need for training or usefulness of the CDSS. On the other extreme, the person with more skills can overestimate the usefulness of the CDSS and can equally reject its need. So, both the group of users need to be motivated to accept the use of CDSS in order to fully implement the CDSS. It is not enough for a basic instruction to be given. There has to be a constant help offered in order to clear doubts at every stage in the use of the CDSS. The users have to trust the CDSS for them to use it [70, 71].

Clinicians who are reluctant to use CDSS have two valid reasons; firstly, they find that the suggested usage does not help them in the patient care and secondly the system distracts them from proper care of the patient [71]. While both the reasons can be improved by proper changes in the system, the overdependency of the clinicians on the system would have to be averted. It can happen only with the proper knowledge of the system and proper training in the use of the system. Abuse of the system can also lead to the faulty implementation of the system. One of the fallouts of the overuse of the CDSS is the 'alert fatigue' that happen either because the system produces too many unnecessary alerts or because the clinicians keep referring to the system too often. Together with the training in the use of the CDSS, the CDSS itself will have to be user friendly in so much as to help the users understand the decision recommendations provided. Such a CDSS will also help in building the trust of users on the CDSS [72].

A subtle but important issue regarding the acceptance of the CDSS is the privacy issue. Privacy and security of data both of the patients and the clinicians are important. Bibliometric analysis and literature reviews on the issue have shown beyond doubt that it is a great concern and is studied by many. Since the databases of all records are stored locally or in cloud servers, the access to it has to be regulated and secure to prevent any leak in the data. The threats to the security and privacy of the data can either be from the people accessing the data or technological failure. Human access can be secured and if there is any leak in the data, it can be found out. Similarly, the danger of the leak from the system should also be prevented. Cybersecurity standards are key indicators to ensure the safety of the data processed in the system. Users have to be familiarised with these standards to prevent unintentional loss or leak in data through the CDSS. The guarantee of security and privacy in the system will help build the confidence in the system and remove reluctance of the users [73, 74]. There may not be any fool-proof systems that can guarantee absolute safety and security of the data, but the level of confidence in the system depends on the importance given to the security and privacy in the system.

4.3.3 Developer's Reluctance to Change

Failure to implement the CDSS can also be as a result of the developer's lack of understanding the need and the inflexibility of the system. CDSS is a complex system. The difficulty in operating the system will lead to its unsuccessful implementation. The more complex the system is, the more involvement should be shown in its acceptance. The stakeholders should feel the necessity to use such complex system for their advantage. Very often the CDSS is rejected because of the lack of flexibility in the system. CDSS is to be incorporated in the workflow of the organisation that is already existing. To overhaul the whole existing workflow to accommodate the CDSS will result in chaos. It will be as good as building a new organisation from the scratch [75]. The problem of interoperability with the existing system can be overcome with the layered structure of the CDSS. In the layered system, the user interface, the database and the inference engine are created in layers and so the incorporation into a particular existing system will only require the recoding at a single layer. However, the layered approach has to follow certain standards. Very often the standards become a bottleneck in the implementation of the CDSS. The code has to follow a particular standard that has to be chosen from many available standards and stick to that standard throughout the system. Such adherence to a standard makes the CDSS rigid. The Service Oriented Architecture (SOA) that uses both the layered architecture using standards and the earlier non-portable architecture is seen as a solution that fits both the problems. Using SOA, the CDSS is created as an application to serve the other existing applications in the system thereby preventing the overhauling of the existing system and reducing the training time for the use of the CDSS [76].

One of the methods for successful implementation of CDSS is to study the threats to the implementation and adopt strategies to overcome the same. For example, the user resistance is a commonly noticed threat to the implementation. The developers of the CDSS should study the threat and adopt methods that will avoid unnecessary disruptions in the normal workflow [77]. Likewise, there are other factors that will lead to successful implementation of CDSS. There cannot be a standardised CDSS for all the organisations since there are differences in the workflow of one organisation with the other. Certain underlying factors will certainly contribute to the success in implementation like, the speed in decision recommendation, anticipation of needs, adaptation to the user's workflow, ease of use, allowing the clinicians to decide, displaying reasons for recommendations, easy user interface and up-to-date knowledge base. All these require a study of the existing system and fitting the CDSS into it [78].

Survey of literature shows that the most common requirement for the acceptability of CDSS is the ease of use, efficiency and compatibility. Poor usability and presentation of relevant information hinder the ease of use of a system. The length of time taken to make the recommendation and minimum user entry contribute to the efficiency of the machine. Compatibility of the CDSS lies in its ability to be integrated with existing software that used for data entry, medicine ordering and so on [63].

Use of models and simulation helps in creating an acceptable CDSS. The models are created and tested in real time to check the acceptance of the real CDSS. The models can be physiological models that simulate the human body, the database model to simulate the datasets and its relationship, the agent-based model to simulate the different scenarios in the hospital, building model to simulate the various buildings and areas of the hospital and the hospital level information model that simulate the various information entry and data links in the hospital [54]. With the various models and simulations, the CDSS developers can handle feedback they receive in order to finally create the CDSS that will be acceptable. It is easier to change or modify a model based on the feedback received during simulation rather than to redo the whole CDSS.

5 Types of CDSS

CDSS can be classified in different ways [79, 80]. The broad classification as knowledge-based and non-knowledge-based CDSS or as active and passive CDSS are the common categorisation of the CDSS. Apart from that there are classifications based on the techniques used, open- or closed-loop systems and system that are event monitors, consultation systems, and clinical guide-lines [80, 81].

5.1 Knowledge-Based and Non-knowledge-Based CDSS

Three interacting components make up the knowledge-based CDSS as shown in Fig. 7.

The Human-Computer Interaction (HCI) in a CDSS is the user interface that allows the entry of data and display of result. It is a quicker and convenient user interface that allows data entry taken at point-of-care that makes it precise [82]. It also helps drug administration entries to be recorded by caretakers [83] and improves decision making processes by physicians [84]. Most of the research publications are on developing and evaluation mobile technology and applications for healthcare [85–89] Visual images on display [90–92] or the use of Natural Language Processing (NLP) tools for data entry help physicians and patients to great extent [93–95]. An effective display on the user interface could also help make clinicians make the right decisions on time [96–98]. Another important factor that helps in the design of the user interface in CDSS is the efficacy of communication of diagnostic results with the patient and their family for further safe decisions [99–102]. For the patient's requirement, there should be a standard user interface incorporating all features and functions [103]. The data available on the electronic health record (EHR) can be efficiently extracted to help CDSS. For the capture of EHR the user interface should be suitably modified to help the data extraction process [104–112]. Such data extraction from EHR will be useful for the decision support to clinicians [113–117]. Care should be taken to avoid information overload on the CDSS when EHR data is passed to the CDSS knowledge base to avoid malfunction of the decision support and to remove constraints [118–120].

Knowledge Engineering (KE) is essential for the creation and development of knowledge-base in CDSS. Software engineering and Artificial Intelligence techniques are used in the KE process. In the early years of KE, the focus was on abstracting knowledge from the expert systems or human experts and transferring it to the knowledge base. Recent modelling techniques used in KE makes use of the problem solving methodology of experts and create a computer model of that kind [121]. KE is a combination of three processes: Knowledge Acquisition (KA),

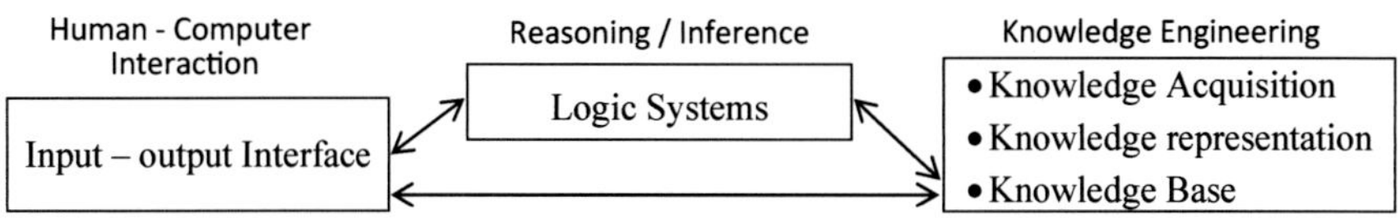

Fig. 7 Interacting components of a knowledge-based CDSS

Knowledge Representation (KR) and Knowledge Base (KB) construction. KA process is an important task in KE. The process consists of extracting information from various sources, structuring the information and organising the information to form knowledge in the knowledge-base. The two great challenges in the KA process are verification and validation of knowledge and the maintenance of the KB [122].

The reasoning or inference system is a crucial link between HCI and KE in the CDSS. The KR schemes could be two layered (disease-symptom) or three layered (disease-symptom-property) or four layered(disease-symptom-property-logic). The logic in the four layered KR could be based on either the declarative or the procedural knowledge. The inference engine works on the KR either using any of the reasoning mechanisms such as rule-based, the model-based, case-based, Bayesian, Bayesian belief networks, decision tree, rough set approach, k-nearest neighbour, fuzzy relations, fuzzy set theory, fuzzy rules, heuristic, semantic network, neural networks, genetic algorithms, support vector machine and so on using the knowledge and an inference strategy [79, 81, 123–127]. The different reasoning mechanisms can be combined or used separately to create an efficient inference. Uncertainty in reasoning is to be expected and therefore the CDSS should be built to handle uncertainty. For this the probabilistic models are used [122].

Nonknowledge-based CDSS on the other hand do not store information in any fixed format or KR scheme. The CDSS learns using training sets using the machine learning algorithms used in artificial intelligence. Two commonly used technologies in nonknowledge-based CDSS are artificial neural network and genetic algorithms. These techniques are also used in knowledge-based CDSS but in a nonknowledge-based CDSS, the techniques are used without predefined knowledge base. The patient's data is analysed and correlation between the symptoms and the disease is established. Unlike the knowledge-based CDSS, the nonknowledge-based CDSS require a higher amount of computing power and are used only in fewer diagnosis. Moreover, the nonknowledge-based CDSS will require more time because of the iterative process that is done in training. Another drawback of the nonknowledge-based CDSS is that the process of reasoning is not known to the user since it uses weights coded in the algorithm. Therefore, there is a trust deficit in the user of the system. Nonknowledge-based CDSS are still popularly used for various image based diagnosis like the analysis of the waveform of an electrocardiogram or electroencephalograms or images of computerized tomography scan or magnetic resonance imaging. The advantages of the nonknowledge-based CDSS includes the elimination of use of different sets of rules for the inference engine, the processing of incomplete information, and it does not require large database [125, 127].

5.2 Active and Passive CDSS

The passive CDSS is one in which the information is made available in the system for the user but the CDSS does not interrupt the workflow. The passive CDSS does not require real time data of the patient. The user needs to make an investigation on the system to find out the possibilities. So, the system is not specific to a particular patient. The passive CDSS is mainly used for cross referencing in order to find relevant data. Such type of a system might only be of very little advantage to the user in comparison with the active CDSS. Passive CDSS are still of use for the users in as much as they are able to present relevant and reliable information and are safe to use. Active CDSS is a more complex system and based on information of a particular patient, the CDSS will be able to present decision recommendations. The active CDSS may be knowledge-based or nonknowledge-based systems unlike the passive CDSS which is only knowledge-based systems. The reasoning mechanisms used for the decision recommendations in an active CDSS vary from system to system. So also, the type of recommendation presentation may vary. While in some active CDSS, the alert or reminder alarm is given, while in some other systems it is silent [128, 129]. An example of the passive CDSS is the clinical pharmacogenetic test interpretation done by an expert and entered in the CDSS. The interpretation of the test can be read by the user before prescribing a treatment. The active CDSS of the same clinical pharmacogenetic test will provide an automated alert to the physician when the details of the patient are entered. The alert can be both pre-test, that is, when a high-risk drug is prescribed for a patient, the physician is altered to do clinical pharmacogenetic test for the patient. The post-test alert is given if the high-risk drug is prescribed for a patient with incompatible pharmacogenetic test results [130].

5.3 *Classification of CDSS Based on Type of Intervention*

There can be different types of CDSS interventions such as alerts and reminders, diagnostic assistance, prescription decision support, information retrieval, image recognition and interpretation and therapy critiquing and planning [131, 132].

Alerts and reminders are active CDSS that provide the clinicians with alerts or reminders at different stages. The alert or reminders could be at the stage of the entry of the patients details when the staff is reminded to enter certain essential details or at the stage of diagnosis when the physician is alerted to look for related symptoms. Real-time alerts are also given from the patients monitors during treatment or when certain high-risk drugs are prescribed. One of the drawbacks of the alerts by CDSS is the alert fatigue that is caused by constant alerts given by the CDSS. A possible solution to prevent alert fatigue is to filter the alerts or order the alerts as per its seriousness.

Diagnostic assistance is provided by the CDSS using the knowledge base of the CDSS or by some machine learning algorithms in case of the nonknowledge-based CDSS. It is useful especially when there is a dilemma in the diagnostic process. Various reasoning and inferencing techniques are used for the CDSS recommendations that may be visible to the user.

Prescription Decision Support is a specialised CDSS recommendation given that is popularly used to check interactions between drugs, possible allergic reactions, error in dosage and contraindications of the drugs that is prescribed. Certain CDSS allow automatic transcript generation and transferring the same to the pharmacy to avoid any errors.

Information retrieval is a passive CDSS application that is used to find relevant and accurate information that could be used for diagnosis or treatment planning. The system works as a search engine to mine to access applicable and important information.

Image recognition and interpretation is a useful CDSS application to study scan reports and provide interpretation automatically. It is very useful to study minute changes over a period of time as in the case of an ultrasound report.

Therapy critiquing and planning is a CDSS alert system used during treatment where the CDSS cautions in case of any errors, omissions, inconsistency or possible contradiction in the treatment. The CDSS compares the prescribed treatment with the guidelines given for such treatment in the knowledge base of the CDSS.

5.4 *Classification Based on Application of CDSS*

The CDSS can be categorised as information management tool, tool for focussing attention and patient-specific recommendation system. The knowledge-based CDSS can also be a tool for acquiring, processing, relating and presenting information. The CDSS as an information tool is a passive CDSS that is useful only for data processing and presentation of the processed data to the user. The decision in the information management CDSS is left to the user. The tool for focussing attention through alerts and reminders are active CDSS that uses the inference engine to keep track of the data that is entered and relate it with information in the knowledge base. In case of discrepancies, the user is alerted. The patient specific recommendation system follows the patient's available information and medical history to prevent any possible error that could happen due to the patient's unique physical features. The pharmacogenetic test results are very important to the patient specific recommendation system as certain drug reactions are unique to genetic makeup. Certain complex reasoning mechanisms like decision theory, rough-set theory, cost-benefit analysis and other such techniques are used for patient specific recommendations [133].

6 Artificial Intelligence (AI) in CDSS

The earliest known work in medicine with AI is probably the 'Dendral experiments' in the late sixties and the early seventies which was a collaborative research to represent and use symbolic logic for expert knowledge. The earliest known CDSS Internist-1, CASNET, and MYCIN were the product of the deep interest in the use of AI in medicine in the seventies [134]. Although, there was reluctance in the use of AI in practical CDSS in the seventies, AI was acknowledged as superior to human reasoning for clinical diagnosis. The first textbook on AI in medicine edited by Szolovits was published in 1982. Among the institutes which began research in AI in healthcare are Stanford, MIT, Pittsburgh and Rutgers and a few other institutes in Europe. Present use of AI in CDSS has resulted from the availability of very large knowledge base that produces Big Data for the machine learning algorithms [135].

The AI algorithms used in CDSS include Support Vector Machine (SVM), Artificial Neural Networks (ANN), Logistic Regression, Discriminant Analysis, Random Forest, Linear Regression, Naïve Bayes, Nearest Neighbour, Decision Tree, Hidden Markov and others. SVM and ANN are the popular CDSS in literature reviews and research. The AI algorithms in Natural Language Processing (NLP) is also popular for data input and knowledge acquisition processes of CDSS [136].

6.1 SVM in CDSS

SVM is used to classify groups of patients based on certain symptoms when the symptoms for more than one disease are similar. For example, if there are two diseases d1 and d2 with symptoms s1 and s2, we classify the patients using SVM as shown in Fig. 8. The maximum margin hyperplane or the optimal separating hyperplane is given by '2/||w||' where the region of the two hyperplanes is 'b/||w||', 'w' is the linear weight connecting the feature with the outcome and 'b' is the offset parameter that allows the increase of margin [136–139].

The goal of the SVM is to find the relationship between the features 'Xi' of a patient and the disease so by classifying patients with similar features and disease in groups. The feature selection is important for the classification to be meaningful. The advantage of SVM is that when the features are large, the SVM is effective. Even if the features are greater than the training samples, the SVM works effectively. It is also very strong model for predictive analysis since it maximises the margin of classification. The choice of the kernel function determines the effectiveness of the SVM. A wrong kernel could lead to increased errors. Also, too many samples can lead to the poor performance of the SVM. Moreover, in the testing phase the SVM is rather slow.

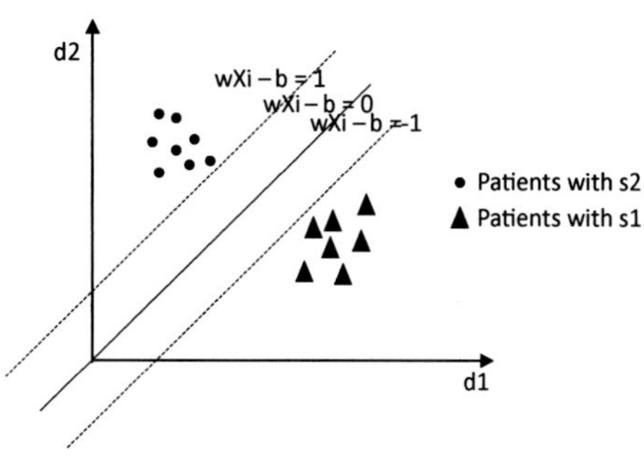

Fig. 8 Illustration of SVM

6.2 ANN in CDSS

ANN is an attempt to mimic the human cognitive process using mathematical representation. The connection between the nodes is indicated by w which represents the weight or the strength of the connection. The set of symptoms S = {s1, s2, …, sn} is used as the input of the ANN for diagnosis and each of which is multiplied with a pre-defined weight and the sum of all the products is together with transforming function forms the input for the next node of the ANN as shown in Fig. 9 where $f\left(\frac{1}{1+e^x}\right)$ is a sigmoidal function.

In a feedforward multilayer neural network which is also called a multilayer perceptron, if the nodes of one layer are connected to all nodes in the next layer, it is called a fully connected network. The output of the first layer serves as input for the second layer. The information is thus propagated from layer to layer till the output is obtained after the last layer. All layers are hidden in the network. Figure 10 is an illustration of a two layered fully connected neural network.

The hidden layers of the neural network are not visible to the user and are considered as a 'black box' since the learning process and the weight adjustments in order to receive n inputs and have m outputs is done according to a mathematical rule called the training algorithm is not visible to the user. The 'black box' reasoning of the ANN makes it unpopular with the users [139–143]. Moreover, the ANN requires immense computational time and power for the training. So, the CDSS with ANN becomes expensive although they are very effective. Clinical applications of ANN based CDSS are in medical diagnosis, image recognition, pattern recognition especially in the analysis and interpretation of waveforms, outcome prediction, identification of pathological specimen and clinical pharmacology [144].

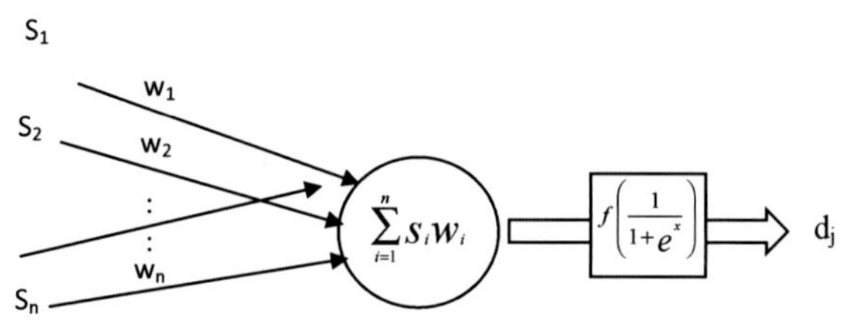

Fig. 9 Illustration of a single node in the neural network

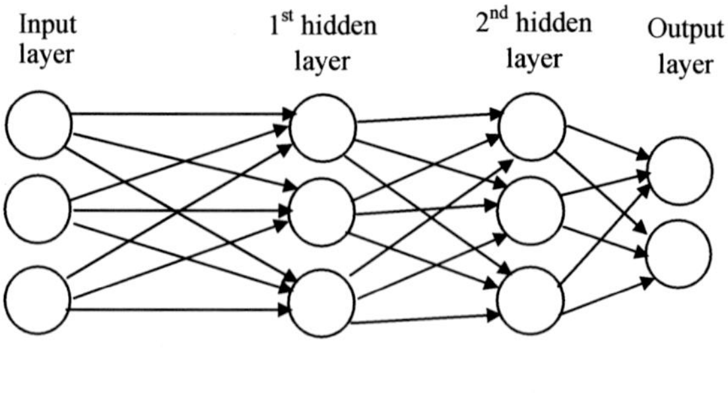

Fig. 10 Illustration of a two layered fully connected neural network

A major drawback of using ANN is the interpretability of the result. It is often noticed that while the training sets are used to get an accurate weight determination, there is no way of knowing when the weight applied is perfect unless the results match the prediction. The iterative testing carries on till there is some sort of guarantee that the weight is the most suitable for the application. In such a case, it is found that the efficiency of ANN is directly proportionate to the number of training sets.

6.3 Natural Language Processing (NLP) in CDSS

NLP are process used for extracting knowledge for unstructured data such as documents, videos, photos, audio files, presentations, webpages and so on. Origins in NLP research dates back to the attempts in machine translation in 1946 when Warren Weaver attempted machine translation using the logic of code breaking used in the second world war. Soon research in machine translation began in various universities and in 1954 IBM and Gerogetown university jointly put up a demonstration of machine translation in a very primitive way. The publication on the concept of syntactic structures by Chomsky 1957, introduced the idea of generative grammar. The statistical information theory used for speech recognition also became a topic of research. In Automatic Language Processing Advisory Committee (ALPAC) submitted a report in 1964 which portrayed a negative progress in machine translation and so funding for it was discouraged. Despite the fall in funding, the Systran system was used by US air-force in 1970 and in 1976 it was commissioned by Commission of European Communities for translation of documents. The university of Montreal developed the Meteo system to translate weather reports the same year. Throughout the 1980s there were a number of prototypes being developed. The growth of NLP has been rapid with the commercialisation of NLP based devices [145–147].

CDSS involves a number of processes like data entry, screening, diagnosis, treatment, monitoring, follow-up and tracking of progress. The users of the CDSS can be the clinicians, patients and developers. The NLP task involves information extraction during any of the CDSS process by any user from free text. There are different applications to the NLP in a CDSS. Information extraction is an important and highly complex application for the extraction of information from documents like the reports and questionnaires that are part of the workflow. Information being crucial to the CDSS, have to be fed into the system with accuracy. It is not enough to identify key terms in the free text that can be mapped to a standardised form but relation between words in context have to be noted. The NLP techniques help gather information precisely from data sheets and medical literature to form the knowledge base. Knowledge acquisition from medical literature is another important application of NLP. The knowledge acquired is converted to the correct format for the knowledge base for later retrieval. Text generation application using NLP techniques allow CDSS to build reports and user understandable recommendations. NLP also allows user friendly human computer interactions. The data can be entered either in free text format or using voice recognition technology. Similarly, the output of the CDSS can be in audio form or in free text form. Machine translation is another popular NLP application in CDSS. Machine Translation allows the use of more than one natural language in the CDSS. The interaction with the CDSS reduces time and increases efficiency of the system [148–150].

NLP has to fulfil three primary tasks in a CDSS. Linguistic knowledge has to be represented, the knowledge has to be used in different applications and new knowledge has to be acquired in computable form. Knowledge representation is done using formalisms which may be symbolic or logical or statistical formalisms. The knowledge representation using formalism is done using a technique called the text parsing and text generation. The other NLP techniques used include morphology, lexicography, syntax, semantics and pragmatics [148].

The NLP algorithms are far from perfect and therefore, the use of the NLP in CDSS is limited as there needs an accuracy in the information for the CDSS to be efficient. If the NLP based CDSS is made to run with poor NLP algorithms leading to the mismatch in the information entered or the knowledge acquired, there could be serious consequences in the recommendations given by the CDSS.

7 Predictive Analytics in CDSS

Predictive analytics is a mathematical method of analysing data in order to find a pattern in the data. Understanding the pattern could help in predicting the position of the newer data in the dataset. Traditional statistical models like multiple regression and logistic regression begins with hypotheses that needs to be tested or a parameter that has to be estimated. If the priori hypothesis is wrong or estimated parameter is incorrect, the whole dataset could be rejected even if there could be an important diagnostic value in the dataset that could help make accurate predictions. Predictive analytics, on the other hand, seeks to find a function model that could predict the outcome from a dataset of predictors. For example, the predictive analysis could predict the cause of a disease from datasets of possible symptoms given for a number of patients [151].

Models for predictive analytics can be of the traditional type in which Bayesian method of statistical inference is used. The model is fit into the validated data. In the data-adaptive model, the data is searched for useful predictors. Here the machine learning algorithms are used and the data-adoptive model are also called statistical learning or data mining approach. In the data-adaptive model, the data determines the technique to be applied. In the model-dependent model, the model is used to create sample data and the sample data is compared with the real data. If the sample data and the real data does not match, then the model is improved and more sample data is generated for comparison with real data. The process continues till the right model is obtained. Often the combination of the different models is used to get the right results [152].

7.1 Data Mining and Predictive Analytics

Data mining is an initial step in the predictive analytics process where the pattern in large datasets is discovered. Predictive analytics extracts the pattern in order to make predictions for later outcomes. Predictive analytics use data mining tools in order to find pattern and make predictions. Predictive models are used in CDSS to make recommendations and alerts. In routine task as in the case of clinical workflow, there is always a possibility of overlooking certain important data. The CDSS alerts the user about these by making a predictive analysis of data in the knowledge base. A well implemented CDSS can make decision recommendation that are more accurate than mere human reasoning. Moreover, the recommendation by a CDSS based on predictive analytics is more consistent that human decision making process as there will not be any user considerations involved in the CDSS [153]. Common data mining tasks used in predictive analytics are done using Classification, Clustering and Association techniques.

7.2 Classification Technique

It is probably the most common data mining task where an event is classified as belonging to a particular class for which target variable is to be given. The classification algorithm first examines the target predictor variables that are already known. From the analysis of the target variable, it learns about its features and form the training sets. When a new variable is introduced, the variable is classified by comparing its feature with the feature of variables in the training set. The most commonly used classification algorithm is the k-nearest neighbour algorithm [154].

In the k-nearest neighbour algorithm, the classification is based on the Euclidean distance between two datasets of predictors. For example, if A = {a1, a2, …, an} and B = {b1, b2, …, bn} are two datasets of predictors then the distance between them is given by $\sqrt{\sum_{i=1}^{n}(a_i - b_i)^2}$. The problem with finding the Euclidean distance is that if one dataset has very large values and the other has very small values, then the dataset with small values will not make any difference even if it is an important feature. Normalisation of values are used to overcome the problem [154, 155].

To elaborate the k-nearest neighbour, an example of prescribing a drug according to the sodium/potassium ratio in the patient's body and the age of the person can be used. The drug X had been prescribed for some patients and drug Y for some other patients as shown in the graph in Fig. 11.

The new patient P will be administered the drug X or Y depending on the k nearest neighbour of P. If k = 1, the algorithm will find the nearest neighbour of P (in this case X) and will be prescribed X. If k = 3, three nearest neighbours will be selected. In the example that would be two X and one Y. So, the drug prescribed to P will be X. k = 2 will be avoided since that would be one X and one Y as the nearest neighbour [154].

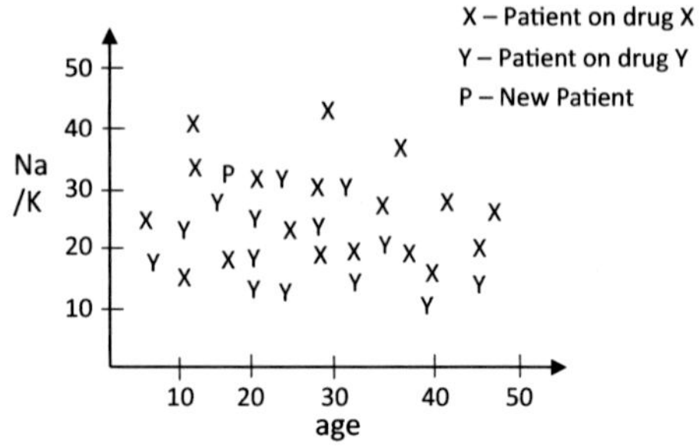

Fig. 11 Classifying observation using k-nearest neighbour

The k-nearest neighbour is a popular classification algorithm that is used since it is simple to use. Because of its simplicity, it is also fast. The k-nearest neighbour allows the use of combined feature and therefore, it is a very flexible algorithm. The k-nearest neighbour is efficient also in large and complicated datasets. The classification ability of the algorithm also helps in filling missing data in some cases. So, if there are some features that are not entered, it will automatically be calculated using the neighbouring features. The ability to predict missing features also makes the algorithm useful in making predictions in uncertainty. The k-nearest neighbour algorithm is a memory based algorithm since the predictor variables and training sets have to be chosen beforehand and committed to memory. The predictions are based on what is memory and it does not find new patterns. Another drawback of the k-nearest neighbour is the value of k by which the number of features for comparison is selected. The value of k is arbitrary and does not follow any rules. There is no way that the algorithm can choose the important features and discard the rest. Moreover, if the value of k is very large, the computing speed and accuracy decreases. The SVM algorithms is an extension of the k-nearest neighbour algorithm and is used where there is a clear separation in the datasets so that the hyperplane could be drawn between them. SVM is considered a learning classification algorithm since the prediction regarding the classification of new data is learned by the SVM based on its earlier classification [151]. SVM are better classifiers that k-nearest neighbour and it does not allow overfitting. ANN are also learning classification algorithms. But SVM are not efficient when too many predictors are used and in case of large datasets, it will require more time for training. ANN have the ability to find relationship between unseen data for generalisation and it has no restriction on the input variable. The drawback of using ANN for classification is that it requires huge computational power and when large datasets are used, the training time is more for better performance [156].

Another classification algorithm is the Decision Trees (Recursive Partitioning Algorithms). The decision tree works on the principle that by splitting the predictors into subgroups recursively with most number of cases belonging to the group until it cannot be split anymore without change in value, the group that classifies the case most accurately can be found. For example, if we take the case of a disease having

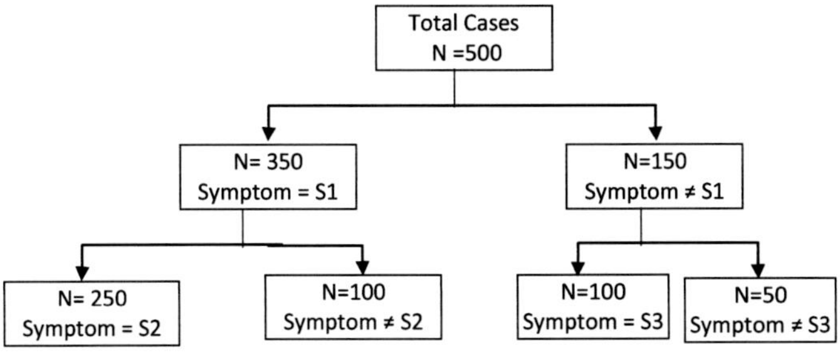

Fig. 12 Illustration of a Decision Trees (Recursive Partitioning Algorithms)

500 patients of which 350 have a particular symptom s1. Then the group is split as shown in Fig. 12 [157].

The decision tree algorithm has the ability to model multifaceted relations in a simple format. It is also flexible and the interpretability of the result is evident. But the problem with decision trees is that it is a solution and not the best possible solution. It is therefore not reliable for all problems [151].

7.3 Clustering Techniques

Clustering algorithms fall in class of unsupervised learning in which the patterns in the data is recognised by the algorithm. The algorithm clusters data with similar patterns. Such clustering of data will allow recognition of the types or 'buckets' of cases. The common clustering algorithms include k-means algorithms, hierarchical clustering and self-organising feature maps. In k-means clustering the distance between the clusters are maximised. At first the k-means algorithm randomly chooses two points which become cluster centre. Then the other predictors are assigned datapoints according to its closeness to the cluster centre. New cluster centres are then computed by taking the mean of the datapoints assigned to each cluster. Computing datapoints and assigning cluster centres are done iteratively until there are no more predictors left for assigning points or an optimal clustering is reached. The k-means clustering will produce datasets that have features far different for each other and is very reliable even for large amount of data. The drawback of k-means algorithm is that initial choice of cluster centre determines the number of iterations that the algorithm has to run.

The hierarchical or tree clustering is done using a distance matrix where each predicator is in the form of a matrix with the relative frequency of the predictor. If two predictors occur very often then the matrices of these two predictors are said to be close to each other. Such pairs of distance matrices are created until all the matrices are paired. Then the pairs that are closest to each other are joined. The tree clustering algorithm evaluates the connection between terms or predictors. The problem with tree clustering is when there are large sets of data, a lot of time will be spent in making matrices and comparing their distances [158].

Self-organising feature maps or Kohonen networks make use of neural networks for clustering. The result of the self-organising feature maps or Kohonen networks are similar to the k-means clustering in the sense that the features in a same cluster are similar compared to the features of a different cluster. The self-organising feature maps or Kohonen network need more computational power and so it is impractical to work with very large datasets [151].

7.4 Association Rule Technique

The aim of association rule is to find the relationship between predictors in a knowledgebase. The outcome of applying an association rule algorithm are pairs datasets (one antecedent and the other consequent) along with the confidence of the rule. Association rules algorithm can be applied in a CDSS to predict the possibility of reaction to a drug among certain group of patients. A common association rule algorithm is the a priori algorithm which makes use of the structure within the rule to reduce the size of the search problem. The a priori algorithm works to create an itemset with a desired maximum size. It begins by counting the frequency of each items greater than the threshold support. It then dose the frequency counting iteratively each time increasing the number of items until a large itemset with a desired maximum size is obtained. The association rule is then created by a binary partition of the frequent items set with those with high confidence. The association rule that is created is called the candidate rule. The a priori algorithm is easy to implement and can be used for large datasets. Computing the candidate rule for very large dataset will require high computational power. Finding the threshold support can also be difficult as the whole database has to be checked [159–161].

8 Conclusion

CDSS is a very vast field of research and study. There has been a lot of literature on CDSS due to the fact that medical decisions are important and critical to lives. The mathematical system for decision began to develop centuries ago and soon after computers came into the scene, the computational systems for decision making was adopted. The CDSS also began its development from the very moment decision making systems was created. Over the years, there has been a steady growth in CDSS with newer technologies being adopted into the system. Although the usefulness of CDSS is widely acknowledged, the adoption of CDSS in the clinical workflow has been remarkably slow particularly due to the various reasons that was discussed in Sect. 4. Yet with newer technologies being used, the reliability and efficiency of CDSS has grown. There are a few fundamental challenges that have to be considered.

CDSS with predictive analysis algorithm will go long way in healthcare especially in medical diagnosis, drug prescription and clinical care. The other uses of predictive analysis such as in user interface and knowledge base creation is also being researched and developed. The big shortcoming of the present predictive analysis algorithms is the computational power that is required to run the algorithms. Although these algorithms will make the CDSS very efficient, they will still be lagging in the hardware requirement. With newer techniques evolving, there is all possibility of the challenges being overcome soon.

In future, CDSS will be implemented in an environment of high connectivity between humans and machines and between devices themselves accelerating the speed of data production and aggregation resulting in highly reliable knowledge discovery. With cognitive aides that will be available in the near future, the clinical data could be acquired directly from the brain resulting in massive data streams and in turn producing higher accuracy in decision making. Methods for organizing clinical knowledge in any form will be part of the CDSS in future. Inference engines will make use of the knowledge base created in a common format in describing and correlating a wide variety of knowledge at multiple levels of inference and decision support. The CDSS with advanced predictive analytics capability will also have a significant impact in precision medicine. The users will have to be trained to make full use of the CDSS.

References

1. A.M. Shahsavarani, E. Azad Marz Abadi, M. Hakimi Kalkhoran, S. Jafari, S. Qaranli, Clinical decision support systems (CDSSs): state of the art review of literature. Int. J. Med. Rev. **2**(4), 299–308 (2015)
2. K. Farooq, B.S. Khan, M.A. Niazi, S.J. Leslie, A. Hussain, Clinical decision support systems: a visual survey (2017). arXiv preprint arXiv:1708.09734
3. D.P. McCallie, Clinical decision support: history and basic concepts, in *Healthcare Information Management Systems*. (Springer, Cham, 2016), pp. 3–19
4. R.A. Greenes, A brief history of clinical decision support: technical, social, cultural, economic, and governmental perspectives, in *Clinical Decision Support* (Academic Press, 2007), pp. 31–77
5. R.A. Miller, Computer-assisted diagnostic decision support: history, challenges, and possible paths forward. Adv. Health Sci. Educ. **14**(1), 89–106 (2009)
6. R.A. Miller, Medical diagnostic decision support systems—past, present, and future: a threaded bibliography and brief commentary. J. Am. Med. Inform. Assoc. **1**(1), 8–27 (1994)
7. V.L. Patel, J.F. Arocha, J. Zhang, Thinking and reasoning in medicine, in *The Cambridge Handbook of Thinking and Reasoning*, vol. 14 (2005) , pp 727–750
8. L.D. Broemeling, An account of early statistical inference in Arab cryptology. Am. Stat. **65**(4), 255–257 (2011)
9. F.N. David, *Games, Gods, and Gambling: A History of Probability and Statistical Ideas* (Courier Corporation, 1998)
10. J.K. Victor, A. Katz, Algebra, geometry, and probability in the seventeenth century, *A History of Mathematics: An Introduction* (Addison-Wesley, Boston, 2009)
11. C.P. Robert, Reading Théorie Analytique des Probabilités (2012). *arXiv preprint* arXiv:1203.6249

12. A. Cantillo, *The Problem of Points* (2011). Retrieved 14 Oct 2019, from https://mpra.ub.uni-muenchen.de/50831/
13. R. Pulskamp, *Summa de Arithmetica, geometria e proportionalita* (2009). Retrieved 14 Oct 2019, from http://www.cs.xu.edu/math/Sources/Pacioli/summa.pdf
14. R. Pulskamp, *Prima parte del General Tratatto Book 16, Section 206* (2009). Retrieved 14 Oct 2019, from http://www.cs.xu.edu/math/Sources/Tartaglia/tartaglia_trattato_2col.pdf
15. R. Pulskamp, *Practica arithmetice et mensurandi singularis* (2009). Retrieved 14 Oct 2019, from http://www.cs.xu.edu/math/Sources/Cardano/cardan_pratica.pdf
16. F.A. Nash, Differential diagnosis, an apparatus to assist the logical faculties. Lancet **266**(6817), 874–875 (1954)
17. M. Lipkin, J.D. Hardy, Differential diagnosis of hematologic diseases aided by mechanical correlation of data. Science **125**(3247), 551–552 (1957)
18. R.S. Ledley, L.B. Lusted, Reasoning foundations of medical diagnosis. Science **130**(3366), 9–21 (1959)
19. C.B. Crumb Jr, C.E. Rupe, The automatic digital computer as an aid in medical diagnosis, in *Papers presented at the December 1–3, 1959, Eastern Joint IRE-AIEE-ACM Computer Conference* (ACM, 1959) , pp. 174–180
20. M. Lipkin, R.L. Engle, B.J. Davis, V.K. Zworykin, R. Ebald, M. Sendrow, C. Berkley, Digital computer as aid to differential diagnosis: use in hematologic diseases. Arch. Intern. Med. **108**(1), 56–72 (1961)
21. H.R. Warner, A.F. Toronto, L.G. Veasey, R. Stephenson, A mathematical approach to medical diagnosis: application to congenital heart disease. JAMA **177**(3), 177–183 (1961)
22. W.V. Slack, P. Hicks, C.E. Reed, L.J. Van Cura, A computer-based medical-history system. N. Engl. J. Med. **274**(4), 194–198 (1966)
23. G.A. Gorry, Strategies for computer-aided diagnosis. Math. Biosci. **2**(3–4), 293–318 (1968)
24. R.E. Bellman, L.A. Zadeh, Decision-making in a fuzzy environment. Manage. Sci. **17**(4), 141 (1970)
25. A.M. Heekin, J. Kontor, H.C. Sax, M.S. Keller, A. Wellington, S. Weingarten, Choosing Wisely clinical decision support adherence and associated inpatient outcomes. Am. J. Managed Care **24**(8), 361–366 (2018)
26. A. Schedlbauer, V. Prasad, C. Mulvaney, S. Phansalkar, W. Stanton, D.W. Bates, A.J. Avery, What evidence supports the use of computerized alerts and prompts to improve clinicians' prescribing behavior? J. Am. Med. Inform. Assoc. **16**(4), 531–538 (2009)
27. J.C. Cox, V. Sadiraj, K.E. Schnier, J.F. Sweeney, Higher quality and lower cost from improving hospital discharge decision making. J. Econ. Behav. Organ. **131**, 1–16 (2016)
28. L.S. Elliott, J.C. Henderson, M.B. Neradilek, N.A. Moyer, K.C. Ashcraft, R.K. Thirumaran, Clinical impact of pharmacogenetic profiling with a clinical decision support tool in polypharmacy home health patients: a prospective pilot randomized controlled trial. PLoS ONE **12**(2), e0170905 (2017)
29. C. Baechle, A. Agarwal, A framework for the estimation and reduction of hospital readmission penalties using predictive analytics. J. Big Data **4**(1), 37 (2017)
30. D. Roosan, M. Samore, M. Jones, Y. Livnat, J. Clutter, Big-data based decision-support systems to improve clinicians' cognition, in *2016 IEEE International Conference on Healthcare Informatics (ICHI)* (IEEE, 2016, October), pp. 285–288
31. D.A. Dang, D.S. Mendon, The value of big data in clinical decision making. Int. J. Comput. Sci. Inf. Technol. **6**(4), 3830–3835 (2015)
32. W. Raghupathi, V. Raghupathi, Big data analytics in healthcare: promise and potential. Health Inf. Sci. Syst. **2**(1), 3 (2014)
33. D.W. Bates, S. Saria, L. Ohno-Machado, A. Shah, G. Escobar, Big data in health care: using analytics to identify and manage high-risk and high-cost patients. Health Aff. **33**(7), 1123–1131 (2014)
34. G. Gürsel, Healthcare, uncertainty, and fuzzy logic. Digit. Med. **2**(3), 101 (2016)
35. E. Chan, H. Zhu, W. Bazzi, *Fuzzy Logic and Probability Theory*. Retrieved 20 Oct 2019, from http://pami.uwaterloo.ca/~sd625/Students/lhan_zhu_bazzi/flpt.pdf

36. P. Hájek, L. Godo, F. Esteva, Fuzzy logic and probability. In *Proceedings of the Eleventh Conference on Uncertainty in Artificial Intelligence* (Morgan Kaufmann Publishers Inc, 1995, August), pp. 237–244
37. D. Dubois, H. Prade, Fuzzy sets and probability: misunderstandings, bridges and gaps, in *[Proceedings 1993] Second IEEE International Conference on Fuzzy Systems* (IEEE, 1993, March), pp. 1059–1068
38. D. Dubois, H. Prade, Fuzzy sets, probability and measurement. Eur. J. Oper. Res. **40**(2), 135–154 (1989)
39. E. Sanchez, Solutions in composite fuzzy relation equations: application to medical diagnosis in Brouwerian logic, in *Readings in Fuzzy Sets for Intelligent Systems* (Morgan Kaufmann, 1993), pp. 159–165
40. A.E. Samuel, S. Rajakumar, On intuitionistic fuzzy modal operators in medical diagnosis. Int. J. Eng. Sci. Math. **7**(4), 313–318 (2018)
41. S.K. De, R. Biswas, A.R. Roy, An application of intuitionistic fuzzy sets in medical diagnosis. Fuzzy Sets Syst. **117**(2), 209–213 (2001)
42. E. Szmidt, J. Kacprzyk, Intuitionistic fuzzy sets in some medical applications. In *International Conference on Computational Intelligence* (Springer, Berlin, 2001, October), pp. 148–151
43. S. Das, D. Guha, B. Dutta, Medical diagnosis with the aid of using fuzzy logic and intuitionistic fuzzy logic. Appl. Intell. **45**(3), 850–867 (2016)
44. N.H. Phuong, V. Kreinovich, Fuzzy logic and its applications in medicine. Int. J. Med. Inform. **62**(2–3), 165–173 (2001)
45. M.A. Madkour, M. Roushdy, Methodology for medical diagnosis based on fuzzy logic. Egypt. Comput. Sci. J. **26**(1), 1–9 (2004)
46. V. Prasath, N. Lakshmi, M. Nathiya, N. Bharathan, P. Neetha, A survey on the applications of fuzzy logic in medical diagnosis. Int. J. Sci. Eng. Res. **4**(4), 1199–1203 (2013)
47. J. Warren, G. Beliakov, B. Van Der Zwaag, Fuzzy logic in clinical practice decision support systems, in *Proceedings of the 33rd Annual Hawaii International Conference on System Sciences* (IEEE, 2000, January), pp. 10-pp
48. J.H. Bates, M.P. Young, Applying fuzzy logic to medical decision making in the intensive care unit. Am. J. Respir. Crit. Care Med. **167**(7), 948–952 (2003)
49. R.W. Leung, H.C. Lau, C.K. Kwong, On a responsive replenishment system: a fuzzy logic approach. Expert Syst. **20**(1), 20–32 (2003)
50. O. Gorgulu, A. Akilli, Use of fuzzy logic based decision support systems in medicine. Stud. Ethno-Med. **10**(4), 393–403 (2016)
51. R. Fullér, Neuro-Fuzzy methods for modeling and fault diagnosis. *Eötvös Loránd University, Budapest VacationSchool* 1–22 (2001)
52. E. Sanchez, C. Toro, A. Artetxe, M. Graña, C. Sanin, E. Szczerbicki, E. Carrasco, F. Guijarro, Bridging challenges of clinical decision support systems with a semantic approach. A case study on breast cancer. Pattern Recogn. Lett. **34**(14), 1758–1768 (2013)
53. A. Donald, R. Milne, Implementing research findings in clinical practice. Getting Res. Find. Pract. 95–106 (2002)
54. S.V. Kovalchuk, K.V. Knyazkov, I.I. Syomov, A.N. Yakovlev, A.V. Boukhanovsky, Personalized clinical decision support with complex hospital-level modelling. Procedia Comput. Sci. **66**, 392–401 (2015)
55. J. Wyatt, Information for clinicians: use and sources of medical knowledge. Lancet **338**(8779), 1368–1373 (1991)
56. A. Berlin, M. Sorani, I. Sim, A taxonomic description of computer-based clinical decision support systems. J. Biomed. Inform. **39**(6), 656–667 (2006)
57. C.P. Friedman, A "fundamental theorem" of biomedical informatics. J. Am. Med. Inform. Assoc. **16**(2), 169–170 (2009)
58. S. Mani, Note on Friedman's 'fundamental theorem of biomedical informatics'. J. Am. Med. Inform. Assoc. **17**(5), 614 (2010)
59. J.S. Hunter, Enhancing Friedman's "fundamental theorem of biomedical informatics". J. Am. Med. Inform. Assoc. JAMIA **17**(1), 112 (2010)

60. S. Medlock, J.C. Wyatt, V.L. Patel, E.H. Shortliffe, A. Abu-Hanna, Modeling information flows in clinical decision support: key insights for enhancing system effectiveness. J. Am. Med. Inform. Assoc. **23**(5), 1001–1006 (2016)
61. R.A. Greenes, D.W. Bates, K. Kawamoto, B. Middleton, J. Osheroff, Y. Shahar, Clinical decision support models and frameworks: seeking to address research issues underlying implementation successes and failures. J. Biomed. Inform. **78**, 134–143 (2018)
62. M.M. Yusof, J. Kuljis, A. Papazafeiropoulou, L.K. Stergioulas, An evaluation framework for Health Information Systems: human, organization and technology-fit factors (HOT-fit). Int. J. Med. Inform. **77**(6), 386–398 (2008)
63. E. Kilsdonk, L.W. Peute, M.W. Jaspers, Factors influencing implementation success of guideline-based clinical decision support systems: a systematic review and gaps analysis. Int. J. Med. Inform. **98**, 56–64 (2017)
64. E.M. Rogers, Elements of diffusion, in *Diffusion of Innovations*, vol. 5, no. 1.38 (2003)
65. M. Fleuren, K. Wiefferink, T. Paulussen, Determinants of innovation within health care organizations: literature review and Delphi study. Int. J. Qual. Health Care **16**(2), 107–123 (2004)
66. M. Pardo del Val, C. Martínez Fuentes, Resistance to change: a literature review and empirical study. Manage. Decis. **41**(2), 148–155 (2003)
67. H.M. Korhonen, I. Kaarela, Corporate customers' resistance to industrial service innovations. Int. J. Innov. Manage. **15**(03), 479–503 (2011)
68. W.H. Delone, E.R. McLean, The DeLone and McLean model of information systems success: a ten-year update. J Manage. Inform. Syst. **19**(4), 9–30 (2003)
69. V. Venkatesh, M.G. Morris, G.B. Davis, F.D. Davis, User acceptance of information technology: toward a unified view. MIS Q. 425–478 (2003)
70. P. Madhavan, R.R. Phillips, Effects of computer self-efficacy and system reliability on user interaction with decision support systems. Comput. Hum. Behav. **26**(2), 199–204 (2010)
71. M.H. Trivedi, E.J. Daly, J.K. Kern, B.D. Grannemann, P. Sunderajan, C.A. Claassen, Barriers to implementation of a computerized decision support system for depression: an observational report on lessons learned in "real world" clinical settings. BMC Med. Inform. Decis. Mak. **9**(1), 6 (2009)
72. K. Alammar, M. Alamrani, S. Alqahtani, M. Ahmad, Organizational commitment and nurses characteristics as predictors of job involvement. Can. J. Nurs. Leadersh. (2016)
73. K.T. Win, W. Susilo, Y. Mu, Personal health record systems and their security protection. J. Med. Syst. **30**(4), 309–315 (2006)
74. C.K. Wang, Security and privacy of personal health record, electronic medical record and health information. Management **13**(4), 19–26 (2015)
75. A.N.H. Zaied, M. Elmogy, S.A. Elkader, Electronic health records: applications, techniques and challenges. Int. J. Comput. Appl. **119**(14) (2015)
76. A. Kumar, Stakeholder's perspective of clinical decision support system. Open J. Bus. Manage. **4**(1), 45–50 (2015)
77. A. Bhattacherjee, N. Hikmet, Physicians' resistance toward healthcare information technology: a theoretical model and empirical test. Eur. J. Inf. Syst. **16**(6), 725–737 (2007)
78. D.W. Bates, G.J. Kuperman, S. Wang, T. Gandhi, A. Kittler, L. Volk, C. Spurr, R. Khorasani, M. Tanasijevic, B. Middleton, Ten commandments for effective clinical decision support: making the practice of evidence-based medicine a reality. J. Am. Med. Inform. Assoc. **10**(6), 523–530 (2003)
79. C. Vaghela, N. Bhatt, D. Mistry, A survey on various classification techniques for clinical decision support system. Int. J. Comput. Appl. **116**(23) (2015)
80. A.T.M. Wasylewicz, A.M.J.W. Scheepers-Hoeks, Clinical decision support systems, in *Fundamentals of Clinical Data Science*, ed. by P. Kubben, M. Dumontier, A. Dekker (Springer, Cham, 2019)
81. E.A. Mendonça, Clinical decision support systems: perspectives in dentistry. J. Dent. Educ. **68**(6), 589–597 (2004)

82. L.A. Wallis, J. Fleming, M. Hasselberg, L. Laflamme, J. Lundin, A smartphone app and cloud-based consultation system for burn injury emergency care. PLoS ONE **11**(2), e0147253 (2016)
83. H. Hagberg, J. Siebert, A. Gervaix, P. Daehne, C. Lovis, S. Manzano, F. Ehrler, Improving drugs administration safety in pediatric resuscitation using mobile technology, in *Nursing Informatics* (2016), pp. 656–657
84. A. Curcio, S. De Rosa, J. Sabatino, S. De Luca, A. Bochicchio, A. Polimeni, G. Santarpia, P. Ricci, C. Indolfi, Clinical usefulness of a mobile application for the appropriate selection of the antiarrhythmic device in heart failure. Pacing Clin. Electrophysiol. **39**(7), 696–702 (2016)
85. K. Blagec, K.M. Romagnoli, R.D. Boyce, M. Samwald, Examining perceptions of the usefulness and usability of a mobile-based system for pharmacogenomics clinical decision support: a mixed methods study. PeerJ **4**, e1671 (2016)
86. U. Sarkar, G.I. Gourley, C.R. Lyles, L. Tieu, C. Clarity, L. Newmark, K. Singh, D.W. Bates, Usability of commercially available mobile applications for diverse patients. J. Gen. Intern. Med. **31**(12), 1417–1426 (2016)
87. B. Brouard, P. Bardo, C. Bonnet, N. Mounier, M. Vignot, S. Vignot, Mobile applications in oncology: is it possible for patients and healthcare professionals to easily identify relevant tools? Ann. Med. **48**(7), 509–515 (2016)
88. N.C. Ernecoff, H.O. Witteman, K. Chon, P. Buddadhumaruk, J. Chiarchiaro, K.J. Shotsberger, A.M. Shields, B.A. Myers, C.L. Hough, S.S. Carson, B. Lo, Key stakeholders' perceptions of the acceptability and usefulness of a tablet-based tool to improve communication and shared decision making in ICUs. J. Crit. Care **33**, 19–25 (2016)
89. A. White, D.S. Thomas, N. Ezeanochie, S. Bull, Health worker mHealth utilization: a systematic review. Comput. Inform. Nurs. CIN **34**(5), 206 (2016)
90. W.Y. Chou, P.T. Tien, F.Y. Lin, P.C. Chiu, Application of visually based, computerised diagnostic decision support system in dermatological medical education: a pilot study. Postgrad. Med. J. **93**(1099), 256–259 (2017)
91. E. Clarkson, J. Zutty, M.V. Raval, A visual decision support tool for appendectomy care. J. Med. Syst. **42**(3), 52 (2018)
92. M. Wagner, D. Slijepcevic, B. Horsak, A. Rind, M. Zeppelzauer, W. Aigner, KAVAGait: knowledge-assisted visual analytics for clinical gait analysis. IEEE Trans. Visual Comput. Graphics **25**(3), 1528–1542 (2019)
93. D. Gavrilis, G. Georgoulas, N. Vasiloglou, G. Nikolakopoulos, An intelligent assistant for physicians, in *2016 38th Annual International Conference of the IEEE Engineering in Medicine and Biology Society (EMBC)* (IEEE, 2016, August), pp. 2586–2589
94. K. Kreimeyer, M. Foster, A. Pandey, N. Arya, G. Halford, S.F. Jones, R. Forshee, M. Walderhaug, T. Botsis, Natural language processing systems for capturing and standardizing unstructured clinical information: a systematic review. J. Biomed. Inform. **73**, 14–29 (2017)
95. S.F. Sung, K. Chen, D.P. Wu, L.C. Hung, Y.H. Su, Y.H. Hu, Applying natural language processing techniques to develop a task-specific EMR interface for timely stroke thrombolysis: A feasibility study. Int. J. Med. Inform. **112**, 149–157 (2018)
96. N. Eskandari, Z.J. Wang, G.A. Dumont, A delayed functional observer/predictor with bounded-error for depth of hypnosis monitoring. J. Clin. Monit. Comput. **31**(5), 1043–1052 (2017)
97. A. Yadollahpour, J. Nourozi, S.A. Mirbagheri, E. Simancas-Acevedo, F.R. Trejo-Macotela, Designing and implementing an ANFIS based medical decision support system to predict chronic kidney disease progression. Front. Physiol. **9**, 1753 (2018)
98. D. Long, M. Capan, S. Mascioli, D. Weldon, R. Arnold, K. Miller, Evaluation of user-interface alert displays for clinical decision support systems for sepsis. Crit. Care Nurse **38**(4), 46–54 (2018)
99. P. Fraccaro, M. Vigo, P. Balatsoukas, S.N. van der Veer, L. Hassan, R. Williams, G. Wood, S. Sinha, I. Buchan, N. Peek, Presentation of laboratory test results in patient portals: influence of interface design on risk interpretation and visual search behaviour. BMC Medical Informatics and Decision Making **18**(1), 11 (2018)

100. H.O. Witteman, B.J. Zikmund-Fisher, Communicating laboratory results to patients and families. Clin. Chem. Lab. Med. (CCLM) **57**(3), 359–364 (2019)
101. J. Liu, C. Li, J. Xu, H. Wu, A patient-oriented clinical decision support system for CRC risk assessment and preventative care. BMC Med. Inform. Decis. Mak. **18**(5), 118 (2018)
102. B. Brown, P. Balatsoukas, R. Williams, M. Sperrin, I. Buchan, Multi-method laboratory user evaluation of an actionable clinical performance information system: Implications for usability and patient safety. J. Biomed. Inform. **77**, 62–80 (2018)
103. D.F. Sittig, A. Wright, E. Coiera, F. Magrabi, R. Ratwani, D.W. Bates, H. Singh, Current challenges in health information technology–related patient safety. Health Inform. J. 1460458218814893 (2018)
104. T. Taft, C. Staes, S. Slager, C. Weir, Adapting Nielsen's design heuristics to dual processing for clinical decision support, in *AMIA Annual Symposium Proceedings*, vol. 2016, (American Medical Informatics Association, 2016), p. 1179
105. M.A. Basit, K.L. Baldwin, V. Kannan, E.L. Flahaven, C.J. Parks, J.M. Ott, D.L. Willett, Agile acceptance test–driven development of clinical decision support advisories: feasibility of using open source software. JMIR Med. Inform. **6**(2) (2018)
106. A. González-Ferrer, M. Peleg, M. Marcos, J.A. Maldonado, Analysis of the process of representing clinical statements for decision-support applications: a comparison of openEHR archetypes and HL7 virtual medical record. J. Med. Syst. **40**(7), 163 (2016)
107. Y. Wang, L. Wang, M. Rastegar-Mojarad, S. Moon, F. Shen, N. Afzal, S. Liu, Y. Zeng, S. Mehrabi, S. Sohn, H. Liu, Clinical information extraction applications: a literature review. J. Biomed. Inform. **77**, 34–49 (2018)
108. D.A. Cook, M.T. Teixeira, B.S. Heale, J.J. Cimino, G. Del Fiol, Context-sensitive decision support (infobuttons) in electronic health records: a systematic review. J. Am. Med. Inform. Assoc. **24**(2), 460–468 (2017)
109. S. Khan, L. McCullagh, A. Press, M. Kharche, A. Schachter, S. Pardo, T. McGinn, Formative assessment and design of a complex clinical decision support tool for pulmonary embolism. BMJ Evid. Based Med. **21**(1), 7–13 (2016)
110. K.D. Lopez, A. Febretti, J. Stifter, A. Johnson, D.J. Wilkie, G. Keenan, Toward a more robust and efficient usability testing method of clinical decision support for nurses derived from nursing electronic health record data. Int. J. Nurs. Knowl. **28**(4), 211–218 (2017)
111. M.M. van Engen-Verheul, L.W. Peute, N.F. de Keizer, N. Peek, M.W. Jaspers, Optimizing the user interface of a data entry module for an electronic patient record for cardiac rehabilitation: a mixed method usability approach. Int. J. Med. Inform. **87**, 15–26 (2016)
112. B.S. Heale, C.L. Overby, G. Del Fiol, W.S. Rubinstein, D.R. Maglott, T.H. Nelson, A. Milosavljevic, C.L. Martin, S.R. Goehringer, R.R. Freimuth, M.S. Williams, Integrating genomic resources with electronic health records using the HL7 Infobutton standard. Appl. Clin. Inform. **7**(3), 817–831 (2016)
113. S.M. Abdel-Rahman, M.L. Breitkreutz, C. Bi, B.J. Matzuka, J. Dalal, K.L. Casey, U. Garg, S. Winkle, J.S. Leeder, J. Breedlove, B. Rivera, Design and testing of an EHR-integrated, busulfan pharmacokinetic decision support tool for the point-of-care clinician. Front. Pharmacol. **7**, 65 (2016)
114. J. Kaipio, T. Lääveri, H. Hyppönen, S. Vainiomäki, J. Reponen, A. Kushniruk, E. Borycki, J. Vänskä, Usability problems do not heal by themselves: National survey on physicians' experiences with EHRs in Finland. Int. J. Med. Inform. **97**, 266–281 (2017)
115. T. Porat, B. Delaney, O. Kostopoulou, The impact of a diagnostic decision support system on the consultation: perceptions of GPs and patients. BMC Med. Inform. Decis. Mak. **17**(1), 79 (2017)
116. S.G. Finlayson, M. Levy, S. Reddy, D.L. Rubin, Toward rapid learning in cancer treatment selection: an analytical engine for practice-based clinical data. J. Biomed. Inform. **60**, 104–113 (2016)
117. U. Guo, L. Chen, P.H. Mehta, Electronic health record innovations: helping physicians–One less click at a time. Health Inf. Manage. J. **46**(3), 140–144 (2017)

118. M.C. Wright, S. Dunbar, B.C. Macpherson, E.W. Moretti, G. Del Fiol, J. Bolte, J.M. Taekman, N. Segall, Toward designing information display to support critical care. Appl. Clin. Inform. **7**(4), 912–929 (2016)
119. P. Chung, J. Scandlyn, P.S. Dayan, R.D. Mistry, Working at the intersection of context, culture, and technology: provider perspectives on antimicrobial stewardship in the emergency department using electronic health record clinical decision support. Am. J. Infect. Control **45**(11), 1198–1202 (2017)
120. M.J. Denney, D.M. Long, M.G. Armistead, J.L. Anderson, B.N. Conway, Validating the extract, transform, load process used to populate a large clinical research database. Int. J. Med. Informatics **94**, 271–274 (2016)
121. K. von Michalik, M. Kwiatkowska, K. Kielan, Application of knowledge-engineering methods in medical knowledge management, in *Fuzziness and Medicine: Philosophical Reflections and Application Systems in Health Care* (Springer, Berlin, 2013), pp. 205–214
122. L. Aleksovska-Stojkovska, S. Loskovska, Review of reasoning methods in clinical decision support systems, in *18th Telecommunications forum TELFOR* (2010)
123. Y. Jiang, B. Qiu, C. Xu, C. Li, The research of clinical decision support system based on three-layer knowledge base model. J. Healthc. Eng. *2017* (2017)
124. K.B. Wagholikar, V. Sundararajan, A.W. Deshpande, Modeling paradigms for medical diagnostic decision support: a survey and future directions. J. Med. Syst. **36**(5), 3029–3049 (2012)
125. M. Alther, C.K. Reddy, Clinical decision support systems, in *Healthcare Data Analytics* (Chapman and Hall/CRC, 2015), pp. 619–656
126. G. Kong, D.L. Xu, J.B. Yang, Clinical decision support systems: a review on knowledge representation and inference under uncertainties. Int. J. Comput. Intell. Syst. **1**(2), 159–167 (2008)
127. E.S. Berner, T.J. La Lande, Overview of clinical decision support systems, *Clinical Decision Support Systems* (Springer, New York, NY, 2007), pp. 3–22
128. S. Hajioff, Computerized decision support systems: an overview. Health Inform. J. **4**(1), 23–28 (1998)
129. G.C. Bell, K.R. Crews, M.R. Wilkinson, C.E. Haidar, J.K. Hicks, D.K. Baker, N.M. Kornegay, W. Yang, S.J. Cross, S.C. Howard, R.R. Freimuth, Development and use of active clinical decision support for preemptive pharmacogenomics. J. Am. Med. Inform. Assoc. **21**(e1), e93–e99 (2013)
130. M. Hinderer, M. Boeker, S.A. Wagner, M. Lablans, S. Newe, J.L. Hülsemann, M. Neumaier, H. Binder, H. Renz, T. Acker, H.U. Prokosch, Integrating clinical decision support systems for pharmacogenomic testing into clinical routine-a scoping review of designs of user-system interactions in recent system development. BMC Med. Inform. Decis. Mak. **17**(1), 81 (2017)
131. J. Osheroff, J. Teich, D. Levick, L. Saldana, F. Velasco, D. Sittig, K. Rogers, R. Jenders, *Improving Outcomes with Clinical Decision Support: An Implementer's Guide* (HIMSS Publishing, New York, 2012)
132. A.B. Al-Badareen, M.H. Selamat, M. Samat, Y. Nazira, O. Akkanat, A review on clinical decision support systems in healthcare. J. Convergence Inf. Technol. **9**(2), 125 (2014)
133. A. De la Rosa Algarın, Clinical decision support systems in biomedical informatics and their limitations (2011)
134. V.L. Patel, E.H. Shortliffe, M. Stefanelli, P. Szolovits, M.R. Berthold, R. Bellazzi, A. Abu-Hanna, The coming of age of artificial intelligence in medicine. Artif. Intell. Med. **46**(1), 5–17 (2009)
135. L.Q. Shu, Y.K. Sun, L.H. Tan, Q. Shu, A.C. Chang, Application of artificial intelligence in pediatrics: past, present and future. World J. Pediatr. WJP **15**(2), 105 (2019)
136. F. Jiang, Y. Jiang, H. Zhi, Y. Dong, H. Li, S. Ma, Y. Wang, Q. Dong, H. Shen, Y. Wang, Artificial intelligence in healthcare: past, present and future. Stroke Vasc. Neurol. **2**(4), 230–243 (2017)
137. V.A. Kumari, R. Chitra, Classification of diabetes disease using support vector machine. Int. J. Eng. Res. Appl. **3**(2), 1797–1801 (2013)

138. N. Barakat, A.P. Bradley, M.N.H. Barakat, Intelligible support vector machines for diagnosis of diabetes mellitus. IEEE Trans. Inf. Technol. Biomed. **14**(4), 1114–1120 (2010)
139. F. Amato, A. López, E.M. Peña-Méndez, P. Vaňhara, A. Hampl, J. Havel, Artificial neural networks in medical diagnosis (2013)
140. S. Viera, W.H. Pinaya, A. Mechelli, Using deep learning to investigate the neuroimaging correlates of psychiatric and neurological disorders. Neurosci. Biobehav. Rev. **74**, 58–75 (2017)
141. A.N. Ramesh, C. Kambhampati, J.R. Monson, P.J. Drew, Artificial intelligence in medicine. Ann. R. Coll. Surg. Engl. **86**(5), 334 (2004)
142. Q.K. Al-Shayea, Artificial neural networks in medical diagnosis. Int. J. Comput. Sci. Issues **8**(2), 150–154 (2011)
143. E. Xhumari, P. Manika, Application of artificial neural networks in medicine, in *RTA-CSIT* (2016), pp. 155–157
144. W.G. Baxt, Application of artificial neural networks to clinical medicine. Lancet **346**(8983), 1135–1138 (1995)
145. S. Joseph, K. Sedimo, F. Kaniwa, H. Hlomani, K. Letsholo, Natural language processing: a review, in *Natural Language Processing: A Review*, vol. 6 (2016), pp. 207–210
146. J. Hutchins, The history of machine translation in a nutshell (2005). Retrieved 20 Dec 2009
147. E.D. Liddy, Natural language processing, in *Encyclopedia of Library and Information Science*, 2nd edn. (Marcel Decker, Inc., New York, NY, 2001)
148. C. Friedman, S.B. Johnson, Natural language and text processing in biomedicine, *Biomedical Informatics* (Springer, New York, NY, 2006), pp. 312–343
149. E. Pons, L.M. Braun, M.M. Hunink, J.A. Kors, Natural language processing in radiology: a systematic review. Radiology **279**(2), 329–343 (2016)
150. D. Demner-Fushman, W.W. Chapman, C.J. McDonald, What can natural language processing do for clinical decision support? J. Biomed. Inform. **42**(5), 760–772 (2009)
151. L. Miner, P. Bolding, J. Hilbe, M. Goldstein, T. Hill, R. Nisbet, N. Walton, G. Miner, Prediction in medicine—the data mining algorithms of predictive analytics, in *Practical Predictive Analytics and Decisioning Systems for Medicine: Informatics Accuracy and Cost-Effectiveness for Healthcare Administration and Delivery Including Medical Research* (Academic Press, 2014), pp. 238–259
152. T.W. Miller, *Modeling Techniques in Predictive Analytics with Python and R: A Guide to Data Science* (FT Press, 2014)
153. S. Finlay, Using predictive models, in *Predictive Analytics, Data Mining and Big Data: Myths, Misconceptions and Methods* (Springer, Berlin, 2014), pp. 21–38
154. D.T. Larose, C.D. Larose, k-nearest neighbor algorithm, in *Data Mining and Predictive Analytics* (Wiley, 2015), pp. 301–316
155. M. Shouman, T. Turner, R. Stocker, Applying k-nearest neighbour in diagnosing heart disease patients. Int. J. Inf. Educ. Technol. **2**(3), 220–223 (2012)
156. D. Pavithra, A.N. Jayanthi, A study on machine learning algorithm in medical diagnosis. Int. J. Adv. Res. Comput. Sci. **9**(4) (2018)
157. A.T. Azar, S.M. El-Metwally, Decision tree classifiers for automated medical diagnosis. Neural Comput. Appl. **23**(7–8), 2387–2403 (2013)
158. D.T. Larose, C.D. Larose, Hierarchical and k-means clustering, in *Data Mining and Predictive Analytics* (Wiley, 2015), pp. 523–541
159. T. Velmurugan, J. Manimaran, Implementing association rules in medical diagnosis test data, in *ICICT2015 Conference: International Conference on Information and Convergence Technology for Smart Society* (2015), pp. 201–205. Retrieved 27 Oct 2019, from https://www.researchgate.net/profile/Dr_Velmurugan_T/publication/287975524_Implementing_Association_Rules_in_Medical_Diagnosis_test_data/links/567aba0508ae051f9adde1ab.pdf
160. A.M. Doddi, S.S. Ravi, C. David, S. Torney, Discovery of association rules in medical data. Med. Inform. Internet Med. **26**(1), 25–33 (2001)
161. D.T. Larose, C.D. Larose, Association rules, in *Data Mining and Predictive Analytics* (Wiley, 2015), pp. 603–621

Dr. Ravi Lourdusamy received his Ph.D. from Bharathidasan University, Tamil Nadu, India in 2012. Currently he is Head and Associate Professor in the Department of Computer Science, Sacred Heart College, Tirupattur, India. His research areas include Software Engineering, Ontology and Knowledge Engineering. He has presented and published over 20 papers in international conferences and journals. He was awarded a couple of UGC funded projects.

Xavierlal J. Mattam is a research scholar under Dr. Ravi Lourdusamy in the Department of Computer Science, Sacred Heart College, Tirupattur, India. He is doing his research on cognitive web architecture for decision support systems. His areas of interest include artificial intelligence for predictive analytics and decision theory.

Yajna and Mantra Science Bringing Health and Comfort to Indo-Asian Public: A Healthcare 4.0 Approach and Computational Study

Rohit Rastogi, Mamta Saxena, Muskan Maheshwari, Priyanshi Garg, Muskan Gupta, Rajat Shrivastava, Mukund Rastogi and Harshit Gupta

Abstract Improving comfort, stress and pollution levels among the fascinating achievements of modern science and technology has become a major challenge for our well-being. The world recognizes that the convenience provided by modern technology does not necessarily make life happy. In fact, apart from stress, fears arise from an increasing number of unknown illnesses, verbal anxiety, and highly polluted environments and environmental imbalances. This created a warning to rethink and change lifestyle and health care. Yajna seems to be a gift from ancient Indian science for this purpose. From the above study and various graphs used we conclude that they benefits through fog company that can be used to used to improve the health of patient through various technical methods like Pattern Classification in the study which proposed schemes GA with nearest neighbor techniques and GA with PNN are the efficient techniques and Expert Design which concludes that it is undeniably reliable in terms of providing reasonable and highly valuable decisions. Knowledge and

---

R. Rastogi (✉) · M. Maheshwari · P. Garg · M. Gupta · R. Shrivastava · M. Rastogi · H. Gupta
Department of CSE, ABESEC, Ghaziabad, India
e-mail: rohit.rastogi@abes.ac.in

M. Maheshwari
e-mail: muskan.18bcs1070@abes.ac.in

P. Garg
e-mail: priyanshi.18bcs1068@abes.ac.in

M. Gupta
e-mail: muskan.18bcs1175@abes.ac.in

R. Shrivastava
e-mail: rajat.18bcs1184@abes.ac.in

M. Rastogi
e-mail: mukund.18bcs1021@abes.ac.in

H. Gupta
e-mail: harshit.18bcs1145@abes.ac.in

M. Saxena
Ministry of Statistics, Govt. of India, New Delhi, Delhi, India
e-mail: saxenamamta@hotmail.com

V. Jain and J. M. Chatterjee (eds.), *Machine Learning with Health Care Perspective*, Learning and Analytics in Intelligent Systems 13,
https://doi.org/10.1007/978-3-030-40850-3_15

experiences from a human expert can lead to the critical decision-making in achieving success. Mental health is happiness. This can disrupt the balance of behavior. Mantra therapy can control stress, depression, anxiety, fear, and promote mental health and well-being. Yanjna and Mantra therapy can be the best, powerful, non-violent choice for the future. Mantras are an important tool for Mental Illness in today's society. The word Mantra is a powerful mind, sound, or vibration tool that can be used to enter a deep state of meditation. Treatment of the Vedic mantra began with the Vedic. This is parallel science with Aveda, also called an alternative medical system. Vedic mantra therapy is based on mantra and evokes the body's natural healing mechanism. This chapter proposes the expert system and methodological framework to control patients with chronic diseases, so that data can be collected and processed effectively. "Yagya and Mantra therapy" is the treatment method which can revolutionize the era of medical science and the way disease are cured will also change significantly by the use of this therapy. It also has the power of curing deadly disease like cancer which now require doing Kemo therapy which is very painful and costly. Thus "Yagya and Matra Therapy" can also be called as treatment of the future.

Keywords Fog computing · Healthcare 4.0 · Big data and IoT · Computational intelligence · Yajna and Mantra science · Mantra therapies · Yajna therapy · Gayatri mantra (GM) · OM chanting · OM symbol · Rudrakash · Elaeocarpus · Yagyopathy · Big data and computational analysis · Swarm intelligence techniques · AI and ML · Ambient computing · Symbolic machine learning · Hybrid intelligent systems · Quantum inspired soft computing · Internet of medical things · Telemedicine · Disease management · e-Health · Remote health monitoring · Pervasive healthcare

1 Introduction

1.1 Status of Healthcare in Indian Context

According to Indian Constitution, Each state government is responsible for improving its Healthcare facilities, raising the level of standard living and nutrition in their own state. In India, Healthcare system under public sector is free for those people who are under poverty line. Healthcare system under private sector consists of 81% doctors, 58% hospitals and 29% of beds in country. According to a survey, Private sector becomes the primary source of healthcare for rural and urban areas. In Medication, it is surveyed that in 2010, India consumes most number of antibiotic per head. But now in 2018, most of the antibiotics are substandard and fake and is not approved. Although there are 1.4 million doctors in India, yet India is not able to reach its Millennium Development Goals (eight international development goals for year 2015) related to health. Initiatives such as The Twelfth Plan, Public-private partnership, PM-JAY and many more are taken for improving access of healthcare in India [1, 2].

1.2 Need of Health Research

There is ninety percent burden of global disease as only ten percent of the total world expenditure is spent on Healthcare conditions; this vast partiality reveals a need to reinforce research in countries through various collaborations. Research outcomes must guide policy and program development as well as the way of providing healthcare services, so it should not only increase knowledge but also lead to action. As Health is a very broad concept involving many determinants so Research is also needed to take an interdisciplinary approach to health problems. As unrivalled amount of resources are invested in Healthcare, so research also help us to know about the impact of these resources. Health research is necessary to remove all cultural, social and logistical barriers that astonish the effort of many health programs.

Zbigniew Fedorowicz experimented in hi research that in resource-poor countries, funding constraints often lead to less healthcare research which has given rise to unfairly held perception that much healthcare research is required in low-resource countries in order to analyze the health data of that region. And the condition is this that research should be of good quality not just for a formality. Some awareness of investigators is also needed in this. This awareness should be increased by knowing about the uncertainties of treatment. To increase healthcare research, diversion of good quality research needs to propagate in low resourced countries [3].

Components Involved in healthcare research are Researchers (MRIs, Universities and Hospitals), Healthcare Professionals (Hospitals, Clinics and others), Investors (Govt., Business and philanthropy) and high quality systems and outcomes [4, 5].

1.3 Computational Intelligence

D. Saxena, S. N. Singh, K. S. Verma experimented in their research that there are five main principles of computational Intelligence. First one is "Artificial Neural Networks", which includes a paradigm to process information. The key element of this is structure of the system that does information processing and consists of neurons (processing elements that are highly interconnected). Second one includes "Swarm Intelligence", which originates from the study of swarms and colonies of social organisms and its application includes the techniques "Ant Colony optimization", "Artificial Bee colony" etc. Third one is "Fuzzy Logic", it is introduced for solving power system problems and is a gist of classical set theory. Fourth one includes "Evolutionary Computation", which is based on principle of Darwin i.e. "survival of the fittest strategy" and is also similar as GA. Fifth one includes "Artificial Immune Systems", these are of the aspect of the concept of NIS (Has a great patter matching ability, used to distinguish between foreign cells) [6].

The Technical Components involved in research are Neural Network, Evolutionary Computation, Swarm Intelligence, Artificial Immune Systems and Fuzzy systems.

1.4 Mist and Edge Computing

Mist computing helps in building IoT systems at large scale as IoT is regarded to have very small things at the very edge of network like less power, limited memory and bandwidth signals etc. Mist computing has some guiding principles such as Network must provide information not only data, Network should deliver only requested information only when it is requested and some more. Edge computing is the practice of developing a system or network which helps us to process the data near the network edge where it is being generated, instead of being processed in a ware house meant for centralized processing. It is basically a open IT architecture that is distributed and have features that enable mobile computing, decentralized IoT technologies and processing power and enables the acceleration of data stream. The cloud works on MIST computing level, foG computing level and Edge Computing level respectively [7].

Yuan Ai, Mugen Peng, Kecheng Zhang experimented in their research that a Cloudlet is a technology which we can say as explanation of Edge computing. As the main problem is of the end to end receptiveness between mobile device and the cloud associated with it. The main purpose of this device is to develop mobile applications which support interaction and is resource intensive by providing advanced computing resources to devices along with low latency. It is a resource—implemented and trusted device that is compatible with the internet and used by only nearby devices [8].

2 Yajna and Mantra Science

Pt. Shriram Sharma Acharya Ji experimented in his research that The Vedic culture describes the fact that Yajna and Mantra are the establishment of the Indian way of thinking and culture. The four Vedas state that Yajna and Mantra plays a conspicuous role in the creation of this divine universe. The Yajurveda depicts parts of performing Yajna as logical and reverential investigations for worldwide welfare. Mantra treats the human framework at physical, mental and profound levels. The actual meaning of Yajna is to sacrifice materialistic things, egotism and adopting logical thinking, burning evil deeds into fire. The thermal energy comes out from Yajna purifies the energy existing in the matter and demolish the sinful energies [9].

2.1 Effect of Mantra and Yajna

Vasanti Gopal Limaye experimented his research that Some scientists researched with some plants which were placed on the terrace for 3 years concerning Yajna, the observation of which were given as cell elongation, initiation of new growth, cell division, induction of flowering and many more. These effects are unbelievable and outstanding. Deep observations of these effects led me to give an important conclusion that Yajna vapor or smoke acted on plants by the generation of a particular class of growth regulatory substances or phytohormones. Yajna is very helpful in keeping the health of human beings. The specific essentialness streams inspired by holy Yajna fire and Mantra Shakti have gigantic mending sway on the messiness and diseases running from cerebral agony, migraine, cold to mental gruffness, scholarly insufficiencies, sorrows, sleep deprivation, lack of restraint, epilepsy, schizophrenia and assortments of lunacies. The purpose of Yajna is to examine the effect of Yajna's smoke on air microbiological quality by bacteria, purification of water and effects on growing root tips [10].

2.2 *Societal Impact of Yajna and Mantra Therapy*

Ruchi Singh and Sunil Kumar Singh experimented in their research that Both indoor and open-air analyses were directed in the general public to see the effect of treatment, some experiments were also done to compare the effect of Yajna and for non-Yajna i.e. just burning the plain wood and burning the plain herbs without any ritual and mantra. Before the Yajna and up to three days after the Yajna, the effect of micro-Petro-bacteria, fungi, and pathogens present in the air was studied. The results were unpredictable. There was a huge lessening in the microorganisms basically the pathogens. These results are in support of the fact that the Yajna makes the atmosphere bacterial and kills harmful microbes in the atmosphere [11].

All the above observations indicate the fact that Yajna process is effective in reducing both vaporous and microbial air contamination and furthermore expels terrible scents. It is generally observed that Yajna stays in the air for quite a long time after it. Yajna is possibly the only solution to the problems of today's environmental solution. Our ancestors have also said that the Yajna should be performed daily in every household and every person. The Ayurvedic (natural medication) lab and its pharmaceutical unit have produced new homegrown medications that have shown great recovering achieves certain examples of tangible framework issues, asthma, heart ailments, diabetes, lung defilements, a wide grouping of skin disorders and ailments of the eyes and ears. Today we have failed to recall this science and the time has come for us to revive this rite enhancement of ourselves and our future generation.

2.3 *Yajna and Mantra Science: Future and Alternate Therapy*

Nowadays, society has become completely dependent on allopath or homeopathy treatment. But the saddest thing is that allopathic medicines can cause undesirable side effects. For example, some antibiotics can cause allergic reaction about 5% of people. Some alternative therapies are completely natural i.e. Sun Therapy, Yajna Therapy, Acupuncture. Sun Therapy treats depression with a seasonal pattern. It is proven that acupuncture helps in case of neck pain, knee torment, migraine, low back agony etc. Yajna Therapy is the most effective therapy which cures humans from diabetes, depression, drug addiction and many more.

Alka Mishra, Lalima Batham, Vandana Shrivastava experimented in their research that cancer is becoming the root cause of death worldwide. India has an infrastructure and facilities to fight diseases. Moreover, the side effects of existing standard therapies such as chemotherapy and radiation therapy pose serious barriers to the quality of life (QOL) of patients. There is also an urgency for alternative adjunct therapy not only to reduce the current therapeutic burden but also reduces side effects. Yagya therapy allows for the ingestion of Yajna's smoke of multi-herb mixture produced by immolation in fire with chanting of Mantras. The study attempted to assess the QOL of three cancer patients (breast cancer, mouth cancer, and chronic myeloid leukemia) by combining specially formulated multi herbs as a part of Yajna Therapy. After scientific research on Yajna science, the study specified that the Yajna therapy supports to improve the QoL of cancer patients taking standard allopathic medication [12].

3 Healthcare 4.0 with Fog Computing on Inhaling Therapies

3.1 Role of Technology in Addressing the Problem of Integration of Healthcare System

There is a huge role of technology in transforming the healthcare industry in many ways. As artificial intelligence is present everywhere so healthcare is no exception. As the technology is advancing day by day, AI could help diagnose heart disease, skin cancer and many more dangerous diseases etc. Virtual healthcare or tele health or telemedicine is a great technological advancement in the field of healthcare. Virtual healthcare allows the interaction of doctors and patients living in remote areas via video conferencing or mobile apps. Then comes 3-D printing. It is capable of building a close and clear 3-Dimensional structure of cells which is layer by layer. Then it can be formed into human tissues and organs eventually for replacements.

Huston, C experimented in his research that now in the era of emerging new technologies, there is a need for changing the practice of nursing in the healthcare industry too. Many technologies have arrived to integrate the healthcare system and technological advancements. One of them is 3-D printing. The benefit of this technology is that in bio printers, a bio-ink is present that is made up of living cell mixtures. It is capable of building a 3-Dimensional structure of cells, layer by layer. Then it can be formed into human tissues and organs eventually for replacements.

3-Dimensional printing which is also called as additive manufacturing. It will be really very beneficial if healthcare industry is combined with new technology. So we should plan for the future and should do research in this area [13].

3.2 Scientific Study on Impact of Yajna on Air Purification

Pushpendra K. Sharma, S. Ayub, C. N. Tripathi, S. Ajnavi and S. K. Dubey experimented in their research that pollution is the most dangerous problem in today's world. Air pollution is going up at a very fast speed day by day. There is great urge to develop some innovative ideas to stop this fast growing air pollution. Scientific study shows that Yajna has a very positive impact on air purification. The Ahuties used in Yajna like cow butter (ghee), Pipal wood (Ficus religiosa), Havan samagri (kapurkachari, gugal, balchhaar or jatamansi, narkachura, illayachi, jayphal, cloves and dalchini etc.) proved very much beneficial in coping air pollution problem. There is no need to add any chemicals. Only these natural products help in air purification. The harmful gases present in the atmosphere like NO_2, CO, SPM and RSPM gets reduced after the Yajna [14].

3.3 Expert System Design

The significance of wellbeing and being medicinally fit is objective for each person. Undesirable circumstances of people are extremely disadvantageous and heartbreaking to our general public. An Expert System is an item that reenacts the show of human authorities in a specific area. The present Expert System has been used in various areas where call for essential initiative. The learning architects assemble the information from human specialists and speak to it into a learning base. The future Expert System will need bigger extent for learning to fathom the further confounded certifiable issues including medicinal and medical issues.

Nana Yaw Asabere experimented in his research that An Expert System is a kind of Artificial Intelligence (AI) that epitomizes the information of at least a specialist. The therapeutic field may utilize the Expert System than some other fields. Many warning projects have been created to enable doctors to analyze a specific disease and sometimes, to recommend treatment. The most established restorative Expert System is called MYCIN. MYCIN is a specialist framework created at Stanford for analyzing blood ailments. It is part of the generally contemplated Expert System on account of prosperity. A noteworthy element of MYCIN is that its standard foundation is independent from its induction scheme. Expert System allows the MYCIN learning part to be disposed of, making void MYCIN. EMYCIN is the pattern after which the cutting edge master improvement shells are designed. Some different MESs(Medical Expert Systems) contain PUFF, Drug Interaction Critic, GUIDON, CASTNET/GLAUCOMA, and ONCOCIN [15].

3.4 Robotics

Robotics is a multifaceted branch of science and engineering that involves the study of engineering of electronics, mechanical, computer, information and others. It deals with whole formation phases i.e. design, construction and operation and use of computer systems and robots (machine programmed by programming language through computer) and also with information processing and feedback for their control. Robots also have some kind of mechanical construction, electric components for providing power and controlling the machinery, and have some level of computer programming code. Some of its components are Power source, Actuators, Motors, Polymers, Nanotubes, Sensors, Grippers, effectors etc [16, 17].

Lorelei Kujat experimented in his research that the field of Robotics has applications in many fields such as in military, labor work, medical, construction, agriculture etc. Application of robotics in medical field include such as in Telemedicine, where recordings and monitors produced by tele health robots shows medical procedures and discussions of a patient, A robot invented by Japan which transport patients who weighs a maximum of 134lbs to their bedsides by using sensors. During the 1990s, Roball robot was invented to assist the child development through providing autistic children with interaction experiences and many more [18].

3.5 Computer Control System Design

Jerzy Moscinski, Zbigniew Ogonowski experimented in their research that the field of adaptive control system design has gathered a lot of interest between researchers and groups over last few years. Many different approaches to computer control system design emerged but almost all of them lacks approach that is appropriate for it. Researchers should do more work in this field. Yagyopathy technology can be benefitted with this technology very much. One should invent some new innovations to inter relate Yagyopathy and computer control system design. It will be very beneficial [19].

Control system design is a device or a set of devices that commands, manages, directs, and regulates the behavior of other connected devices. This functionality can be used in the field of Yagyopathy to manage different devices that are connected and it can also be helpful in case of patients living in remote area. There are many types of control system design namely open loop control system, close loop control system. Computer control system design will be very helpful in the field of advancements in Yagyopathy.

3.6 *Machine Learning Algorithms*

G. E., Krizhevsky, A., Srivastava, N., Sutskever, I., and Salakhutdinov, R. experimented in their research that the most important sub field of machine learning is deep learning. It is just like the backbone of machine learning. Deep learning has many amazing advantage and many more which is yet to be discovered. In Yagyopathy machine learning can be used widely. So innovations in machine learning and deep learning will be very beneficial for the field of Yagyopathy. Yagyopthy is one of the advancement that will make the world more advanced definitely [13].

3.7 *Expert System*

An expert system is an application of artificial intelligence which can take decision depending on previous instances and experience. It has large set of data and statistics and uses complex machine learning algorithm to analyze it and show the result when new data set is given as input to it. The main aim of the expert system to is make data availability easy. The logic required to work can be reviewed by any person of a particular field rather than an IT expert as in case of traditional system. The expert system is a robust software that can be used in "Yagya and Mantra therapy" for cure of diseases. The expert system can be provided with large set of data containing details about various diseases and their symptoms. Along with this it will also contain data about the therapy required for treatment of various diseases. The expert system can then hear patient's problem and provide the necessary ailment to the disease. This will reduce the work and need of the doctor. The expert system can be deployed at places where availability of doctors is difficult.

The paper by Sami S. Abu Naser designed an expert system to diagnose feeding problem in children. The expert system was set so that parents can know what problem a child is facing by recognising and gets recommendations about how to deal with that problem of child. The system is given data related to what all reaction a child gives and also when does it give it. For example if a child is crying is morning then most probably he is hungry. The data has been collected from research on many children. Now the expert system can give suggestion to parents to deal better with their children [20].

The same method can also be applied in case of 'Yagya and Mantra Therapy'. The patient can tell the system his problem and system will provide him the required recommendations and diagnosis of the disease. The diagnosis will be based on the data given to the system previously. The system will analyze the symptoms with various other present in its data set and provide the necessary therapy required to heal the disease. This Expert system can be deployed a remote places and at those villages where easy availability of doctors is not possible.

3.8 *Ambient Computing*

Ambient Computing is a term which is a mix of many things. At the basic level it consists of hardware, software, human interaction with machine and machine learning. The main idea is to use a computer or any device which is connected to internet without actually concisely using it. To make it simpler ambient computing refer to all the devices around us, in our home, in our workplace which are connected to internet and we are using them anywhere. We now don't need to sit in front of a desktop or laptop to get our things done. The IOT technology is used to connect various house hold devices under one network and further to the internet. Every device is then in control from just our mobile or now even by just voice. Let's take an example, the advance lightning systems. The light, fans and other electronic devices are connected over Wi-Fi to some intelligent system. We now don't need to physically go and press the switch to turn it on rather we can control them from our mobile phone or it voice recognition is enabled we can directly operate them from our voice command. The technology can be used in "Yagya and Mantra Therapy" to train a device which can be operated from a smartphone and can do required therapy. The need of an expert or doctor is minimized [21, 22].

Alexiei Dingli have developed a method called PINATA to aid doctor and nurses in better delivery of healthcare services. It focuses on passive devices and thus help doctor and nurses to focus on patient and thus improving the quality of healthcare [23].

In the similar manner the ambient computing can be used to build a device which can perform the required therapy in case of our "Yagya and Mantra Therapy". The requirement of doctor will be minimized if such a machine is operational.

3.9 *Computer Application in Information and Communication Technology*

The computers are now used everywhere. There is no a single domain left where the computer are not playing an important role. Be it from education to finance, government to judiciary, communication and aviation sector. The use of computer has revolutionized the communication world. The information and communication technology is a wide term that covers any service which help people to reach out to others. The communication via internet started from e-mail services which travelled from the age of instant message then to voice call and now have come to the era of video call. The information is now available to everyone who has access to internet. In the medical sector the advent of internet and recent technology like machine learning and artificial intelligence have completely changed the way of diagnosing diseases. In early days computers were just use to do MRI or an X-Ray or any other type of scans. But now Robots are able to do complex operations with better accuracy. The use of expert system and ambient computing can further provide health care too remote areas where reach of professional doctor is not easy. In many hospitals around the world the robots have proved to be a better companion with doctors doing major operations.

Peter Idowu in his research work talked about how he used the Information and Communications technology to deliver better health care facility in Nigeria. They review the present and past health care reports of people of Nigeria and compared it with people living in the UK as an example of underdeveloped and developed nation. They also analyzed the problem faced by people in getting better health care facility. So with the use of ICT they were able to give better health diagnosis about people of Nigeria [24].

3.10 Hybrid Intelligent Systems

Hybrid intelligent system denotes a software system which employs a combination of methods and techniques from artificial intelligence subfields. In other words, it combines at least two intelligent technologies. For example, combining a neural network with a fuzzy system results in a hybrid neuro-fuzzy system. It is independent, non-interacting components. A hybrid system that combines a neural network and a rule-based expert system is called a neural expert system.

The Architectural diagram of hybrid intelligent system shows that Graphical User Interface is connected with Clinician and rule based reasoning and Case Based Reasoning. These two reasoning systems are connected with knowledge base which is intermediate between clinician and knowledge acquisition followed and used by knowledge engineers. Rule based reasoning consists of working memory and inference engines while the case based reasoning consists of retrieve to reuse and retain to revise processes [25].

It has been developed for modeling expertise, decision support, complicated automation tasks etc. The use of intelligent systems for stock market predictions has been widely established. It is based on an artificial neural network trained using scaled conjugate algorithm and a neuro-fuzzy system. It can be easily implemented and factual results are very promising [25].

Ajith Abraham Baikunth Nath P. K. Mahanti experimented in their research that it has been developed for modeling expertise, decision support, complicated automation tasks etc. The use of intelligent systems for stock market predictions has been widely established. It is based on an artificial neural network trained using scaled conjugate algorithm and a neuro-fuzzy system. It can be easily implemented and factual results are very promising.

Block diagram of hybrid intelligent systems in process, it can easily be identified that stock index values are input to data preprocessors and it inputs to neural network which trains the data using scaled conjugate algorithms. It outputs to both the stock forecasting as well as to neuro fuzzy systems which further outputs for stock trend analysis purposes [25].

3.11 Symbolic Machine Learning

Symbolic Machine Learning is the term for the collection of all methods in artificial intelligence research that are based on high-level "symbolic" (human-readable) representations of problems, logic and search. The history of Artificial Intelligence (AI) began in antiquity, with myths, stories and rumors of artificial beings endowed with intelligence or consciousness by master craftsmen. Research into general intelligence is now studied in the sub-field of artificial general intelligence. Machines were initially designed to formulate outputs based on the inputs that were represented by symbols. Symbols are used when the input is definite [26, 27].

Goldberg, D. E. and Holland experimented in their research that the field of Machine Learning depends upon nature‘s bounty for both inspiration and mechanism machine learning must borrow from nature. It consists of well-defined algorithms, data structures, and theories of learning, without once referring to organisms, cognitive or genetic structures, and psychological or evolutionary theories. Machine Learning is devoted to papers concerning genetic algorithms and genetics-based learning systems. Genetics-based machine learning have often been attacked on the grounds that natural evolution is simply too slow to accomplish anything useful in an artificial learning system [28].

Commonalty used symbolic machine learning algo for material sciences are

A. Regression
   SVM, ANN, LR and MLR, KRR(Kernal Ridge Regression) etc.
B. Classification and Clustering
   SVM, ANN, Naïve Bayes, KNN, Decision Tree, K Means, DBScan etc.
C. Probability estimations
   EM, Naïve Bayes etc. [28].

3.12 Neuro-Fuzzy System

A neuro-fuzzy system is a fuzzy system that uses a learning algorithm which is derived from neural network theory and is used to determine its parameters (fuzzy sets and fuzzy rules) by processing data samples. It is represented as special multilayer feed forward neural networks. A neuro-fuzzy system can be always (i.e. before, during and after learning) interpreted as a system of fuzzy rules. It is based on a fuzzy system that is trained by a learning algorithm which is derived from neural network theory. It should not be seen as a kind of (fuzzy) expert system, and it has nothing to do with fuzzy logic in the narrow sense.

Akbar Esfahanipour Werya Aghamiri experimented in their research that Neuro-Fuzzy Inference System adopted on a Takagi–Sugeno–Kang (TSK). It is a type of Fuzzy Rule Based System which is developed for stock price prediction. It is so difficult to forecast stock price variation. It has six layers. The main steps are as below

1. Training data.
2. Factor selection.
3. FCM clustering on output.
4. Assigning Membership degree to the output (using various clusters).
5. Projection membership degree to the input(which uses clusters) used for identification of membership functions of input variables.
6. Adapted fuzzy inference systems are used which takes care of setting up a TSK fuzzy rule base which inputs for tuning system parameters by ANFIS.
7. Outputs are used for stock price forecasting [29].

It's goal is to predict the trend of the price variation which includes various influential factors such as macro-economic change, political reasons, fundamental analysis and the technical index etc. Neuro Fuzzy system uses multiple layers where in first layer, inputs are provided, inlayer-2, membership functions are used, and in layer-3, Fuzzy rules are applied. Layer-4 uses output membership functions and layer-5 uses de-fuzzification process [29, 30].

3.13 Nature Inspired Computing

Nature inspired computing (NIC) is a very new discipline that strives to develop new computing techniques. It solve complex problems in various environmental situations. They are used in the fields of physics, biology, engineering, economics and even management. It consists of two concepts independently i.e. effectors and detectors. Multiple detectors that receive information regarding neighboring agents and the environment whereas multiple effectors that exhibit certain behaviors and cause changes to their internal state and drive changes to the environment. Effectors facilitate the sharing of information among autonomous entities. It can also be seen that nature inspired computing takes variety of inputs from Artificial Neural computations, Evolutionary Computations, Swarm Intelligence and Artificial immune systems [31].

Nazmul Siddique Email author Hojjat Adeli experimented in their research that the nature-inspired computing prototype is fairly large. It helps us when problem is complex and nonlinear and involves a large number of variables or potential solutions or has multiple objectives. It also helps us when problem cannot be suitably modelled. It refers to a class of multiobjective algorithms that imitate some natural phenomena which is explained by natural sciences. Nature inspired computing can be classified into three classes: physics-based algorithms (PBA), chemistry-based algorithms (CBA) and biology-based algorithms (BBA) [31].

4 Literature Survey

Dr. Rashi Sharma has described that Ayurveda is the divine science of life which deals with Ayurvedic study. Therefore, the Ayurvedic study also emphasizes treatment-related health prevention. In Ayurveda people are classified. According to Prakruti, there are two types of prakruti: Sharir prakruti and tamas prakruti, where Sharir prakruti belongs to Vataj, Pittaj, kaphaj, Manas prakruti belongs to Satva, Rajas, and tamas prakruti.

In terms of people's stress, Charac Asharia says he tries to kill an elephant if he performs more than his ability to put excessive pressure on slaughter such as milk. This means that overstressing the system can have a devastating effect on the body.

According to modern science, stress is a situation involving demand for physical and mental energy. This is a condition that can impair physical or mental health. Observations in current scenarios indicate that psychological stress and rapid life stress are increasingly contributing to health risks. Manas Prakruty. Raising his mind with the help of his psychoanalysis and positive thoughts, he can achieve a stronger and higher spiritual quality.

She has revealed, Upanishad, allied literature, and even the character of Samita, describe various tools for overcoming rabia and tama. One such effective tool is the Gayatri mantra. Gayatri also cleans the stimuli. It strengthens Satba and thus triggers spiritual enlightenment and the progress of life at the Tamoguna Refinery that creates fear and patience gives the courage to fight injustice.

Thus, individuals can follow the path of complete social well-being and influence minimal tension [17, 32].

According to Dr. Selvamurthy, amidst the modern and fascinating achievements of science and technology to improve our comfort level, stress and pollution have posed great challenges to our well-being. The world recognizes that the convenience provided by modern technology does not necessarily make life happy. In fact, apart from stress, unknown illnesses, untreated diabetes, anxiety, and fear from highly polluted environments and environmental imbalances. This created a warning to rethink and change lifestyle and health care. He said about Yajna, which seems to be a gift from ancient Indian science for god to achieve this goal. This paper highlights potential drug applications elsewhere in light of recent research results.

The author has revealed the following reported experiments and case studies. Some physiological studies: In his study, Dr. Slamoumerti observed the mantra-neurophysiological effects of certain types of agnihotra at sunrise and sunset. He described his experimental study in which eight healthy men were selected as subjects. They reported for two consecutive days. The first day was to record the controls when performing the Agnitra ritual, but instead of the default mantra, some unrelated syllables were spoken at specific times. The next day, Agnihotra was performed with the appropriate mantra. Record physiological parameters. Heart rate, blood pressure, etc. were measured every 2 days.

As per Dr. Selvamurthy, The results showed that the heart did not change during the first day of agnihotra, but significant changes occurred after proper agnihotra. This includes:

1. G.S.R. significantly higher during appropriate agnihotra.
2. ECG showed baseline DC change.
3. EEG showed increased alpha and delta suppression over 15 min.

Diabetes treatment: Some acute diabetic patients do not experience a complete increase in urinary glucose levels, and blood sugar levels have been found to return to normal immediately after 2–3 weeks of a daily pyranometer.

Gas tragedy of Bhopal and Agnihotra: This tragic incident occurred on the night of December 3, 1984, when MIC poisoning has emerged from Bhopal's Union Carbide. Hundreds of people died and thousands were hospitalized, but there were two families, Shrizohanral S. Kushwaha and Shri M. L. Rasole lives a mile away from the plants that came out of the harbor. These families regularly played Agnihotra (Havana). Despite being in an area affected by toxic gas leaks, no one died in these families. These observations suggest that agnihotra is a proven antidote to infection.

The laboratory has a glass chamber and gas analysis wing for collecting and analyzing Yajna smoke and vapor. The effectiveness of various herbal ingredients in havishya and the quality of Samida are evaluated in a Phytochemical laboratory equipped with devices such as gas-liquid chromatography.

They explain the purpose of first analyzing the raw content and what these substances remain after full steaming [33].

If blood samples are filled with smoke or vapor during daily exercise, they are stored in a glass chamber and changes in the blood biochemistry and hematological parameters of these samples are recorded.

The overall conclusion of the current results is that Yajna's performance greatly enhances the vitality and resistance of metro's harmful biological changes and the invasion of viruses and deadly bacteria. Mental relaxation, emotional stability, and creative mind development are common observations of psychological analysis.

The possibilities of treatment of mental illness is even more encouraging. Diagnosis and treatment of mental illness is just beginning in modern treatment systems. There are no good diagnostic aids, and there are no known systems for the treatment of diseases such as neurosis, psychosis, schizophrenia, depression, tension, depression, mania disease, hysteria. On the other hand, mental illness is more common than physical illness.

5 Methodology

5.1 Instruments Required

During Healthcare Experiments, we have done experiment on a set of patients suffering with pre-diabetic stage, for that we have used checkers and testers for measuring levels of various parameters in blood to provide patient with correct prescription. We have used Glucose checker to measure the amount of glucose present in blood, Haemoglobin checker to measure the amount of red blood cells present in blood and Blood Sample checkers to measure the various parameters in blood to identify disease.

While we are collecting the data of patients and pollution level too, then we need to compare the data to check whether we are able to improve the conditions or not. So to achieve this, we need a machine learning software or an expert system which can store the data of last 10 years of pollution so that we compare our current data to the stored data. Similarly, we need expert system in field of Healthcare too such that by taking the blood sample of patient, it is able to tell the stage of diabetes with which the patient is suffering and to tell the patient about the diet pattern to follow to control it.

5.2 Experimental Setup of an Expert System

While we are doing the experiment for Pollution Checking, we have done Yagya on the places on crossroads. We have taken the reading before doing Yagya and then after doing Yagya. Then with the help of checkers and testers we installed, able us to compare the pollution level.

Mamta Saxena, Sushil Kumar Sharma, Sulochana Muralidharan, Vijay Beriwal, Rohit Rastogi, Parul Singhal, Vishal Sharma, Utkarsh Sanga experimented in their research that while we are performing the experiment on patients suffering with diabetes, we have performed yagya in all its three Sessions i.e. Pre-Yagya Therapy Session, Main Yagya Therapy Session and Post Yagya Therapy Session in a close environment. In Pre-Yagya Therapy Session, all requisites of therapy like CHS and SHS, Utensils, Flowers etc. are prepared. In Main-Yagya Therapy Session, yagya is performed with some Vedic mantras with Ahutis of Cow ghee and mixture of CHS and SHS in 2:1. In Post-Yagya Therapy, Patients are advised to do meditation, breathing exercises and yogasan in the same close environment in order to inhale the fumes of the Yagya. Then the patients are advised to perform this daily at their homes and have also prescribed with healthy diet pattern and they are also advised not to take any processed meal, potatoes, packed juices etc. During this analysis, the Body weight, Fasting Blood Glucose Level (FBS), Post Prandial Glucose Level (PPBS), Hb Alc Level and their overall general health assessment are recorded periodically [34].

5.3 *Flow Chart of Healthcare Expert System*

The figure (as per Fig. 1) demonstrates the progression of the expert system, the flow of this chapter starts with a discussion about Healthcare 4.0 with Fog Computing on inhaling Therapies. Then go through to the deep knowledge of Healthcare and methodology used (i.e. instruments required, experimental setup and factors and parameters). After that some significant outcomes are drawn from research and discussion on Healthcare 4.0. At that point critical outcomes are examined and deciphered.

5.4 *Parameters and Factors Under Study*

Different clinical parameters of diabetes are emphatically connected with both fatigue and depression. Fatigue itself has noteworthy connection with depression in type 2 diabetes. Standard observing of biochemical parameters are central to anticipate the advancement of fatigue and depression in type 2 diabetes. A research conducted for youth in which 21 youths focuses who were assessed through the Diabetes Quality of Life (DQOL) survey, It reasoned that reduced HbA(1c) was related with decreased sway, less stress, more prominent fulfillment and better wellbeing discernment for teenagers. These are some factors which describes the whole essence of healthcare 4.0 [35, 36].

6 Results and Discussion

6.1 *Results, Interpretation and Analysis on Healthcare Experiments*

Yagyopathy Shivir was conducted in Navchetna vistar Kendra Durgapura, Yagyopathy Research Centre, 43 Srijinagar, Durgapura, Jaipur, Rajasthan 302018, nckdurgapura@gmail.com, in May 2019 and following results were found for different experiments on different subjects suffering with various physical problems.

6.2 *Asthma Related Experiments*

Asthma/Allergy-Patient Description-Symptoms were checked on 1st day, after 10, 30 and 60 days.

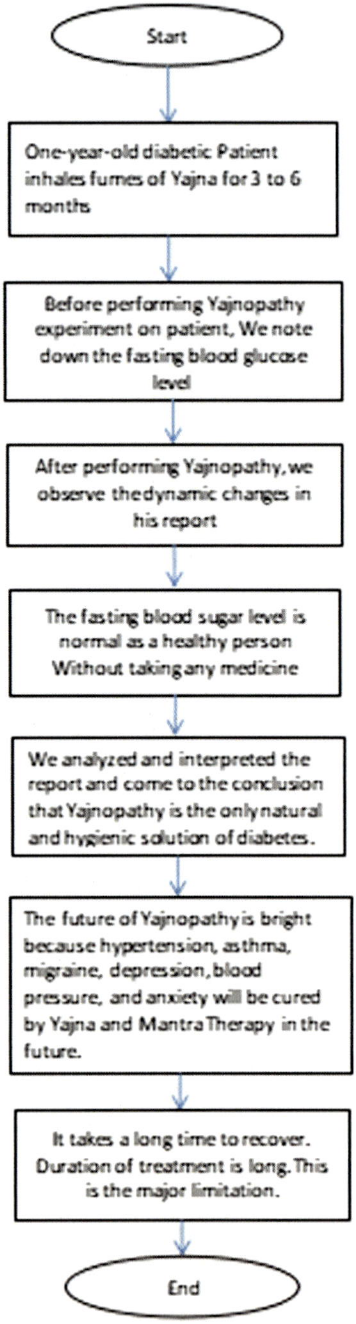

Fig. 1 Flow chart of diabetes analysis of subjects through Yajna and Mantra Therapy

Data was recorded under date, weight, B.P., other records, Limb dysfunction/extreme tiredness or weakness, Inhaling problem while lying, Unable to breathe deeply, Sneezing, Cold, Restlessness, Breathlessness and other issues.

6.2.1 Case Study 1 (As per Fig. 2)

Report on Clinical Trials of Effect of Yagyopathy on Asthma
Shri Baidyanath Shukla

1. I have been diagnosed for Bronchial Asthma since past three years. However, with regular pranayam, I have been able to contain it to a large extent and need medication only occasionally.
2. **Yagyopathy**. During the meeting on Yagyopathy chaired by Shraddhey Dr. Sahab at Shantikunj on 23 Nov 18, I volunteered to experiment its effects on myself. I consulted Dr. Vandana Srivastava and was advised following treatment:
   (a) Yagyopathy twice a day, at sunrise and sunset, with Havan Samagri for asthma and normal samagri in 3:1 ratio, recitation of Surya Gayatri mantra 24 times and nadi-shodhan pranayam for 30 min.
   (b) Kwath of mixed Hawan Samagri twice a day.
3. **Clinical Trials**. In order to measure the efficacy of Yagyopathy, I undertook Lung Function Test (LFT) on 18 Dec 18 and started Yagyopathy as above instructions (once a day) since 24 Dec 18. Following three important parameters are measured during LFT:
   (a) Forced Vital Capacity (FVC): measure of lung capacity/volume of air that can be exhaled after deep inhalation.
   (b) Forced Expiratory Volume-One Second (FEV1): measure of rate of exhalation.

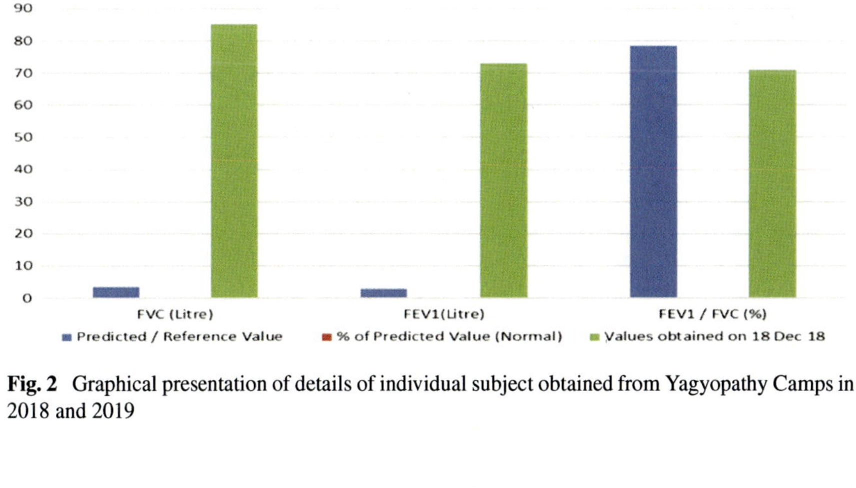

Fig. 2 Graphical presentation of details of individual subject obtained from Yagyopathy Camps in 2018 and 2019

Table 1 Predicted value and comparative data for FEV1 and FEC

Parameter (unit)	Predicted/reference value	% of predicted value (normal)	Values obtained on 18 Dec 18	Values obtained on 05 Mar 19
FVC (L)	3.5	≥80	85	94
FEV1(L)	2.86	≥80	73	80
FEV1/FVC (%)	78.4	≥70	70.9	89

(c) FEV1/FVC: Percentage of lung size (FVC) that can be exhaled in one second.

4. **Comparison of Results**. After continuing yagyopathy and kwath treatment once a day, LFT was repeated on 05 Mar 19 and following comparative results have been obtained (as per Table 1).
5. **Conclusion/Findings**. Results obtained after undertaking yagyopathy treatment once a day for approximately 2½ months indicated considerable improvement in lung function parameters. Results can be further improved with increasing treatment doses to twice a day.

6.2.2 Case Study 2 (As per Fig. 3)

(TYPE) - TYPE-2
Patient Name - SreeJgadish Prasad Sharma, Age - 67 years.
Address - 181 Suryanagar, Chamkoot, Tonkrod, Jaipur.
Diabetes Type (TYPE) - TYPE-2
Patient's background - Mr. Sharma has been suffering from heart problems for 10 years, his heart is also stunted. He has been taking diabetes, B.P. medicines

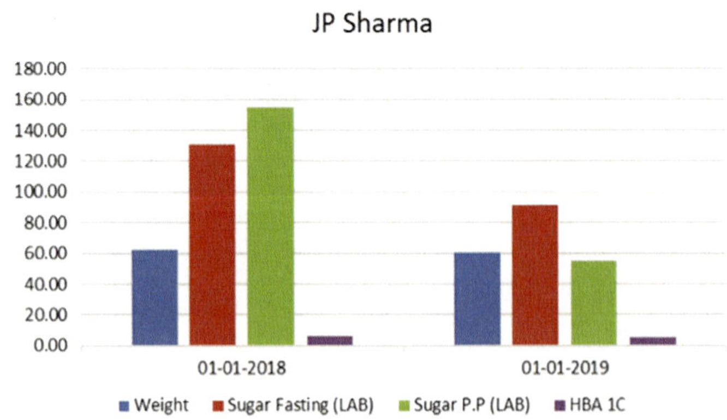

Fig. 3 Graphical presentation of details of individual subject obtained from Yagyopathy Camps in 2018 and 2019

Table 2 Mr. Sharma data through Yajna therapy

Jagdish Prasad Sharma					
Date	Weight	Sugar fasting (LAB)	Sugar PP (LAB)	HBA 1C	Medicine status
12-03-18	62.40	130.9	154.6	6.17	Same medicine
04-04-19	60.60	91.40	55	5.56	Same medicine

for a long time but despite this, diabetes could not be stabilized and fasting PP both ever also used to touch the 200 level. Despite taking the diabetes medicine at a high level, there was no significant improvement in the sugar level. List of prescription and drugs are given below (as per Table 2).
Mr. Sharma had a lot of complaints of knee pain before the camp but that complaint became almost negligible from the forest for 1 month of Yagyopathy medicine.
The Yajnopathy therapy camp ran from December 3 to January 2, in which daily 20 min VanaushadhiYajna, 10 min meditation, 30 min Pranayama and 30 min Yogasan program were regularly conducted which was regularly attended by Mr. Shameena. The details of daily routine are given below.

Daily Yagyopathy Routine
Forest practitioner 20 min, Meditation 10 min
Pranayama
>PranakranPranayam 5 min >Bhastrika Pranayama 5 min >Anulom Vilom Pranayam 5 min >Kapalbhati Pranayama 5 min >Fire extinguishing 5 min >Bandha - Moolbandh, Udianbandh, Jalandharbandh (3 times) >Bhramari Pranayama (3 times) >Omkar 5 times Yogasana
A. microbiology
B. Sitting
*Mandukasan, *Gomukhasana, *Yogamudrasana, *Shashankasan *Paschimottasan
C. Commercable
*UttanapadaHastasana, *Pawanmuktasan, *Mercutasan
D. On Stomach
* Bhujangasana
E. Standing
*Engine racing, *Tadasana, *Katichkrasan

Diet Routine

Morning juice intake, Abundant only fruits consumed till 11:00 am, Abundant salad along with food, Fruit intake again at 4:00 pm, Eat plenty of salad with evening meal, After 7:00 pm, the next day till 11:00 am the next day, only vallikvid/fruits should be taken 30 min of sunshine daily.

Avoid them—Fried things, processed food, packaged food, potatoes and things made from it.

Routine after Sacrificial camp

1. Daily Havan and Aushadhi Yajna was performed daily.
2. Ahar routine partially followed.
3. Pranayam was performed daily after the yajna.
4. Yoga and Morning Walk were done irregularly.

6.2.3 Case Study 3 (As per Fig. 4)

(TYPE) - TYPE-2
Patient Name - Mrs. Manju Kanwar, Age - 55 years.
Address - 55-56 Suresh Nagar, Durgapura, Jaipur.
Type of diabetes (TYPE) - TYPE-2

Patient's background - Mrs. Manju Kanwar had a problem in her lungs, she was breathless when climbing stairs and walking and there was a lot of swollen on her face, due to which she approached the Yajnopathy camp. It was found out that he is a patient of diabetes and also a patient of B.P. His sugar was found to be PP 162 and HBA 1C 6.31 and B.P. was also found 169/102. Gone.

Medicine Status - Diabetes was not taken in the past and was still under diabetes control without any medication. Lung treatment was taken by her in 2017 and the blood report and the MRI report are given below (as per Table 3).

Daily sacrificial routine—Same as Mr. Sharma in case Study 7

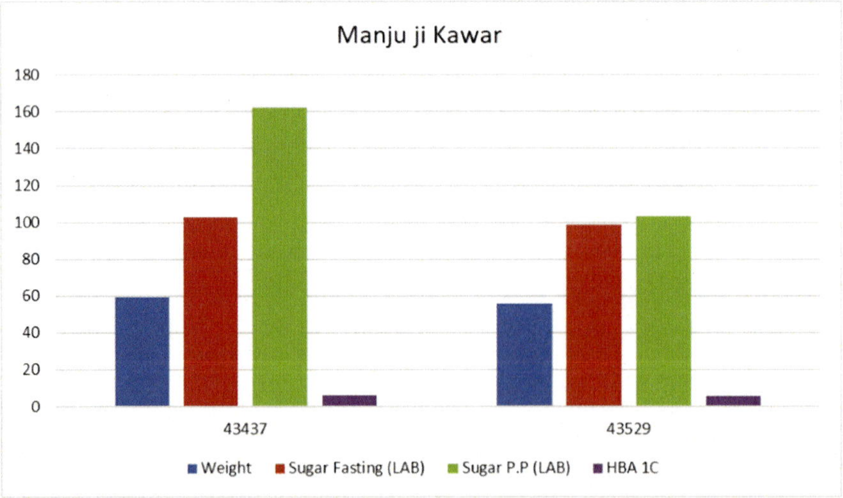

Fig. 4 Graphical presentation of details of individual subject obtained from Yagyopathy Camps in 2018 and 2019

Table 3 Mrs. Manju Kanwar data through Yajna therapy

Date	Weight	Sugar fasting (LAB)	Sugar PP (LAB)	HBA 1C	Medicine status
3/12/2018	59.20	103	162	6.31	No medicine
5/3/2019	55.70	98.6	103.6	5.88	No medicine

Mrs. ManjukanvarJi performed Yogopathy Medicine Camp JOIN on 3 December and on completion of 3 months on 5-3-2019, her sugar level was rechecked. Comparative analysis of both is being given.
NOTE - The above report is given in the separate folder.

1. Sugar fasting has come down from 103 to 98.6.
2. Sugar has come down from P162 to 103.6.
3. HBA 1C has come down from 6.31 to 5.88.
4. This improvement is after taking any medicine for 3 months.
5. Weight is reduced by 3.50 kg.
6. Fatigue and heaviness is over and energy level has increased.
7. There has been an amazing improvement in the working capacity of their lungs, now they can climb up the ladder comfortably and they have been cured of breathing problems while walking.
8. Her B.P. is normal without any medication. 31-3-2019 His B.P. was 101/66.
9. Their acidity problem has been fixed.

Yagonopathy medical camp was held from 3rd December to 2nd January, in which 20 min VanaushadhiYagya, 10 min meditation, 30 min Pranayama and 30 min Yoga Yasanas were regularly organized which was regularly attended by Mrs. Manjukanwarji. The details of daily routine of the Gyneopathy Camp conducted for patients are mentioned in Fig. 4.

6.2.4 Case Study 4 (As per Fig. 5)

Yagya-Therapy provides pulmonary inhalation of medicinal-smoke of multiple herbs (generated through oblation in fire along with chanting of Vedic hymns), which have the potential for seizure treatment. A case study had conducted in Chetna Kendra Durgapur, study is being reported wherein Yagya-Therapy was prescribed to a prostate, joint pain, cholesterol, high blood pressure, allergy, headache, depression and asthma seizure patient. Before the start of Yagya-Therapy, the patient (Male/65 years) had been suffering from above mentioned diseases (~8–10 episodes annually) since ~1 years (pre-observation). Subsequently, the patient has been doing Yagya-Therapy since past ~1 years, wherein only 2–3 episodes occurred during the first year, that too during sleep only, and after that he was relieving 75% no have been experienced of joint pain, cholesterol and prostate. All this time, the patient continued to take the joint pain medication that he was taking earlier. Thus, Yagya-Therapy can be an effective treatment option for prostate and other diseases patients (as per Table 4).

Table 4 Many Pateint's reports as below

Name of patient	Problem	Progress
M.l Gupta	Prostate, Joint pain, Cholesterol	75% relief
Premdevipareek*	Joint pain, High B.P.	75% relief
GovindNaraynPareek*	Prostate, B.P., Joint Pain	50% relief
RoshaniDevi*	B.P., Allergy, Headache	90% relief
Premdevichaudhry	B.P., Depression	50% relief
L.R. Sharma	Asthma	50% relief

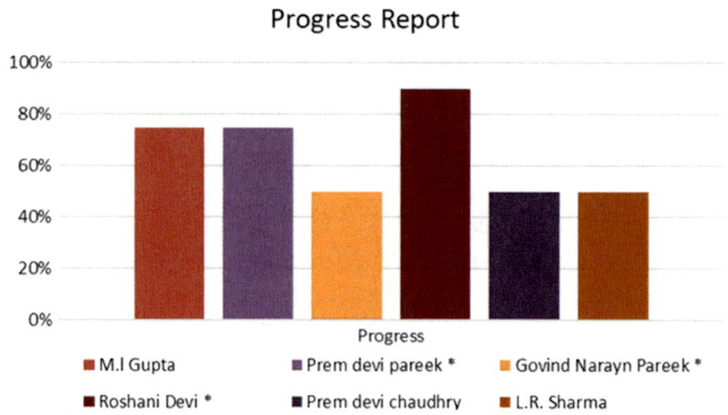

Fig. 5 Graphical presentation of details of individual subject obtained from Yagyopathy Camps in 2018 and 2019

6.2.5 Case Study 5 (As per Fig. 6)

Name of Patient - Mr. S. K. Agarwal, Age - 63 years.
Address - 111 Shriji Nagar, Durgapur, Jaipur.
Type of diabetes (TYPE) - TYPE-2
Patient's background - Between 5/3/2019 to 14/3/2019, a second camp of Yagyopathy was organized in which the same patient of diabetes, Shri SK Agarwal participated. Mr. Agarwal has been a patient of diabetes and B.P. for 32 years. Despite taking high doses of his allopathic medicine, diabetes was not coming under control. Even before coming to Yajnopathi Shivar, his sugar level was running between 200 and 300 and even after three to four months, the retina was injected into the retina after the diabetes affected the retina. Their HBA 1C was 10.71 on 5/6/2018 to 9.29 on 5/3/2019, thus the diabetes level was running out of control. Shri Aggarwal walks 10 km daily, performs yoga and takes a balanced diet, but even then, diabetes was running out of control. Before joining Yagya camp, they have given the inquiry report which is given below (as per Table 5).
Medicine status:
DIABITIES METFORMIN 1000 M.G IN MORNING AND EVENING.

Daily sacrificial routine—Same as Mr. Sharma in case Study 7

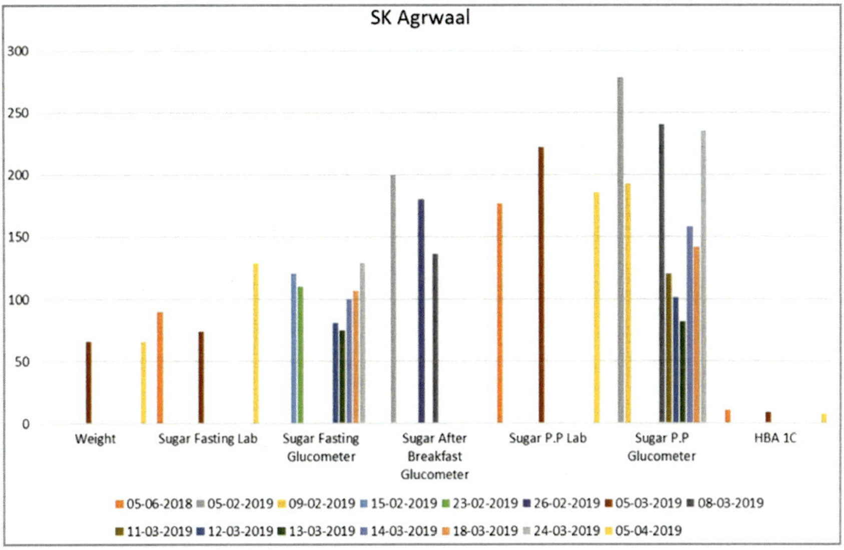

Fig. 6 Graphical presentation of details of individual subject obtained from Yagyopathy Camps in 2018 and 2019

GLIMPRIDE 4 M.G IN MORNING AND 4 M.G IN EVENING.
ZITA PLUS 20 M.G ONCE A DAY.
DOSE REDUCED TO 50% W.E.F 13/3/2019. EVEN THEN HBA 1C CAME DOWN FROM 9.29 TO 8.05 IN 1 MONTH.
B.P.
LOSARH 50 M.G ONCE A DAY.
MEDICINE STOPPED ON 5/3/2019 EVEN THEN B.P. IS WELL WITHIN CONTROL.
Yajna Medical Camp was joined by Shri SK Aggarwal on 5 March 2019 and after 4 months before the date of completion and on 4 April, he was tested from the same lab and on completion of one month, test again on 5/4/2019. Were made. Analysis of Yagyopathy before and after 1 month of experiment is given in Table 5.

Table 5 Mr. SK Aggarwal data through Yajna therapy

Date	Weight	B.P./Pulse	Sugar fasting		Sugar after breakfast		Sugar PP		HBA 1C	Remark
			Lab	Glucometer	Lab	Glucometer	Lab	Glucometer		
Prior to yagyopathy treatment										
5/6/2018			90				177		10.71	
5/2/2019						200		278		
9/2/2019		147/84/92						193		
15/2/2019		148/80/93		121						
23/2/2019				110						
26/2/2019		142/84/84				180				
After Joining Yagyopathy										
5/3/2019	66	141/68/82	74.4				222		9.29	
6/3/2019		146/73/92								B.P. medicine was discontinued
7/3/2019		149/71/91								
8/3/2019		144/68/86				136		240		
9/3/2019		140/87/84								
10/3/2019		141/80/87								

(continued)

Table 5 (continued)

Date	Weight	B.P./Pulse	Sugar fasting		Sugar after breakfast		Sugar PP		HBA 1C	Remark
			Lab	Glucometer	Lab	Glucometer	Lab	Glucometer		
11/3/2019		145/77/85						120		
12/3/2019		130/85/79		81				101		
13/3/2019		144/82/81		75				82		Doses reduces by 50%
14/3/2019		135/76/78		100				158		
18/3/2019		134/78/76		107				142		
23/3/2019		137/81/70								
24/3/2019				129				235		
5/4/2019	66		129				186		8.05	

Note The above report is given in the Separate folder
B.P. Normal is still running despite discontinuation of B.P. medicine on 6/3/2019
Mr. Agarwal has been a regular diet-feeding person and despite taking allopathic medicine of high level, he was running HBA 1C level 9.29 but HBA 1C has come down to 8.05 despite reducing the medicine by 50% on 13/3/2019. And SUGAR fasting and SUGAR PP level has also decreased

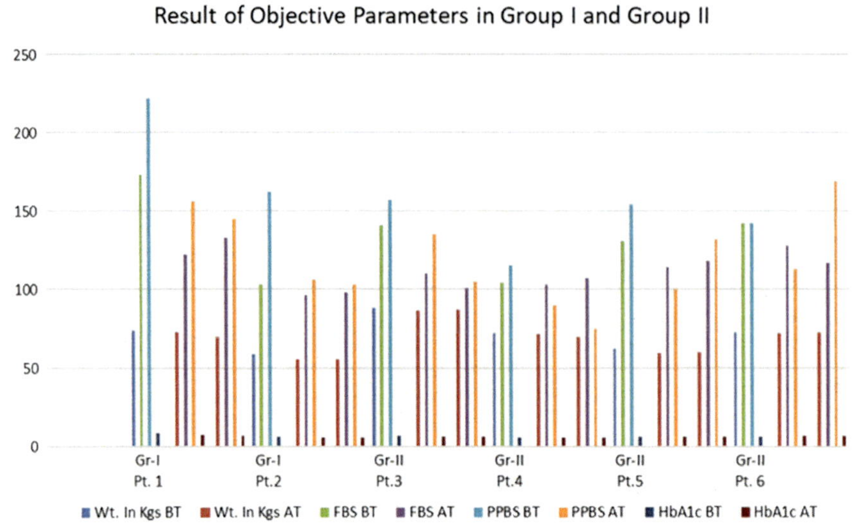

Fig. 7 Graphical presentation of details of individual subject obtained from Yagyopathy Camps in 2018 and 2019

6.2.6 Case Study 11 (As per Fig. 7 and Table 6)

As experiment conducted on many subjects

7 Novelty in Our Work

It neither requires eating of costly and chemical rich medicine nor it require injecting of chemicals in the body of the patient. The disease is cured by the fumes which rise from doing a special type of "Yagya" in a closed room. The material for the "Yagya" is prepared from extensive research. The fumes have some feature which when inhaled effect the disease causing bacteria or virus and thus cure the disease. It is also not a slow process the considerable result is seen at the end of the "Yagya" in the patient. The different experiment conducted at different places also has proven the usefulness and power of the "Yagya and Mantra therapy". As the method doesn't involve consumption of medicine or any type of injection, the cost to perform the therapy is too low. It also don't require expensive machine to perform operations. Thus the "Yagya and Mantra therapy" is the treatment method which can revolutionize the era of medical science and the way disease are cured will also change significantly by the use of this therapy. It also has the power of curing deadly disease like cancer which now requires doing Kemo therapy which is very painful and costly. Thus "Yagya and Mantra Therapy" can also be called as treatment of the future [37, 38].

Table 6 Different case data for FBS and PPBS attributes

Case S. No.	Date of lab test	Wt. in kg		FBS		PPBS		HbA1c	
		BT	AT	BT	AT	BT	AT	BT	AT
Gr-I Pt. 1	3-Dec-18	74.1		173		222		8.27	
	2-Jan-19		72.8		122		156		7.04
	5-Mar-19		**70**		133		145		6.61
Gr-I Pt. 2	3-Dec-18	59.2		103		162		6.31	
	2-Jan-19		55.6		96		106		5.83
	5-Mar-19		55.7		98		103		5.88
Gr-II Pt. 3	3-Dec-18	88.3		141		157		6.47	
	2-Jan-19		86.7		110		135		6.17
	5-Mar-19		86.8		101		105		6.13
Gr-II Pt. 4	3-Dec-18	72		104		115		5.77	
	2-Jan-19		71.6		103		90		5.69
	5-Mar-19		70		107		75		5.74
Gr-II Pt. 5	3-Dec-18	62.4		130.9		154		6.17	
	2-Jan-19		59.7		114		100		6.2
	5-Mar-19		60		118		132		6.2
Gr-II Pt. 6	3-Dec-18	73		142		142		6.39	
	2-Jan-19		71.9		128		113		6.52
	5-Mar-19		72.5		117		169		6.51

8 Recommendations

The "Yagya and Mantra therapy" is very useful for patient suffering from Diabetes. The current form of treatment available is to eat the medicine lifelong which will just control disease form increasing. But no complete cure is of diabetes is there. The "yagya and mantra therapy" has been found to be very useful while dealing with diabetes. The experiment done on various patients has proved that it has significantly affected the disease. The "Yagya and Mantra therapy" is thereby a strong medium of treatment especially in case of diabetes [39, 40].

For other diseases the research is still going to find similar method of treatment. The therapy is also good for lowering the pollution content from atmosphere.

9 Future Scope, Limitations, Applications of Yagya and Mantra Therapy

9.1 Future Scope

There are many scopes of Yagya Therapy and Mantra Therapy. Some are mentioned below

1. Diseases like Hypertension, Migraine, Depression, Asthma, Arthritis, Blood pressure, Anxiety etc. will be cured by yagya therapy and mantra therapy.
2. Water can be purified through yagya therapy.
3. Fertility of the soil can be improved by these therapies.
4. Different skin diseases can also be cured.
5. Growth of organic farming can be increased through Yagya Therapy.
6. It also helps to increase yield in the field which helps farmers.
7. It also helps in increasing rain in the drought regions.
8. Strep throat, urinary tract infections and tuberculosis which is caused by bacterias can easily be cured.
9. Ringworm and athlete's foot which is caused by fungi can also be cured and malaria, swine flu can also be cured through these therapies [41, 42].

9.2 Limitations

1. Faith Issue—Many people or patients have belief on homeopathy and allopathy that they will be cured easily and feel better within few days. So, people have less belief on Yagya and Mantra Therapy. So, they treated it as unfaithful and have less profit.

2. Knowledge—The number of subjects is less within patients. Patients don't know about this therapy.
3. Treatment—Duration of treatment is long. It takes long time to recover.
4. Duration—Patient arrives late for their treatment.

9.3 Applications

In a Yagya, medicines and herbs are vaporized by offering them into the sacrificial fire, and they enter the human body in a vaporous form through the nose, lungs and the pores of the skin. This might be proved to be easiest, least toxic, less risky and most effective method of administrating a medicine to reach every single cell of the body. It is used for the treatment of physical and mental disease. It consists of various Samidhas which create desired effects. It may lead to the development of a scientifically established Yagnopathy and also in other therapies of the world such as Allopathy, Homeopathy, Chromopathy, Naturopathy, etc. It is used in herbal/plant medicinal preparation is used in anti-tuberculosis yagya [43, 44].

10 Conclusion

We can design the automatic expert system which can be helpful to analyse the data or instruction. We can cure many diseases through Artificial Intelligence. It is at the centre of a new enterprise to build computational models of intelligence. It uses many technical methods in yagya therapy. The main assumption is that intelligence (human or otherwise) can be represented in terms of symbol structures. From Machine Learning, it can be concluded that it is a technique of training machines to perform the activities a human brain can do, even though bit faster and better than an average human-being so that any long term disease can be cured. From Swarm Intelligence Techniques, it can be concluded that basically it is a set of algorithms and can be used in the context of forecasting problems [44].

From quantum inspired soft computing, we can conclude that it explores the use of a hybrid soft-computing paradigm for the prediction of the adsorption capacity of an environmentally-friendly and low-cost adsorbent. From intelligent control, we can conclude that it is a class of control techniques that use various artificial intelligence computing approaches like neural networks, Bayesian probability, fuzzy logic, machine learning, reinforcement learning, evolutionary computation and genetic algorithms. From Applications and experience with deployed systems, it can be concluded that it is the set of contents in the box. An application needs at least one deployment type, as it determines how to install the app. From ambient learning, it can be concluded that it is the idea that we don't need to interact directly with any devices.

References

1. C.D. Mathers, D. Loncar, Projections of global mortality and burden of disease from 2002 to 2030. PLOS Med. **15**, 1 (2006). Available: https://journals.plos.org/plosmedicine/article?id=10.1371/journal.pmed.0030442
2. M. Marjani, F. Nasaruddin, A. Gani, A. Karim, I.A.T. Hashem, A. Siddiqa, I. Yakoob, Big IoT data analytics: architecture, opportunities, and open research challenges. IEEE Access **5**, 5247–5261 (2017). Available: https://ieeexplore.ieee.org/document/7888916
3. Z. Fedorowicz, More or less healthcare research or, healthcare research 'more or less'?. Bahrain Med. Bull. **30**, 2 (2008). Available: https://www.academia.edu/692355/More_or_Less_Healthcare_Research_or_Healthcare_Research_More_or_Less
4. A. Kasthuri, Challenges to healthcare in India—the five A's. Indian J. Community Med. 141–143 (2018). Available: https://www.ncbi.nlm.nih.gov/pmc/articles/PMC6166510/
5. C. Mouradian, D. Naboulsi, S. Yangui, R.H. Glitho, M.J. Morrow, P.A. Polakos, A comprehensive survey on fog computing: state-of-the-art and research challenges. IEEE **20**, 416–464 (2017). Available: https://ieeexplore.ieee.org/abstract/document/8100873/authors#authors
6. D. Saxena, S.N. Singh, K.S. Verma, Application of computational intelligence in emerging power systems. Int. J. Eng. Sci. Technol. **2**, 1–7 (2010). Available: https://www.ajol.info/index.php/ijest/article/view/59166
7. Z. Pang, G. Yang, R. Khedri, Y.-T. Zhang, Introduction to the special section: convergence of automation technology, biomedical engineering, and health informatics toward the healthcare 4.0. IEEE **11**, 249–259 (2018). Available: https://ieeexplore.org/document/8421122
8. Y. Ai, M. Peng, K. Zhang, Edge computing technologies for internet of things: a primer. Digit. Commun. Netw. **4**, 77–86 (2018). Available: https://www.sciencedirect.com/science/article/pii/S2352864817301335
9. S.S. Acharya, The integrated science of yagna. Shantikunj **01**, 16–17 (2001)
10. V.G. Limaye, Agnihotra (The Everyday Homa) & Production of Brassinosteroids: a scientific validation. Int. J. Mod. Eng. Res. **08**, 44 (2018)
11. R. Singh, S.K. Singh, Gayatri mantra chanting helps generate higher antimicrobial activity of yagya's smoke. Interdiscip. J. Yagya Res. **1**, 11–12 (2018)
12. S.S. Acharya, The scientific basis of yajnas along with its wisdom aspects. Shantikunj **01** (2001)
13. C. Huston, The impact of emerging technology on nursing care: warp speed ahead. OJIN: Online J. Issues Nurs. **18**(2), 10–46 (2013)
14. P.K. Sharma, S. Ayub, C.N. Tripathi, S. Ajnavi, S.K. Dubey, AGNIHOTRA—a non conventional solution to air pollution. Int. J. Innov. Res. Sci. Eng. (IJIRSE) **7**(2), 34–74 (2017)
15. N.Y. Asabere, mMES: a mobile medical expert system for health institutions in Ghana. Int. J. Sci. Technol. **02**(6), 334–336 (2012). ISSN 2224-3577
16. R.K. Barik, A.C. Dubey, A. Tripathi, T. Pratik, S. Sasane, R.K. Lenka, H. Dubey, K. Mankodiya, V. Kumar, Mist data: leveraging mist computing for secure and scalable architecture for smart and connected health. Procedia Comput. Sci. **125**, 647–653 (2018). Available: https://www.sciencedirect.com/science/article/pii/S187705091732851X
17. A. Mishra, L. Batham, V. Shrivastava, Yagya therapy as supportive care in cancer patients improved quality of life: case studies. Interdiscip. J. Yagya Res. **01**, 1 (2018)
18. L. Kujat, How have robotics impacted healthcare?. Fish. Digit. Publ. **12**, 6–8 (2010). Available: https://fisherpub.sjfc.edu/cgi/viewcontent.cgi?referer=https://scholar.google.co.in/&httpsredir=1&article=1055&context=ur
19. J. Moscinski, Z. Ogonowski, Computer aided adaptive control system design. Silesian Univ. Technol. Gliwice, Poland **2**(3–4), 560 (1994)
20. S.S. Abu Naser, M.W. Alawar, An expert system for feeding problems in infants and children. Int. J. Med. Res. **1**, 79 (2016)
21. S.S. Acharya, The integrated science of yagna. Shantikunj **01**, 04 (2001)
22. A. Kumari, S. Tanwar, S. Tyagi, N. Kumar, Fog computing for Healthcare 4.0 environment-opportunities and challenges. Comput. Electr. Eng. **72**, 1–13 (2018)

23. A. Dingli, C. Abela, I. D'Ambrogio, Pervasive nursing and doctoral assistant—PINATA. In: International conference on pervasive computing technology, p. 189 (2008)
24. P. Idowu, D. Cornford, L. Bastin. Health informatics deployment in Nigeria. J. Health Inform. Dev. Countries **2** (2008)
25. A. Abraham, B. Nath, P.K. Mahanti, Hybrid intelligent systems for stock market analysis. In: V.N. Alexandrov, J.J. Dongarra, B.A. Juliano, R.S. Renner, C.J.K. Tan (eds.) Computational Science—ICCS 2001. ICCS 2001. First Online: 17 July 2001; ICCS 2001: Computational Science—ICCS 2001. LNCS, vol. 2074, pp. 337–345. Available: https://link.springer.com/chapter/10.1007/3-540-45718-6_38
26. G.S. Thakur, YAJÑA-A vedic traditional technique for empirical and transcendental and achievement. Indian Streams Res. J. **04**, 5 (2014)
27. D.P. Acharjya, P. Kauser Ahmed, A survey on big data analytics. Int. J. Adv. Comput. Sci. Appl. **7**(2), 1–13 (2016)
28. D.E. Goldberg, J.H. Holland, Genetic algorithms and machine learning. Mach. Learn. **3**(2–3), 95–99 (1988). Kluwer Academic Publishers. https://doi.org/10.1023/A:1022602019183
29. A. Esfahanipour, W. Aghamiri, Adapted neuro-fuzzy inference system on indirect approach TSK fuzzy rule base for stock market analysis. Expert Syst. Appl. **37**(7), 4742–4748 (2010). https://doi.org/10.1016/j.eswa.2009.11.020
30. S.S. Acharya, The integrated science of yagna. Shantikunj **01**, 14 (2001)
31. N. Siddiqu, H. Adeli, Nature inspired computing: an overview and some future directions. Cogn. Comput. **7**(6), 706–714 (2015). Available: https://doi.org/10.1007/s12559-015-9370-8
32. R.R. Nair, A. Yajna, J. Acupunct. Meridian Stud. **10**(2), 143–150 (2017)
33. S. Makhanov, W. Antaipaiboon, Advanced numerical methods to optimize cutting operations of five axes milling machines. **49**(3–4), 395–413 (2007)
34. M. Saxena, S.K. Sharma, S. Muralidharan, V. Beriwal, R. Rastogi, P. Singhal, V. Sharma, U. Sangam, Statistical analysis of efficacy of yagya therapy on type-2 diabetic mellitus patients on various parameters. In: Proceedings of 2nd International Conference on Computational Intelligence in Pattern Recognition (CIPR-2020), Institute of Engineering and Management, Kolkata, West Bengal, India, 4–5 Jan 2020
35. P. Mahajan, Application of pattern recognition algorithm in health and medicine: a review. Int. J. Eng. Comput. Sci. **05**(5), 16580–16583 (2016)
36. C. Gunavathi, K. Premalatha, A comparative analysis of swarm intelligence techniques for feature selection in cancer classification. Sci. World J. **2014**, 12 (2014). Available: https://www.hindawi.com/journals/tswj/2014/693831/
37. K.-J. Kim, L.L. Tagkopoulos, Application of machine learning rheumatic disease research. Korean J. Int. Med. **34**, 2 (2019)
38. P. Ray, G. Patian, A. Srinivasan, D. Rodbard, D. Price, Systems and methods for pattern recognition in diabetes management. Patent No. US8758245B2, Year of Patent, 24 June 2014. Available: https://patents.google.com/patent/US8758245B2/en
39. G.E. Hinton, A. Krizhevsky, N. Srivastava, I. Sutskever, R. Salakhutdinov, J. Mach. Learn. Res. **15**, 1929–1958 (2014). (cited 2084 times, HIC: 142, CV: 536)
40. S. Bhattacharyya, Quantum inspired soft computing for Binary image analysis. Res. Gate **1** (2016)
41. J.A. Mendez, Artificial intelligence in medicine. Sci. Direct **84**, 159 (2018)
42. R.L. Glass, I. Vessey, V. Ramesh, Research in software engineering: an analysis of the literature. Commun. ACM **44**(8), 491–506 (2002). https://doi.org/10.1016/S0950-5849(02)00049-6
43. J.H. Holmes, P.L. Lanzi, W. Stolzmann, S.W. Wilson, Learning classifier systems: new models, successful applications. Inf. Process. Lett. **82**(1), 23–30 (2002). https://doi.org/10.1016/S0020-0190(01)00283-6
44. D. Subarna, Construction of an expert system: 4 tools. Artificial Intelligence (2018). Retrieved from http://www.engineeringenotes.com/artificial-intelligence-2/expert-systems/construction-of-an-expert-system-4-tools-artificial-intelligence/35582

Rohit Rastogi received his B.E. C. S. S. Univ. Meerut, 2003. Master's degree in CS of NITTTR-Chandigarh from Punjab University. Currently he is pursuing his a doctoral degree. From the Dayalbagh Educational Institute in Agra, India. He is an associate professor in the CSE department of ABES Engineering College, Ghaziabad, India. He has won awards in a various of areas, including improved education, significant contributions, human value promotion, and long-term service. He keeps himself engaged in various competition events, activities, webinars, seminars, workshops, projects and various other educational learning forums.

Mamta Saxena is Additional Director General in ministry of Statistics, GoI and has completed her Ph.D. In Yajna Science with CPCB (Central Pollution Control Board). She has keen interest to revive our ancient culture and science through modern instruments. She is scientist by thought and working on the study of effect of Yajna, Mantra and Yoga on mental patients, patients suffering with various diseases like diabetes, stress, arthritis, lever infection and hypertension etc. with joint collaboration with different organizations AIIMS, NIMHANS, NPL etc.

Muskan Maheshwari is B.Tech. Second Year student of CSE in ABESEC, Ghaziabad. She is a brilliant student in the CSE Department of ABES Engineering. Ghaziabad, India. She is working presently on data mining (DM) and machine learning (ML). She is also working on Yagyopathy. She has keen interest in Google surfing. Her hobbies is playing badminton and reading books. She is young, talented and dynamic.

Priyanshi Garg is B.Tech. Second Year student of CSE in ABESEC, Ghaziabad. She is a brilliant student in the CSE Department of ABES Engineering. Ghaziabad, India. She is working presently on data mining (DM) and machine learning (ML). She is also working on Yagyopathy. She has keen interest in Google surfing. Her hobbies is playing badminton and reading books. She is young, talented and dynamic.

Muskan Gupta is B.Tech. Second Year student of CSE in ABESEC, Ghaziabad. She is a brilliant student in the CSE Department of ABES Engineering. Ghaziabad, India. She is working presently on data mining (DM) and machine learning (ML). She is also working on Yagyopathy. She has keen interest in Google surfing. Her hobbies is playing badminton and reading books. She is young, talented and dynamic.

Rajat Shrivastava is B.Tech. CSE Second Year student of ABESEC, Ghaziabad, He is working presently on data mining (DM) and machine learning (ML). He is currently working of TTH and stress. His hobbies is playing badminton and reading books. He is talented, young, and dynamic.

Mukund Rastogi is B.Tech. CSE Second Year student of ABESEC, Ghaziabad, He is working presently on data mining (DM) and machine learning (ML). He is currently working of TTH and stress. His hobbies is playing badminton and reading books. He is talented, young, and dynamic.

Harshit Gupta is B.Tech. CSE Second Year student of ABESEC, Ghaziabad, He is working presently on Deep Learning and Neural networks and Machine learning (ML). He is currently working of TTH and stress. His hobbies is playing badminton and reading books. He is talented, young, and dynamic.

Identifying Diseases and Diagnosis Using Machine Learning

K. Kalaiselvi and D. Karthika

Abstract Machine learning is once a computer system has been trained to identify patterns by provided it with information and a set of rules to help recognize that data. We demand the process of knowledge 'training' and the output that this method produces is called a 'model'. The possible of ML in this area is meaningfully from top to bottom meanwhile it delivers us with computational approaches for accruing, altering and informing information in smart medical duplicate understanding systems, and, in specific, knowledge machines that will benefit us to tempt information from instances or information. This Chapter is selected to describe an inconsistency of groupings calculated to describe, rise, and approve multi-disciplinary and multi-institutional machine learning exploration in healthcare Perceptive. While the healthcare part is certainty transformed by means of the ability to uppermost huge dimensions of data about separate patients, the enormous volume of information certainty for human beings to scrutinizes. To identify and diagnosing diseases using Machine learning, brings a method to mechanical discovery of outlines and aim about information, which permits healthcare experts to interchange the adapted care known as precision medicine.

Keywords Machine learning (ML) · Learning healthcare systems (LHS) · Machine learning algorithms · Population health management · Disease diagnostics

1 Introduction

Machine learning is once a computer system has been trained to identify patterns by provided it with information and a set of rules to help recognize that data. We demand the process of knowledge 'training' and the output that this method produces is called

K. Kalaiselvi · D. Karthika (✉)
Department of Computer Science, VELS Institute of Science Technology & Advanced Studies, Chennai, India
e-mail: d.karthi666@gmail.com

K. Kalaiselvi
e-mail: kalairaghu.scs@velsuniv.ac.in

V. Jain and J. M. Chatterjee (eds.), *Machine Learning with Health Care Perspective*, Learning and Analytics in Intelligent Systems 13,
https://doi.org/10.1007/978-3-030-40850-3_16

a 'model'. Machine Learning (ML) studies algorithms which can absorb from data to increase acquaintance from knowledge and to make conclusions and predictions. Machine learning representations govern a set of commands using massive quantities of computing influence that a human brain would be unable of dealing out. The additional data a machine learning model is nourished, the extra multifaceted the guidelines—and the additional accurate the predictions. Whereas a statistical prototypical is probable to consume an integral logic that can be unwritten by maximum people, the guidelines formed by machine learning are repeatedly beyond human knowledge since our intelligences are unable of processing and investigating massive data sets. Machine Learning (ML) delivers approaches, methods, and tools that can benefit resolving investigative and predictive difficulties in a variability of medical areas. ML is actuality secondhand for the study of medical boundaries and their mixtures to estimate, e.g. prediction of disease growth, removal of medicinal information for the consequence research, treatment planning and provision for the general persistent organization.

ML is also an actuality, used for data examinations such as discovery of symmetries in the figures dealing with defective data, clarification of nonstop data used in the Exhaustive Care Unit, and brainy disturbing resultant with well-organized monitoring. It is also contended that the positive claim of ML methods can help the addition of computer-based systems in the healthcare atmosphere by providing the chances to enable and increase the effort of health. The possible of ML in this area is meaningfully from top to bottom meanwhile it delivers us with computational approaches for accruing, altering and informing information in smart medical duplicate understanding systems, and, in specific, knowledge machines that will benefit us to tempt information from instances or information.

Machine Learning (ML) Fig. 1 brings practices, methods, and devices that can assistance deciding logical and prognostic problems in a collection of healing areas.

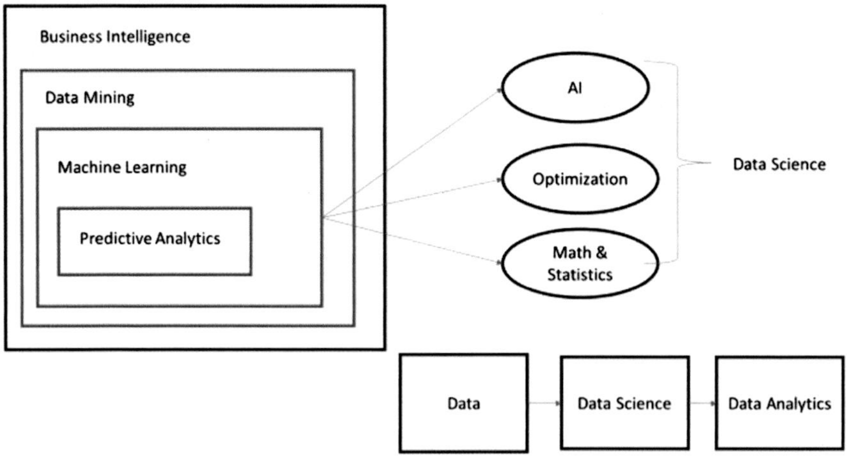

Fig. 1 Machine learning with health care perspective

In Machine learning Business intelligence, Data mining, Machine Learning and Predictive Analysis plays a major role in Healthcare Perceptive. Which processes data through data science and analytics and also combine with three more Data sciences as AI, Optimization and Math Statistics.

Specifically in nominally aggressive imaging events that smear novel imaging philosophies, such as fluorescence imaging or laser perusing microscopy, ML approaches can be valuable since algorithmic answers are not obtainable, there is absence of official replicas, or the information about the request domain is unwell clear due to absence of preceding knowledge and/or medicinal skill in the clarification of the learned images. Where the problem faced in all these areas are focused and find an appropriate solution when bringing ML in Medical identification. Machine Knowledge is an existence, used for the review of skillful limits and their combinations for prediction e.g. prediction of disease growth. Exclusion of therapeutic data for a status examination which conduct supervision for delivery and also for the general lasting group.

Machine knowledge is also used for numbers inspection, such as detection of extents in the information by correctly trading with faulty data, explanation of nonstop information used in the Active Care Component and brainy worrying which are following in the actual and well-ordered treatment. It is resisted, that the positive performance of ML arrogances can assist the count of computer-based structures in the healthcare location as long as probabilities to comfort and improve the effort of medical cases which ultimately used to improve the capability and brilliance of therapeutic restoration.

2 Inevitability of Machine Learning in Healthcare

Originally, ML systems were intended and cast-off to perfect and examine enormous medicinal datasets. By way of these ML procedures developed additional dependable, they remained accepted into gears for supplementary medicinal analysis. Unique of the initial requests of ML in medicinal analysis yield residence in the initial 1970s by the growth of Internist-1 which is a professional advisor package for analysis in over-all interior medication. The scheme usages a probabilistic flawless which controlled to the growth of Bayesian system and estimated implication. Over the centuries, the ML and data mining preceding to produce, additional progressive in addition with influential procedures such as K-nearest neighbor, decision trees, and artificial neural network, were the practical problems to resolve additional complex and particular medicinal analysis difficulties.

Knowledge from enduring information unique meetings numerous problems, meanwhile datasets remain considered by incompleteness (lost parameter principles), erroneousness (methodical or chance sound data), thinness (insufficient in addition/or else non-representable enduring archives obtainable), and the imprecision (unsuitable assortment of limits for the assumed job).

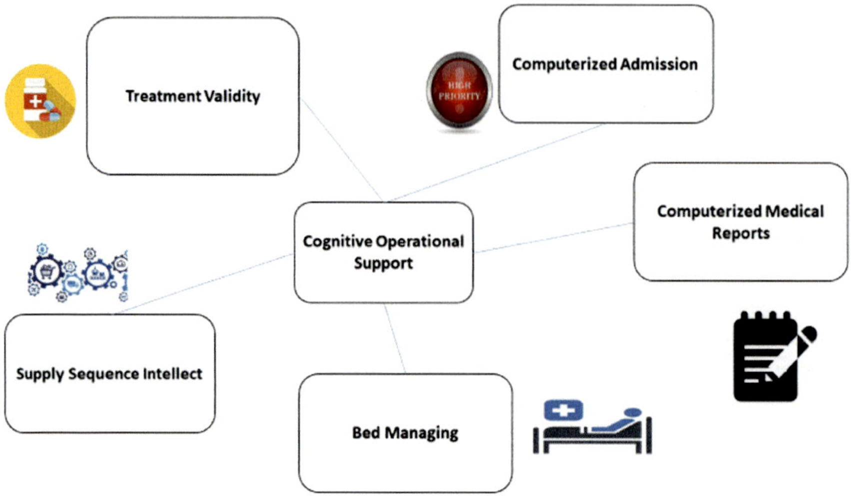

Fig. 2 Cognitive operational support

Healthcare Business in Fig. 2 is revolving available to be additional consumer-driven. By the increasing obtainability of fitness data, the breadwinners are observing for healthier habits to appreciate their patrons and goal them through detailed contributions. Cog nub is emerging machine learning representations that participate the uninterruptedly graceful clinical information from numerous data foundations. ML delivers mechanisms for making with these features of medicinal datasets. Represent typical knowledge approaches, particularly the concept of neural networks which are used to switch these datasets and are typically cast-off their design identical aptitudes and their humanoid comparable features (simplification, strength to sound), in instruction towards recovery of medicinal choice creation.

Additional ground of request is biomedical sign dispensation. Subsequently our considerate of biotic systems is not wide-ranging, there are vital structures and data unseen in the physical signs that are not eagerly deceptive. Also, the belongings amid the dissimilar subsystems are not unique. Genetic signals are considered by extensive inconsistency, produced moreover by impulsive interior devices or by outside incentives. Relations amid the dissimilar limits might be too multifaceted to be resolved with conservative methods.

ML approaches usage these groups of information, which container remain shaped informal, and container assistance to perfect the nonlinear relations that be amid this information, and excerpt limits and structures which can recover medicinal attention. Computer-based medicinal image clarification schemes include a main request area uncertainty important support in medical detecting. The medical information is incessantly curated by means of machine education methods which in try consequence in additional precise conclusion assembly particularly in the zones of healthcare representations, hazard documents, medical processes, etc.

The unbiassed is to rise the specialist's aptitude to classify hateful areas although lessening the essential for interference, and upholding the aptitude for precise finding.

Also, it might be likely to inspect a greater part, learning alive tissue in vivo, perhaps at an objectivity, and, therefore, minimalize the failings of surgeries, such as distress for the enduring, stay in analysis, and imperfect amount of muscle illustrations. The essential for additional actual approaches of initial discovery, such as persons that processer assisted medicinal analysis schemes aim to deliver is clear.

3 Learning Healthcare System (LHS)

Learning Healthcare Systems consume main, personnel, structural, controlling and financial insinuations for humanity. Learning health systems (LHS) are healthcare schemes in which information group procedures are fixed in everyday repetition to yield repeated recover in attention.

Learning Health Care cycle Fig. 3 which brings three process together to form a cycle. The information which are processed through Data to Knowledge (D2K), Knowledge to Performance (K2P) and Performance to Data (P2D) based on these cycles the information is gained from ML applications. Which exchanges all the collected data in the cycle to give an optimal medical report which is the formation of medical community.

The impression was primary abstracted in a 2007 workshop prearranged by the US Institute of Medicine, structure on thoughts about evidence-based medication and "practice-based evidence".

LHS can be labeled as consuming four significant essentials:

1. An administrative planning that provisions the creation of groups of patients, healthcare specialists and investigators who cooperate to yield and practice "big data";

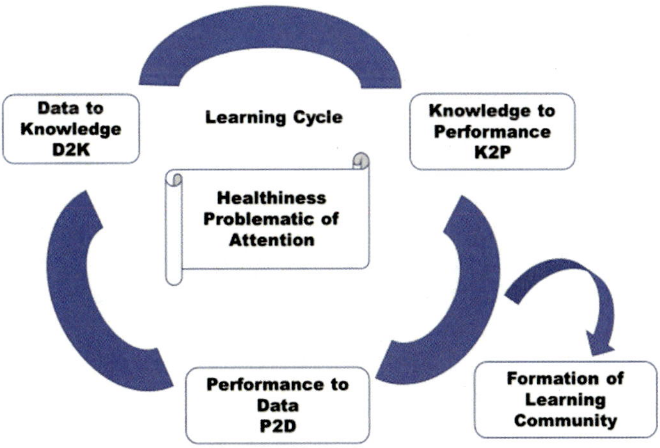

Fig. 3 Learning healthcare systems

2. Huge automated health data sets, i.e. "big data";
3. Excellence development at the idea of maintenance for respectively insistent by means of original information created by investigation;
4. Investigation wide-ranging in repetitive healthcare locations.

They are therefore in need of on the usage of electronic health records (EHRs) and consume hereditary the acceptance tests of EHRs.

LHS involve a scientific lifespan. Enduring information is composed, it is combined crossways numerous patients and a problematic is clear. These are doings mainly ambitious by healthcare specialists. With the provision of knowledge, an examination is done, which proceeds indication, from which information is produced, which mains to altered clinical practice, and therefore to novel enduring information existence together.

McLachlan and colleagues (2018) propose a classification of nine LHS sorting types:

- Unit identification appearances for patients with comparable features.
- Confident aberration discoveries instances of healthier maintenance in contradiction of a standard.
- Bad nonconformity discoveries instances of sub-optimal maintenance.
- Prognostic persistent hazard demonstrating usages designs in information to discovery collections at better danger of opposing proceedings.
- Prognostic maintenance hazard and consequence replicas classify circumstances that are at better danger of unfortunate upkeep.
- Scientific choice provision schemes use enduring procedures practical to persevering information to style exact action approvals.
- Relative efficiency investigation regulates the greatest real actions.
- Brainy help usage of information to mechanize tedious procedures.
- Investigation displays information for illness occurrences or additional action problems.

4 Population Health Management

Population health characterizes an enormous change in the social groundwork of healthcare. Popular its widest intelligence, population health is the health consequences of a collection of persons, but it also comprises observing at the delivery of those consequences inside the cluster. Individuals collections can be strongminded through numerous demographics as well as,

- Topography
- Competition/Society
- Gender
- Socioeconomic position
- Nations/Communities

- Corporation of service
- Incapacity
- Association with the punishing scheme

Individual of the comprehensive areas of populace health as a organization inventiveness is to remove changes of health inside collections, or at smallest decrease them considerably. It's greatest significant to message the difference amongst the general fitness inside these inhabitants and the delivery of fitness inside individuals' comparable collections.

Additional significant difference to type is among that of populace health and community health. Community health mentions the dangerous purposes of government and home-grown community health objects (e.g. widespread calculation, repression of ecological dangers, the inspiration of strong performance.) That supposed, wider meanings that comprise the assurance of the fitness of the community can interconnect with the idea of populace fitness. Whether it interconnects with community health or not, populace fitness determination comprises manifold investors, knowledges, and values in its request. Populace health organization dynamically includes all healthcare investors, and earners must be prepared to style variations to efficiently achieve their enduring inhabitants.

By way of its defiance, greatest performs are under-prepared for the period of population health management with workflows motionless planned about visit-based maintenance replicas and recruitment constructions positioned about worker requirements in its place of patients. Re-engineering determination need that medical doctor no lengthier understand themselves as the midpoint of the workplace, nonetheless as share of a team that is placed everywhere the persistent.

5 The Goal of Using ML Algorithms in Healthcare

It's continued supposed earlier that the greatest machine learning implement in healthcare is the doctor's brain. Might here remain a propensity for doctors to interpretation machine learning as an unwelcome additional estimation? At unique opinion, autoworkers be afraid that automation would remove their occupations. Also, near might be doctors who anxiety that mechanism learning is the start of a procedure that might reduce them outdated.

Nonetheless it's the sculpture of medication that container not ever be substituted. Patients determination continuously essential the humanoid touch, and the thoughtful and sympathetic association with the persons who bring care. Neither machine learning, nor slightly additional upcoming technologies in medication, determination remove this, but determination develop utensils that clinician's usage to improve continuous maintenance.

The emphasis must be on in what way to usage machine learning to supplement patient care. In Fig. 4 machine learning in medicine states that diagnosing, imaging and monitoring of patient data which helps to predict valuable results. The clinical

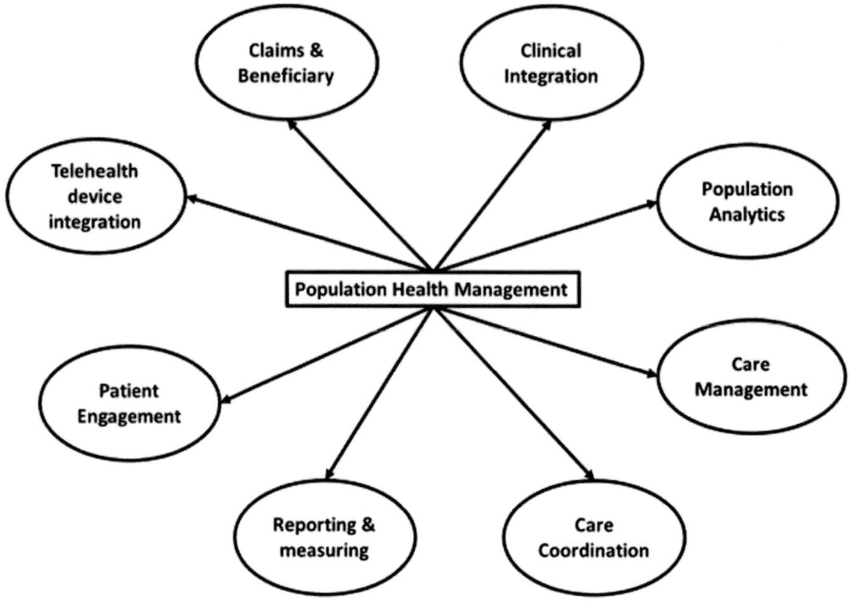

Fig. 4 Population health management

research which identifies all personalized data and diagnosis and make the analysis by monitoring the results. For instance, if I'm challenging a patient for cancer, formerly I poverty the highest-quality surgery consequences I container maybe become (Fig. 5).

A machine learning algorithm that container appraisal the pathology photos and effect the pathologist by an inspection, is valued. If its container produces the standards in a serving of the period with an unclear notch of exactness, before, lastly, this is optimistic to improve lasting upkeep and gratification. Machine Learning (ML) Fig. 6 brings practices, methods, and devices that can assistance deciding logical and prognostic problems in a collection of healing areas. In Machine learning Information interoperability, Budget, Change Control, Information Interchange, Hazards Credentials and Commitments play a vital role in Medical diagnosis. Where the problem faced in all these areas are focused and find an appropriate solution when bringing ML in Medical identification.

Healthcare requirements to change from rational of machine learning as an innovative idea to sighted it as a real-world instrument that can remain organized nowadays. Uncertainty machine learning is to consume a part in healthcare, formerly, necessity to take an incremental method. This determination remains a step-by-step path to joining additional analytics, machine learning, and predictive procedures into ordinary medical repetition.

Originally, our goalmouths essential to competition our competences. Exercise a machine learning procedure to classify membrane cancer after a great traditional of skin cancer imageries is somewhat that maximum persons realize. This necessity remains linked completed period. Radiologists won't always develop outdated,

Machine Learning in Medicine

Diagnostic Testing	• Predict viral failure in AIDS patients • Personalized diagnosis
Oncology	Clinical Research: Identify which gene is associated with breast cancer Predict probability of survival
Medical Imaging	Clinical research: Deep Learning Cellular image analysis
Remote Patient Monitoring	Real time Predictions Medical Monitoring

Fig. 5 Machine learning in medicine

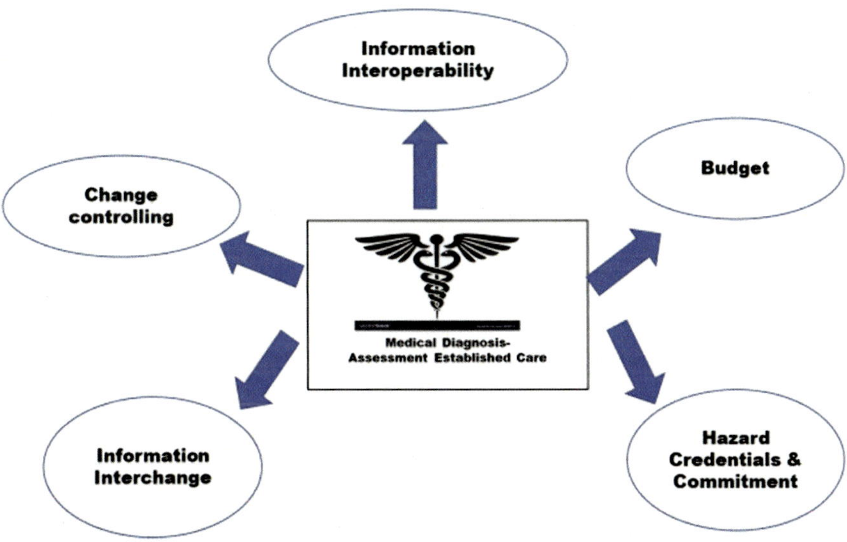

Fig. 6 Machine learning with health care perspective

nonetheless radiologists of the upcoming determination oversee and appraisal interpretations that consume remained originally recite by a mechanism. They determination employ machine learning similar a cooperative significant other that classifies exact parts of emphasis, illumines noise, and benefits emphasis on high chance zones of concern.

5.1 ML in Medical Imaging

Trendy medical imaging, Computer Aided Diagnosis (CAD) is a fast-rising active zone of research. Machine learning is significant in Computer Aided Analysis. Afterward by means of an informal equation, substances such as tissues may not remain designated precisely. So, design recognition basically includes knowledge from instances. In the arena of bio-medical, design recognition and machine learning potential the better correctness of insight and analysis of illness. They likewise indorse the impartiality of decision-making procedure. For the examination of high-dimensional and multimodal bio-medical information, machine learning proposals a well-intentioned method for creation elegant and involuntary procedures.

5.2 Healthcare Accuracy Medication

The methodical collection and examination of hereditary information in mixture with illnesses, treatments, and consequences consumes the possible to melodramatically recover the assortment of the finest actions, evading the hurting of patients, and the usage of unsuccessful treatments. The obtainability of ancient longitudinal patient information about ecological contact and existence would likewise assistance to healthier control the (ensemble of) reasons activating the start of an illness national.

5.3 Assembling Patient Testified Consequences and Full Pathway Prices for Value-Based Healthcare

In healthcare, reimbursement for presentation, is a perfect that proposals monetary inducements to healthcare benefactor for refining excellence and efficiency of healthcare by conference sure presentation events. Proceeding the other arrow, it is likewise a test to regroup managerial care schemes to be talented to attach all the complicated prices of exact care-paths in instruction to consume a precise estimate of the full charges complicated. As quickly as maintenance procedures have remained related and upkeep trails can be sketched, choices for precise treatments can be founded on experiential indication, as reinforced by a enormous catalog of 'enduring described consequences' of patients through alike illnesses and the related entire price of behaviors and treatments. It is vital that the approaches to gather enduring stated well-being consequences and charges per treatment/maintenance pathway are consistent and authenticated.

5.4 *Enhancing Workflows in Healthcare*

The industrial business includes procedures which are in numerous belongings expectable. Though, circumstances inside a hospital remain highly self-motivated and frequently in need of on a huge amount of inter-related issues straddling the patients themselves and their requirements, manifold sections, staff memberships and possessions. This instable condition brands slightly procedure of workflow instrumentation to recover efficiency highly stimulating unless hospital supervise and managers consume an appropriate impression of the hospital's process. This varieties it indispensable for a healthcare earner to have the essential tools to integrate manifold information watercourses such as actual place tracking systems, microelectronic medical records, attention information systems, persistent monitors, research laboratory information and mechanism logs to mechanically classify the present working state of a hospital in instruction to let for additional real decision-making that consequences in healthier reserve operation and thus advanced output and excellence.

5.5 *Contamination Inhibition, Estimate and Control*

Contagion control is the punishment concerned with stopping hospital acquired (HAI) or healthcare associated infection. Rendering to the European Centre for Sickness Deterrence and Control 24, 100,000 patients are projected to obtain a healthcare-associated infection in the EU apiece day. The amount of deceases happening as a straight importance of these contagions is projected to remain at least 37,000 and these contagions are supposed to donate to an extra 110,000 demises each year. It is projected that about 20–30% of healthcare-associated contagions are avoidable by concentrated cleanliness and switch agendas. Actual and big data knowledges are wanted to assimilate genomic through epidemiology information in instruction to not fair regulator, but also stop and forecast the feast of contagions inside a healthcare location.

5.5.1 Social-Clinical Care Pathway

Healthcare is touching near a combined care method, which rendering to the description of the World Health Organization (WHO) is "an idea carrying organized inputs, delivery, organization and society of services connected to analysis, treatment, care, reintegration and fitness advancement".

5.5.2 Patient Provision and Participation

In adding to gathering patient stated health consequences here are additional occasions for enduring empowerment and participation. The patient controls for handling health information must provision dissimilar levels of numerical/well-being literateness and permit following patient agreement of choosing in/out for medical investigation educations. For instance, web opportunities of patient governments production a significant role in swapping data about illness, medication and managing approaches, balancing to unvarying patient meeting data.

5.6 Public Decision Support

By means of underlining the patient's participation inside decision procedures, patients are able to improvement a healthier sympathetic of all the health-related subjects. In this intelligence, generous patients switch ended and vision in their individual fitness information can assistance to reinforce patient-centered upkeep afterward periods of a disease-centered perfect of upkeep, and letting the informal customization of healthcare and exactness medicine. Rationally, existence information composed and combined into expressive data should stimulate patients to attain advanced obedience charges and inferior medicinal prices. Expressive data disapprovingly be contingent on the aptitude of schemes to enumerate the characteristic indecision complicated in the analysis and also the indecision with admiration to the consequences of action replacements and related hazards.

5.7 Home-Based Care

Specialized following and footage of medicinal information as well as individual information must not be incomplete to individual infirmaries and registrars. Outstanding to demographic variations, novel replicas for home care or casualty care (facilities) have to be industrialized. Big Data technologies can provision the general ICT founded alteration in this zone. Through uniting smart home knowledges, wearables, medical information and episodic vital sign capacities, home care breadwinners will be in the least reinforced by a prolonged healthcare substructure, while persons are authorized to conscious lengthier on their individual.

5.8 Scientific Research

The addition and examination of the enormous capacity of capability information impending from numerous dissimilar capitals such as microelectronic healthiness accounts, social media surroundings, drug and toxicology records and all the 'omics' data such as genomics, proteomics and metabolomics, is a important motorist for the alteration after (populace level) indication founded medicine near precision medication.

6 Spread Over Machine Learning Towards Health Care

Machine learning in medication consumes lately complete headlines. Google consumes an advanced machine learning algorithm which is used to assist and classify cancerous growths through mammograms. Stanford is a deep learning algorithm to classify skin cancer. A new JAMA article states that the consequences of a deep machine-learning algorithm that remains intelligent results to diagnose the diabetic retinopathy in the retinal imageries. It's strong that the machine learning, places an additional projectile in medical conclusion making.

Silent, machine learning advances the aforementioned to approximately procedures healthier than others. Procedures can deliver instant advantage to punishments with procedures which are reproducible or consistent. Likewise, the persons with great image datasets, such as radiology, cardiology, and pathology, are treated and called as robust applicants. Machine learning can remain skilled and appearance with imageries, identify irregularities, and opinion to the parts of essential care. Thus, by refining the correctness of all these procedures the long period, machine learning determination and the advantage of domestic doctor or internist at bedside. Machine learning can be the proposal of an objective and the estimation to recover competence, dependability, and correctness.

At Health Catalyst, usage an exclusive platform to examine information, and twist it posterior in real period to medical doctor to aid in scientific conclusion making. At the similar period a surgeon understands an enduring and arrives indications, information, and examination consequences into the EMR, there's machine learning behindhand the divisions observing at the whole thing around that enduring, and encouragement the medic with valuable data for creation an analysis, collation a examination, or signifying a defensive broadcast. Extended term, the competences will spread into all features of drug as get more practical, healthier combined information.

6.1 *Machine Learning Algorithms*

Its definition is as tracks: the field of edu̧cation that stretches processers the aptitude to study deprived of existence openly automatic. Numerous practical glitches can remain demonstrated through the machine learning algorithms. Approximately the instances are: classification, regression, clustering, dimensionality reduction, structured prediction, face detection, decision making, speech recognition, signal de-noising, anomaly detection, deep learning and reinforcement learning.

Machine learning can be categorized into four types as given below as,

a. un-supervised learning
b. supervised learning
c. semi-supervised learning and
d. reinforcement learning.

6.1.1 Un-supervised Learning

In realism, near happen numerous unlabeled information which can remain deliberate lost label or accidental misplaced tag. The previous information is typically branded originally; one might eliminate the tag and express the problematic as relative or association examination amid examples. Un-supervised knowledge contracts with the problematic of project of perfect to relate the concealed design and association of unlabeled data by means of the machine learning algorithms. The characteristic of methods used in unsupervised learning are un-supervised collecting and un-supervised anomaly discovery. In an Un-supervised clustering, the goals to establish a similar unlabeled information into groups which are also called as named groups.

Consequently, an information within a same cluster can consume the similar characteristic and which remains the dissimilarity to the data in supplementary clusters. Here, are three elementary methods used in clustering for nearness measure (similarity or dissimilarity measure), criterion function for clustering evaluation, and algorithms for clustering. It can also be situated for an additional separable clustering into hierarchical clustering (divisive or agglomerative), partitional clustering (centroid, model based, graph theoretic, or spectral) and Bayesian clustering (decision based or non-parametric). Concealed infectious illnesses are appalling since medical specialists are usually not acquainted with their topographies.

6.1.2 Supervised Learning

In machine learning, the outstanding request in classification and regression are the greater collection of investigators to absorb the supervised learning as well. The datasets, comprises separate matching of information which can consume input and output principles. Normally, an algorithm is design to make the inner relatives based on the physical activity information and to simplify the hidden information (Fig. 7).

There are five over-all steps for supervised knowledge model:

- Information gathering of exercise and stimulating datasets;
- Feature abstraction;
- Assortment of machine knowledge algorithm;

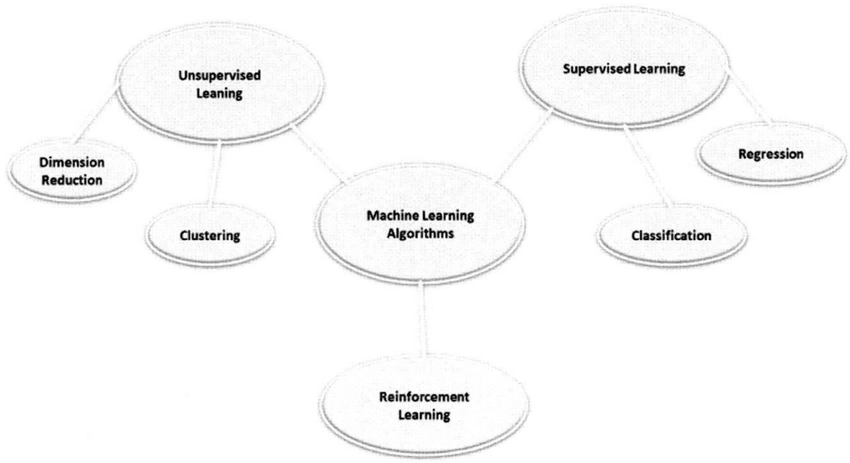

Fig. 7 Machine learning algorithms

- Prototypical building by means of the designated procedure; and
- Procedure assessment, in addition contrast to supplementary algorithms.

Additional distinguishing of SVM is that it contracts with curved optimization problematic so that slightly resident answer is a worldwide best resolution.

6.1.3 Semi-supervised Learning

Semi-supervised learning remains comparatively new likened to supervised and un-supervised learning. Probabilities are that here is numerous unlabeled information nonetheless lone insufficient labeled information are obtainable. This remains the aim for the designation semi-supervised by way of it lies among supervised learning (pairwise labeled inputs and outputs) and un-supervised learning (completely unlabeled data). Nonetheless, it is period overwhelming and expensive to alteration unlabeled information into considered data. Then, the instructions are modified to comprise unlabeled information. The toughness of such method be contingent on the dependability of classification, i.e., it consumes deprived heftiness if it consumes unadorned classification mistake.

When building semi-supervised knowledge representations, at the minimum unique to subsequent expectations are finished:

- Evenness supposition is an idea that examples (in the forms of feature vector) which are nearby to each additional have an advanced chance have the similar production label as they part similar distinguishing in feature space;
- The cluster supposition entitlements that the datasets and
- Alike to the first supposition, but mentioning to gathering, various supposition income that information untruth on a low-dimensional various entrenched in a

higher-dimensional interplanetary. It is renowned that various is a topological interstellar that nearby look like Euclidean space nearby individually argument.

6.1.4 Reinforcement Learning

Reinforcement learning is connected to managers annoying to exploit the entire prize below communication by indeterminate and multifaceted setting. In the field of switch and process investigation, it is too named estimated lively software design. Varied after normal administered learning, reinforcement learning fixes not hold correct input/output pairs and sub-optimal movements. Essentially, two plans are usually secondhand to resolve reinforcement learning difficulties.

A reinforcement knowledge founded neuro prosthesis supervisor was qualified to appraise target-oriented mission achieved by means of humanoid arm. Though the consequences reflects that the humanoid plunders are real events to partition the supervisor. As soon as it originates to moveable health submissions, medical audio-visual streaming via adaptive degree switch algorithm was deliberate. Reinforcement learning remained functional to fulfill the obligation of high excellence of facility.

7 Machine Learning in Health Care Diagnostics

Machine Learning in health care is an exclusive determination to characterize a variability of classifications designed to characterize, increase, and authorize multi-disciplinary and multi-institutional machine learning research in healthcare informatics. The combined, panoramic assessment of data and machine learning techniques can deliver a chance for original clinical understandings and detections. Health Informatics (HI) trainings the operative use of probabilistic information for conclusion making.

The grouping of both Machine Learning (ML) and Health Informatics (HI) has extreme possible to increase quality, effectiveness and competence of treatment and care. Although the healthcare area is actuality altered by the capability to highest enormous volumes of data about discrete patients, the huge capacity of data actuality composed is unbearable for human beings to scrutinizes. Machine learning delivers a way to robotically discovery of patterns and reason about data, which allows healthcare specialists to move to modified care known as precision medicine.

The quick growth of Artificial Intelligence, particularly Machine Learning (ML), in addition Datamining authorizations the knowledge and healthcare innovators towards produce intellectual schemes in the direction of improve and development the present measures. Nowadays, Machine language consumes remained purposeful fashionable a multiplicity of zone in the healthcare manufacturing such through income of judgement, modified behavior, medication detection, experimental research, smart electronic health records and epidemic outbreak prediction.

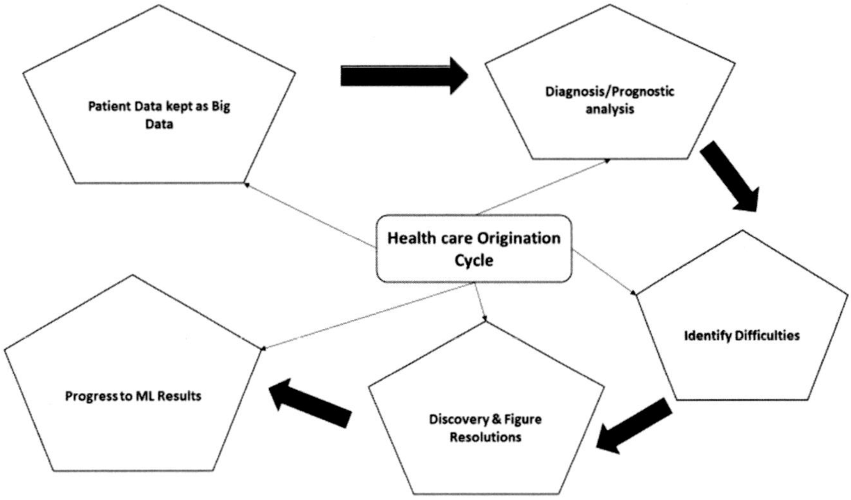

Fig. 8 Healthcare origination cycle

This is a troublesome, particularly once persons exist are at stick. An analysis of a illness or a disorder trusts on data which covers issues that brands receiving it precise test. These issues include uncertainty, doubt, battles, and reserve and structural restraints. In Fig. 8 carries us to a novel test- in what way to style expressive usage of such huge capacities of information. A review discloses that nearly 44% of healthcare managements are powerless to use all the obtainable information, estimate our healthcare manufacturing a vast $350 billion per year.

This can be located powerless to use patient information in a all-inclusive way to style medical development. For example, serving Workers with opportune admission to precise enduring data is unavoidable for evidence-based conclusion creation and positive analysis. Though, the information took finished manifold schemes are not combined or public, which is a barricade for perceptive examination. Also, meanwhile it is calm finished manifold schemes, healthcare information is multifaceted and needs composite analytics to confirm expressive usage.

Altogether these steeplechases are subsequent of information foundations occupied in storage tower and their incapability to attain interoperability amongst themselves. Medical diagnosis is one of the procedures for decisive hopefully, the reason of neither a patient's disease nor a disorder through examining the data learnt from the numerous bases such as counting corporal examination, enduring meeting, workshop tests and current medical information of the reason of experiential symbols as indications. Receiving a precise analysis is the greatest vital step for giving a patient to make the surgeons to discover the finest action for the patient's disorder.

The fast growth in the grounds of Artificial Intelligence, particularly Machine Learning (ML), and Datamining permits the knowledge of healthcare modernizers to make intelligent systems to enhance and recover the present processes. Today, ML has stood functional in a diversity of part within the healthcare manufacturing.

Such that the analysis, modified treatment, medication discovery, scientific experimental investigation, radiology and radiotherapy, clever electronic health best, and widespread outbreak calculation. In medicinal diagnosis, ML and Datamining procedures are mainly valuable and they rapidly imprisonment unexpected outlines inside the multifaceted and bulky datasets.

The ground of medicinal diagnosis for medicinal systems is fairly ironic with potentials and its compensations such as saving, early analysis and possibly redeemable humanoid life. It has numerous limits such as confidentiality. Since, the confidentiality issues relate the patient's delicate data which makes those data widespread about the information and cannot be providing to repetition ML algorithms. Additionally, inadequacy is not a portion of doctors/surgeons but are conscious of ML tackles which obtainable in the marketplace. The determination of involvement suitable to keep fit when comprehend to novel increasing technical requests. When it originates to information distribution, it must not remain unnoticed to the individuals laboring on ML requirements and procedures. An essential to comprehend multifaceted medicinal information and association among patient's consequence with attention.

7.1 Resolving Medical Diagnostic and Prognostic Difficulties Using Machine Learning System

Machine Learning (ML) affords approaches, methods, and apparatuses which are benefited in resolving analytic and predictive difficulties in the diversity of medicinal fields. Machine learning (ML) are the wildest rising field in computer science, and Health Informatics (HI) is between the utmost request challenges, as long as upcoming aids in better-quality medical diagnoses, disease analyses, and pharmaceutical development. Medical diagnosis remains the sequence of conclusive the reason of a patient's disease or complaint by investigative information are cultured.

Since frequent fundamentals counting corporeal inspection, enduring conference, workshop examinations, patient's and the patient's personal therapeutic greatest record, besides present therapeutic information of the reason of experimental cryptograms and indications. Still, it is difficult procedure and necessitates lots of humanoid effort and time. An analysis of a disease or a disorder depend on information which comprises issues that makes receiving it precise contest. A portion of indications are general and adjustable, dependent scheduled an individual. Numerous analytical examinations are exclusive, but not often done. The arena of medical analysis in therapeutic organizations remains impartially frustrating through possibilities and its reimbursements.

Machine learning methods can be positioned used for the examination of medical information and it is cooperative in medicinal analysis for detecting dissimilar particular diagnostic difficulties. By means of Machine learning, organizations take the patient information like indications, research laboratory information and approximately of the significant characteristics as a contribution and makes a precise analysis consequence.

Constructed on the correctness of the consequence, machine determination resolve which information will remain functioned as exercise and skilled dataset for the upcoming orientation. In present situation, medic is gathering all the best of the persistent and founded on that will stretch drugs to patients. Through this situation, enormous quantity of time is misused due to numerous explanations which approximately time formed adversity in slightly when lifespan. By means of machine learning classification algorithms, for any precise disease, we can advance the accuracy, rapidity, consistency and presentation of the analytic on the present system.

7.2 Disease Prediction Using Machine Learning

Numerous healthcare administrations about the nation have previously ongoing refining consequences and convertible lives by joining with Health Catalyst and by means of its catalyst ai-driven analytics. Machine learning is portion of ordinary life for greatest Americans, since triangulation apps to internet shopping, and extensively secondhand in additional businesses, such as trade and investment. Nonetheless it be situated monotonous in healthcare since of the difficulty and incomplete obtainability of information—and the absence of admission to extremely accomplished information researchers and sides obligatory to go that information into expressive developments. Greatest businesses as long as machine learning answers need patrons to number out in what way to attach as numerous as 100 dissimilar data bases to brand the knowledge work. Through difference, Health Catalyst's Healthcare Analytics Platform mixes 120 dissimilar data bases, counting the electronic health record (EHR), rights, important economic, working and enduring gratification schemes.

Greatest current machine learning answers merchants deliver academically-appealing, separate replicas deprived of a sympathetic of in what way to interpret them addicted to expressive, climbable consequences. As a consequence, here are insufficient practical instances of extensive machine-learning assisted consequences development in healthcare. The lowest line is that health systems can excluding additional survives and recover more maintenance, though saving currency at the similar period.

Currently numerous international organizations and actions crossways the biosphere, similar e.g. the World Health Organization (WHO), brand usage of semantic knowledge-bases in health care schemes to:

- Advance accurateness of identifies by as long as actual time associations of indications, test outcomes and separate medical antiquities;
- Assistance to shape additional influential and additional interoperable facts schemes in healthcare;

- Provision the essential of the healthcare procedure to convey, re-use and portion persistent information;
- Deliver semantic-based standards to provision dissimilar arithmetical combinations for dissimilar drives;
- Transport healthcare schemes to provision the addition of information and information;
- Aggressive information organization systems in level on healthcare can ease the movement of data and consequence in healthier, more-informed conclusions.

8 Applications of Machine Learning in Medical Diagnosis

The progressively rising quantity of requests of mechanism knowledge in healthcare permits us to foretaste at an upcoming anywhere information, examination, and novelty effort hand-in-hand to assistance uncountable patients deprived of them always understanding it. Rapidly, its determination be fairly shared to discovery ML-based claims entrenched with actual enduring data obtainable after dissimilar healthcare schemes in manifold republics, thus cumulative the effectiveness of novel action choices which remained unobtainable earlier.

8.1 Identification of Diseases and Diagnosis

Unique, the principal of ML applications in the medical healthcare is the documentation and analysis of illnesses which are then careful hard-to-diagnose. This can be comprised with whatever from the tumors. Tumors are threatening to fastening growth through the early stages to additional hereditary illnesses. IBM Watson Genomics is a major instance, in what way the mixing and reasoning of calculation with genome-based tumor growth can assist in the creation of fast analysis. Berg, the biopharma massive is leveraging an AI to grow the therapeutic actions in the zones such as oncology.

8.1.1 Typhoid Fever Diagnosis

Typhoid disease is a unique for most of the major life-threatening diseases and secretarial for the death of millions of people each year. Rapid and precise diagnosis is a foremost key in the medical field, where the large number of deaths are related with typhoid fever. It is a consequence of many factors which includes poor diagnosis, self-medication, shortage of medical experts and insufficient health institutions. These provoked are the growth of a typhoid diagnosis system that can be cast-off by anybody as normal intellect to this will involvement in quick diagnosis of the disease. Despite,

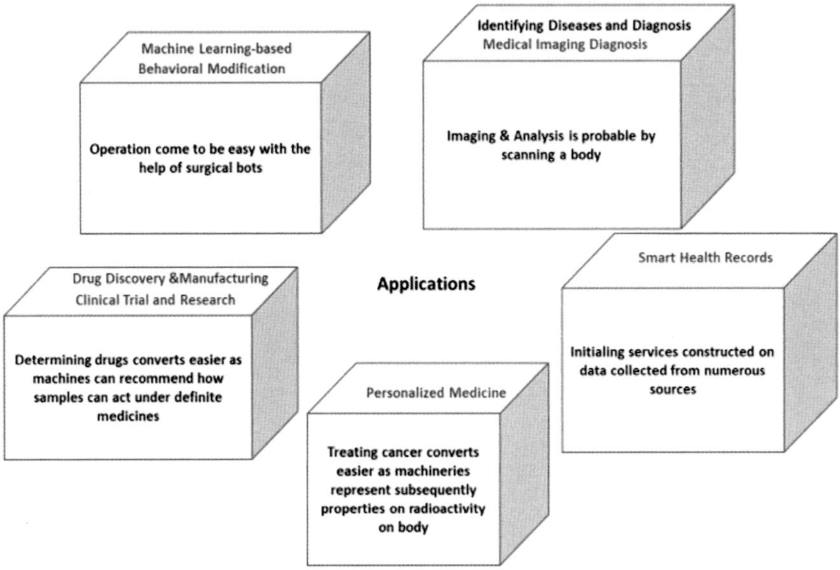

Fig. 9 Machine learning applications

the lack of health institutions and medical experts. A machine learning method was used on the labelled set of typhoid fever in which the provisional variables are used to produce the understandable instructions for the diagnosis of typhoid fever (Fig. 9).

8.1.2 Breast Cancer Diagnosis Using ML

Breast cancer consumes and industrialized a foremost foundation of expiry in women nowadays. Therefore, the common awareness of the conceivable aids of main detection of breast cancer consumes enlarged powerfully. A genuine and trustworthy visionary prototypical aimed at preliminary diagnosis can expressively reduction the weakening to women, the previous neutral of the estimate stands to numeral out whether patients stay owed into a benevolent collection (non-cancerous) or malevolent collection (cancerous). In the existing case, the calculations can remain gaped as organization difficulties. In the learning of various ML/Data Mining methods have stayed upcoming to provision the breasts cancer primary discovery and calculation. In malignance exploration zone, the keeping of trained data is typically significant, as an appropriate dataset afterwards scrubbing up and overhauling irrelevant or inacceptable information assistances in demonstrating a additional multifaceted interpreter of cancer analysis.

8.2 Medication Discovery and Manufacturing

Distinct, the main medical applications of machine learning falsehoods are in early-stage of medication detection procedure. This process also comprises R&D knowledges in next-generation arrangement and exactness drug that can assist in discovery other trails for the treatment of multifactorial illnesses. Now, these mechanisms knowledge methods which includes the unverified knowledge that can classify the designs of information. Scheme Delivery manufacturing, by of ML-based skills for numerous creativities, where the counting of emerging AI-based knowledge for the tumor to conduct and personalizing a painkiller mixture for AML (Acute Myeloid Leukemia).

8.3 Health Imaging Diagnosis

Machine learning and deep learning be located together as an answerable for the advance technology which are named as Computer Vision. This consumes instigate in the Internal Eye inventiveness industrialized by the Microsoft, which the whole thing on duplicate analytic apparatuses for duplicate examination. As mechanism learning develops additional available and as they raise in their descriptive volume, suppose to understand additional information bases from diverse medicinal images develop a portion of this AI-driven analytical procedure.

8.4 Personalized Medicine

Adapted handlings can not individual remain additional actual by combination separate fitness by prognostic analytics nonetheless is too ready remain for additional investigation and healthier illness valuation. Now, surgeons are incomplete to selecting after an exact usual of identifies or approximation the risk to the enduring founded on his indicative past and obtainable hereditary information.

8.5 Smart Health Records

Preserving in the know well-being histories is a thorough procedure, and although knowledge consumes occupy yourself its share in facilitation the information admission procedure, the fact is that smooth today, a popular of the procedure's income a ration of period to whole. The chief part of mechanism knowledge in healthcare is to comfort procedures to but period, exertion, and cash. Text organization approaches by means of course machineries and ML-based OCR obligation methods are gradually meeting steam.

8.6 *Clinical Trial and Research*

Machine learning consumes numerous possible requests in the field of scientific prosecutions and investigation. As anyone in the pharma business would express you, scientific hearings price a ration of period and cash and container income ages to whole in numerous possessions. Smearing ML-based prognostic analytics to classify possible scientific experimental applicants can assistance investigators attraction a pond after an extensive diversity of data ideas, such by way of preceding administrator calls, communal media, etc.

9 Conclusion

Machine learning structures are one of the principle methods for emerging sophisticated, automatic, and objective algorithms for analysis of high-dimensional and multimodal biomedical data. This assessment is mostly concentrating on numerous advances those are in the state of the art. Important in the progression has been focused on the growth of in-depth sympathetic and theoretic analysis of critical issues which connected to algorithmic building and knowledge theory. These comprise trade-offs for exploiting simplification presentation, which uses the bodily realistic restrictions, and combination of prior knowledge with uncertainty.

This Chapter is an exclusive determination to characterize a variability of classifications designed to characterize, increase, and authorize healthcare informatics. The rapid development in the arenas of Artificial Intelligence, particularly Machine Learning (ML), and Data mining authorizations the knowledge and healthcare innovators to create intelligent systems to improve and development in the present procedures. Nowadays, ML has remained purposeful in a variety of zone within the healthcare manufacturing such as analysis, modified behavior, drug detection, clinical trial investigation, radiology and radiotherapy, computerized microelectronic well-being utmost, and widespread eruption forecast. In medical diagnosis, Machine Learning and Data mining procedures are largely valuable. They can quickly detention to unforeseen outlines inside compound and great datasets that are stored. Through neutral and self-possessed datasets, machine learning processes can comfort the aforementioned reasoning partiality challenging and produce advanced accurateness.

References

1. G.D. Magoulas, A. Prentza, Machine learning in medical applications, 11 Sept 2015. https://doi.org/10.1007/3-540-44673-7_19. Source: DBLP https://www.researchgate.net/publication/225171947_Machine_Learning_in_Medical_Applications
2. F. Campion, G. Carlsson, in *Machine Intelligence for Healthcare*. The full book can be obtained from Amazon.com, bit.ly/MIforHealthcare 2017
3. A. Smola, S.V.N. Vishwanathan, in *Introduction to Machine Learning*. Published by the press syndicate of the University of Cambridge, United Kingdom, 1 Oct 2010, svn://smola@repos.stat.purdue.edu/thebook/trunk/Book/thebook.tex
4. D. Karthika, K. Kalaiselvi, ISVS3CE: incremental support vector semi-supervised subspace clustering ensemble and enhanced bat algorithm (ENBA) for high dimensional data clustering. Int. J. Recent Technol. Eng. (IJRTE) **8**(2) (2019). ISSN: 2277-3878
5. http://www.cnn.com/2016/06/28/us/epa-lead-in-u-s-water-systems/index.html
6. K. Ross, Applying machine learning to healthcare. Precision Driven Health. precisiondriven-health.com @healthprecision, 2016
7. John Wiley & Sons, Machine Learning for Dummies, IBM Limited edn., Published by, Inc. 111 River St. Hoboken, NJ 07030-5774 www.wiley.com Copyright © 2018
8. A. Holzinger, Machine learning for health informatics. HCI-KDD, Institute for Medical Informatics, Statistics and Documentation, Medical University Graz, Graz, Austria, Springer International Publishing AG 2016
9. F. Jeanquartier, C. Jean-Quartier, M. Kotlyar, T. Tokar, A.C. Hauschild, I. Jurisica, A. Holzinger, Machine learning for in silico modeling of tumor growth, in *LNCS (LNAI)*, vol. 9605 (Springer, Heidelberg, 2016)
10. M. Burr, Building foundations to build better care. Pure Storage's Flash, Blade Business Unit, mburr@purestorage.com @purehealthcare, vol. 2, no. 1 (2019)
11. C.M. Bishop, Pattern recognition and machine learning. Library of Congress Control Number: 2006922522, ISBN-10: 0-387-31073-8, ISBN-13: 978-0387-31073-2, Printed on acid-free paper. © 2006 Springer Science + Business Media
12. F. Jiang, Y. Jiang, H. Zhi, et al., Artificial intelligence in healthcare: past, present and future stroke and vascular neurology, 2017. https://doi.org/10.1136/svn-2017000101, http://creativecommons.org/licenses/by-nc/4.0/
13. Int. J. Pure Appl. Math. **114**(6), 1–10 (2017). ISSN: 1311-8080 (printed version); ISSN: 1314-3395 (on-line version). http://www.ijpam.eu
14. S. Vinitha, S. Sweetlin, H. Vinusha, S. Sajini, Disease prediction using machine learning over big data. Comput. Sci. Eng.: Int. J. (CSEIJ) **8**(1) (2018)
15. M. Fatima, M. Pasha, Machine learning-based prediction of heart failure readmission or death: implications of choosing the right model and the right metrics (2019)
16. D. Raval, D. Bhatt, M.K. Kumhar, V. Parikh, D. Vyas, Medical diagnosis system using ML. Int. J. Comput. Sci. Commun. **7**, 177–182 (2016). https://doi.org/10.090592/ijcsc.2016.026. Nirma University, Ahmedabad, India
17. Machine learning classification techniques for heart disease prediction: a review (2016)
18. Best treatment identification for disease using machine learning approach in relation to short text. IOSR J. Comput. Eng. (IOSR-JCE) **16**(3), 05–12. e-ISSN: 2278-0661, p-ISSN: 2278 8727, Ver. VII (May-June 2014). www.iosrjournals.org

Dr. K. Kalaiselvi has received her M.Sc., Computer Science degree from Periyar University, M.Phil. degree from Bharathidasan University, Tamil Nadu, India and Ph.D. degree in Computer Science from Anna University, Chennai, Tamil Nadu, India. She is currently working as Professor and Head in the Department of Computer Science, School of Computing Sciences, Vels Institute of Science, Technology and Advanced Studies, Chennai, India which is well known university.

She has more than 16 years of teaching experience in both UG and PG level. Her research interests include Knowledge Management, Data Mining, Embedded Systems, Big Data Analytics and Knowledge Mining. She has produced four M.Phil. Scholars. Currently, she is guiding Ph.D. scholars and M.Phil. Scholars in VISTAs. She has published more than 32 research Papers in Various International and two papers in National Conferences. She has published a book titled "Protocol to learn C Programming Language". She has received the Best Researcher Award from DK International Research Foundation, 2018 and received Young Educator and Scholar award—6th women's day Awards 2019, from the National Foundation for Entrepreneurship Development NFED 2019. She is a professional Member of CSI. She serves as Editorial Board Member/Reviewer of various reputed Journals. She has been invited a resource person for various International National conferences and seminars organized by various Institutions. She has completed the mini project funded by VISTAS.

Ms. D. Karthika Completed her Master of Computer Applications (M.C.A.), in PSG College of Arts & Science, Coimbatore, M.Phil. from Hindusthan College of Arts & Science, Coimbatore. She is currently Pursuing her Research Work (Ph.D.) in Vels Institute of Science, Technology and Advanced Studies under the guidance of Dr. K. Kalaiselvi, Head and Associate Professor in the Department of Computer Science, School of Computing Sciences, Vels Institute of Science, Technology and Advanced Studies, Chennai.

She has more than three years of teaching experience in both UG and PG level. She is Interested in Machine Learning, Artificial Intelligence and Big Data Analytics. Broad area of research is Big Data Analytics focusing on High Dimensional Data Clustering. She has published two research articles in Elsevier SSRN Digital Library. She has Published a research article in IJRTE Scopus indexed Journal. Published a book chapter titled as "Women & Leadership: Leading under Extreme Diversity" on a book "Women Empowerment: Leadership & Socio-Cultural Dimension" in Association with NFED (National Foundation for Entrepreneurship Development), held on March 8th, 2019 at Coimbatore. She has also presented papers at International conferences.

Made in the USA
Las Vegas, NV
16 September 2026

bf7d8807-b06d-461a-92e1-342f1709da53R02